ENVIRONMENTAL IMPACT ASSESSMENT

McGraw-Hill Series in Water Resources and Environmental Engineering

---

Ven Te Chow, Rolf Eliassen, and Ray K. Linsley
Consulting Editors

Bailey and Ollis Biochemical Engineering Fundamentals
Bockrath Environmental Law for Engineers, Scientists, and Managers
Canter Environmental Impact Assessment
Chanlett Environmental Protection
Graf Hydraulics of Sediment Transport
Hall and Dracup Water Resources Systems Engineering
James and Lee Economics of Water Resources Planning
Linsley and Franzini Water Resources Engineering
Linsley, Kohler, and Paulhus Hydrology for Engineers
Metcalf and Eddy, Inc. Wastewater Engineering: Collection, Treatment, Disposal
Nemerow Scientific Stream Pollution Analysis
Rich Environmental Systems Engineering
Schroeder Water and Wastewater Treatment
Tchobanoglous, Theisen, and Eliassen Solid Wastes: Engineering Principles and Management Issues
Walton Groundwater Resources Evaluation
Wiener The Role of Water in Development: An Analysis of Principles of Comprehensive Planning

ENVIRONMENTAL IMPACT ASSESSMENT

LARRY W. CANTER
University of Oklahoma

McGRAW-HILL BOOK COMPANY
New York St. Louis San Francisco Auckland Bogotá Düsseldorf
Johannesburg London Madrid Mexico Montreal New Delhi Panama
Paris São Paulo Singapore Sydney Tokyo Toronto

This book was set in Press Roman
by Hemisphere Publishing Corporation.
The editor was B. J. Clark and
the production supervisor was Milton J. Heiberg.
The Kingsport Press was printer and binder.

ENVIRONMENTAL IMPACT ASSESSMENT

1 2 3 4 5 6 7 8 9 0 KPKP 7 8 3 2 1 0 9 8 7

Library of Congress Cataloging in Publication Data

Canter, Larry W.
Environmental impact assessment.
(McGraw-Hill series in water resources and
environmental engineering)
Includes bibliographies and index.
1. Environmental impact statements. I. Title.
TD194.5.C36 333.7 76-42480
ISBN 0-07-009764-X

To Donna, Doug, Steve, and Greg

Contents

Preface

The National Environmental Policy Act requires environmental impact considerations to be included in project planning along with technical and economic concerns. The action-forcing mechanism is that environmental impact statements must be prepared to describe the environmental consequences of proposed actions and various alternatives. Over 6,000 statements have been prepared since the effective date of the act, January 1, 1970, and even more will be prepared in the future.

New terminology has been developed as a result of the environmental impact statement process or components thereof. For example, environmental impact statements are also referred to as impact statements, environmental statements, and 102 statements. One component of the process is the preparation of an environmental inventory. Many agencies prepare environmental impact reports or environmental analysis reports as a prelude to the actual preparation of impact statements. A companion term is environmental impact assessment, with definitions ranging from a cursory preliminary evaluation of potential impacts of projects to the preparation of a document that is more extensive than an impact statement.

This author views the process culminating in an environmental impact statement as consisting of five components: basics, environmental setting, prediction and assessment of impacts, selection of proposed action, and preparation of impact statement in accordance with extant guidelines. This textbook is organized according to these components. Chapters 1 and 2 encompass the basic requirements and framework of the process. Chapter 3 describes environmental factors that must be considered in defining the study area environmental setting. Chapters 4 through 9 address the steps for impact prediction and assessment for the physical-chemical (air, water, and noise), biological, cultural, and socioeconomic environments. Chapter 10 presents various impact assessment methodologies that can be utilized in evaluation of alternatives and proposed actions, and Chapter 11 describes public participation in the environmental impact assessment process, particularly as related to selection of proposed actions. Chapter 12 discusses pertinent considerations in writing environmental impact statements. Finally, Chapter 13 presents information regarding future trends and needs in the environmental impact assessment/environmental impact statement process.

This book is intended for use in upper division or graduate level courses dealing with environmental impact assessments/statements. The orientation is primarily for science and engineering majors; however, individuals trained in other disciplines such as planning can utilize this text. This book can also be used by professionals working on environmental impact studies. The attempt has been to assemble pertinent information both for the classroom and for practice. It is intended that all users gain familiarity with the following topics:

1 Requirements and implementation of the National Environmental Policy Act.
2 A conceptual framework for conducting environmental impact studies.
3 Approaches for defining environmental settings.
4 Technological methods that can be used to predict changes in environmental characteristics.
5 Criteria and considerations involved in interpreting significance of predicted impacts.
6 State of the art of environmental impact assessment methodologies.
7 Techniques for accomplishing public participation.
8 Practical considerations for writing environmental impact statements.

This book is not meant to encompass every possible consideration in the environmental impact assessment/environmental impact statement field. This is a dynamic field, and proper use of this textbook is as a reference for a point in time, with the understanding that it must be supplemented as additional information and technology become available.

The author expresses his gratitude to the College of Engineering, University of Oklahoma, for its support during the preparation of this textbook. He is also grateful to the U.S. Civil Service Commission and Mr. Marvin H. Brannon for the many opportunities afforded through conduction of workshops in this subject area. Particular thanks are due to Dr. Loren Hill and Dr. Paul Risser, University of Oklahoma, for their cooperative efforts on several environmental impact assessment studies. Appreciation is also extended to Mrs. Edna Rothchild, Mrs. Pat Huddleston, and Mrs. Darlene Scallon for their tireless typing efforts. Finally, the author thanks his wife for her encouragement in the process of developing this textbook.

Larry W. Canter

ENVIRONMENTAL IMPACT ASSESSMENT

Chapter 1

National Environmental Policy Act and Its Implementation

The past several years have been characterized by passage of major federal legislation dealing with the environment, including specific legislation on control of water and air pollution (1, 2). Perhaps the most significant legislation is the National Environmental Policy Act (NEPA) of 1969 (PL 91-190), which became effective on January 1, 1970. This act was the first signed in the 1970s (3). The thrust of this act, as well as of subsequent executive orders, Council on Environmental Quality (CEQ) guidelines, and numerous federal agency procedures, is to ensure that balanced decision making occurs in the total public interest (4). Project planning and decision making should include the integrated consideration of technical, economic, environmental, social, and other factors. Prior to NEPA, technical and economic factors dominated the decision-making process. The NEPA is reproduced in Appendix A.

TERMINOLOGY

New terminology has arisen in conjunction with the process of complying with the requirements of NEPA. Three of the most significant new terms are "environmental inventory," "environmental assessment," and "environmental impact statement."

Environmental Inventory

Environmental inventory is a complete description of the environment as it exists in an area where a particular proposed action is being considered. The inventory is compiled from a checklist of descriptors for the physical, biological, and cultural environment. The physical environment includes such major areas as geology, topography, surface-water and groundwater resources, water quality, air quality, and climatology. The biological environment refers to the flora and fauna of the area, including species of trees, grasses, fish, herpetofauna, birds, and mammals. Specific reference must be made to any rare and/or endangered plant or animal species. General biological features such as species diversity and overall ecosystem stability should also be presented. Items in the cultural environment include human population trends and population distributions, historic and archeological sites, and economic indicators of human welfare.

The environmental inventory serves as the basis for evaluating the potential impacts on the environment, both beneficial and adverse, of a proposed action. It is included in an impact statement in the section referred to as "description of the existing environment" or "description of the environmental setting without the project" (5). Development of the inventory represents an initial step in the environmental impact assessment process.

Environmental Assessment

The environmental assessment represents the key step in meeting the requirements of NEPA. In essence, it is an attempt to evaluate the consequences of a proposed action on each of the descriptors in the environmental inventory. The essential steps in an environmental impact assessment are

1 Prediction of the anticipated change in an environmental descriptor.
2 Determination of the magnitude or scale of the particular change.
3 Application of an importance or significance factor to the change.

Many of the current assessment approaches embody the steps of prediction, scaling, and significance interpretation, although the methods use many terms to describe these particular steps.

The scientific validity of the technology available for the prediction of impacts varies depending upon the particular environmental descriptor. For example, extensive research and sound scientific methods have been developed for prediction of air quality impacts (6), at least with regard to anticipated concentration levels of pollutants in the ambient air; however, impacts on flora or fauna as a result of the calculated concentration levels are less quantifiable. Thus it is possible to utilize sound technology for some impact predictions, whereas other predictions must be primarily based on professional judgment.

In order to accomplish an environmental assessment, as well as to prepare an inventory and write an impact statement, it is necessary that the approach used be interdisciplinary, systematic, and reproducible. Requirements for an interdisciplinary approach indicate that the environment must be considered in its broadest sense; thus the input of persons trained in a number of technical fields needs to be included (7). The disciplines represented in a specific environmental assessment must be oriented to the unique features of the proposed action and the environmental setting; however, at a minimum it is necessary to have input from a physical scientist or engineer, a biologist, and a person who can address cultural and socioeconomic impacts. Requirements for a systematic and reproducible approach indicate that a degree of organization and uniformity should be utilized in the assessment process. In this regard several assessment methodologies have been developed since 1970 (8, 9), and these will be discussed in more detail later.

Environmental Impact Statement

The environmental impact statement (EIS) is a document written in the format as specified by NEPA, CEQ guidelines, and specific agency guidelines. The EIS represents a summary of the environmental inventory and the findings of the environmental assessment. Environmental impact statements are also referred to as "environmental statements," "impact statements," "environmental impact reports," or "102 statements" (10). The term 102 statement refers to the section in NEPA that spells out the requirements for the preparation of an EIS.

There are two categories of EISs: draft statements and final statements. The draft statement is the document prepared by an agency proposing an action; it is circulated for review and comment to other federal agencies, state and local agencies, and public and private interest groups. Specific requirements with regard to timing of review are identified in the CEQ guidelines (the 1973 CEQ guidelines are included here as Appendix B). The final statement is the draft statement modified to include a discussion of problems and objections raised by reviewers. The final statement must be on file with CEQ for at least a 30-day period prior to initiation of construction on a project.

NATIONAL ENVIRONMENTAL POLICY ACT

The act is divided into two basic parts: title I, which is a declaration of a national environmental policy, and title II, which establishes the CEQ. The national goals as specified in section 101 of the act are as follows (4):

1 Fulfill the responsibilities of each generation as a trustee of the environment for succeeding generations.

2 Assure for all Americans safe, healthful, productive, and aesthetically and culturally pleasing surroundings.

3 Attain the widest range of beneficial uses of the environment without degradation, risk to health or safety, or other undesirable and unintended consequences.

4 Preserve important historical, cultural, and natural aspects of our national heritage and maintain, where possible, an environment that supports diversity and variety of individual choice.

5 Achieve a balance between population and resource use that will permit high standards of living and a wide sharing of life's amenities.

6 Enhance the quality of renewable resources and approach the maximum attainable recycling of depletable resources.

Section 102 of NEPA has three primary parts related to the environmental impact assessment process. Part A specifies that all agencies of the federal government shall utilize a systematic, interdisciplinary approach, which will ensure the integrated use of the natural and social sciences and environmental design arts in planning and in decision making that may have an impact on the human environment. Part B requires agencies to identify and develop methods and procedures that will ensure that presently unquantified environmental amenities and values may be given appropriate consideration in decision making along with economic and technical considerations. This part has provided impetus for the development of several environmental assessment methods. Part C indicates the necessity for preparing environmental statements and identifies basic items to be included. It also indicates that agencies should include in every recommendation or report on proposals for legislation and other major federal actions significantly affecting the quality of the human environment a detailed statement that covers five major areas (4):

1 The environmental impact of the proposed action.

2 Any adverse environmental effects that cannot be avoided should the proposal be implemented.

3 Alternatives to the proposed action.

4 The relationship between local short-term uses of the human environment and the maintenance and enhancement of long-term productivity.

5 Any irreversible and irretrievable commitments of resources that would be involved in the proposed action should it be implemented.

The requirement for preparing an EIS was not a part of the original proposed legislation that subsequently became NEPA (11). Detailed histories of the legislative background of NEPA have been presented by Andrews (12) and Yannacone and Cohen (13). Section 102 requirements were added late in the legislative review process, just prior to final action on the part of Congress. These particular requirements have been called the "action-forcing mechanism" of NEPA (14), indicating that agencies must prepare a draft statement, which is then subject to review and critique by other federal agencies as well as state and local governmental and private groups. It is an understatement to say that the NEPA requirements for preparing impact statements have been controversial (15, 16), and many court cases have resulted from this section of NEPA (17-20). Perhaps one of the problems with NEPA is that it does not contain a provision for ensuring compliance (21).

One section of NEPA that has received very little attention is section 103, which requires that all agencies review their present statutory authority, administrative regulations, and current policies and procedures for the purpose of determining whether there are any deficiencies or inconsistencies therein that prohibit full compliance with the purposes and provisions of NEPA. Very few written responses have been recorded with regard to action taken in conjunction with section 103 (11, 22).

MAJOR ACTIONS SIGNIFICANTLY AFFECTING THE QUALITY OF THE HUMAN ENVIRONMENT

Section 102 of NEPA requires that environmental statements be prepared for "major Federal actions significantly affecting the quality of the human environment." The Corps of Engineers uses the acronym MASAQHE for this phrase. Definitions are not included in NEPA for what constitutes a major action or what constitutes a significant effect on the quality of the human environment. Concern with these definitions is relevant since the preparation of an EIS requires both human and economic resources (23). One of the negative results of the passage of NEPA has been the preparation of impact statements on projects that perhaps should not require much attention, such as installation of traffic control signals and minor roadway resurfacing work. On the other hand, many major actions of the federal government, such as peace-time military activities and space activities, have not had EISs filed.

To attempt to define a "major action significantly affecting the quality of the human environment" involves many quantitative and qualitative considerations. The simplest way of defining a major action is to compare a predicted impact with an environmental quality standard for a given parameter. It is possible to do this for many substances found in air and water, for example, suspended particulates in the atmosphere and dissolved oxygen in water. However, there are many environmental parameters for which only subjective standards are available, such as scenic vistas and archeological sites. Agencies can best define MASAQHE by project type, indicating that certain projects require impact statements because they are major actions, and others do not because they are minor actions. The Federal Highway Administration has developed guidelines of this type (24). Major actions include a highway section entirely or generally on a new location and major upgrading of an existing highway section that requires extensive right-of-way acquisition and construction. Highway sections that may have a significant effect on the quality of the human environment include those

1 That are likely to have a significantly adverse impact on natural ecological, cultural, or scenic resources of national, state, or local significance.

2 That are likely to be highly controversial regarding relocation housing resources.

3 That divide or disrupt an established community; disrupt orderly, planned development; are inconsistent with plans or goals that have been adopted by the community in which the project is located; or cause increased congestion.

4 That involve inconsistency with any national, state, or local standard relating to the environment; have a significantly detrimental impact on air or water quality or on ambient noise levels for adjoining areas; involve a possibility of contamination of a public water supply system; or affect groundwater, flooding, erosion, or sedimentation.

Negative declarations can be prepared on the following types of highway improvement actions since they are not likely to have significant impacts:

1 Signing, marking, signalization, and railroad protective devices.

2 Acquisition of scenic easements.

3 Modernization of an existing highway by resurfacing; less-than-lane-width widening; adding shoulders; auxilliary lanes for localized purposes.

4 Correcting substandard curves.

5 Reconstruction of existing stream crossings where stream channels are not affected.

6 Reconstruction of existing highway/highway or highway/railroad separations.

7 Reconstruction of existing intersections including channelization.

8 Reconstruction of existing roadbed, including minor widening, shoulders, and additional right-of-way.

9 Rural two-lane highways on new or existing location that are found to be generally environmentally acceptable to the public and local, state, and federal officials.

BASIC CONTENTS OF AN IMPACT STATEMENT

Section 102, part C, of NEPA identifies five points that need to be adhered to in an EIS. The first one is to describe "the environmental impact of the proposed action." In the early years of the preparation of impact statements, attention was primarily focused on the negative or detrimental impacts associated with a given proposed action. To be complete both beneficial and detrimental impacts should be delineated. The basic thrust of NEPA is that it is a "full disclosure law," implying that both the positive and negative ramifications of a given proposed action should be explored in complete detail (25). In addition, attention must also be directed toward the primary and secondary impacts associated with a proposed action (5). Primary and secondary impacts are also referred to as direct and indirect consequences. Table 1-1 includes a partial listing of the direct and indirect impacts of a sewage treatment plant, with only adverse impacts being identified (26). In general, agencies have developed methods and procedures to respond in part to direct impacts, both beneficial and adverse. However, the major impact of a project is often from secondary or even tertiary effects, and these are much more difficult to assess due to the dearth of predictive techniques available.

The second item required by NEPA is an identification of "any adverse environmental effects which cannot be avoided should the proposal be implemented." If a thorough approach has been utilized in describing the environmental impact of the proposed action, this section should basically be an abstract of the negative impacts, both direct and indirect, of the proposed action. New information is not included in this section.

The third point focuses on a discussion of "alternatives to the proposed action." This section has caused a great deal of difficulty, and many court cases have resulted from inadequate treatment of this section by the proposing agency (17-20). Kennedy and Hanshaw (27) reported on an analysis of the alternatives sections of 200 randomly selected environmental statements from several agencies. Of the 127 actions that listed adverse environmental effects, a total of 214 alternatives were listed, all of which were rejected: 130 were rejected for economic

Table 1-1 Direct and Indirect Impacts of Sewage Treatment Plant[a]

	Primary effects	Secondary effects	Tertiary effects
		Adverse impacts—Direct	
Short term	Erode soil during construction	Degrade aquatic habitat of stream	Decrease fisheries
Long term	Periodic releases of noxious gases	Decrease surrounding property values	Change socioeconomic composition of neighborhood[b]
		Adverse impacts—Indirect	
Short term	Construction employment	Temporary housing	Do not occur
Long term	Permit/encourage residential development within service area[b]	Increase traffic on local streets	Traffic congestion, noise, smog

[a] Impacts are described for illustrative purposes only; a complete matrix for a sewage plant would contain many more impacts within each of the respective cells of the matrix. A separate matrix could be constructed for beneficial impacts.

[b] Irreversible impact.

reasons, 47 for environmental reasons (that is, the alternative would do more harm than the proposed action), and 37 because of engineering problems. Implicit in the alternatives section is the idea that the alternatives to the proposed action should be compared on a common basis, presumably their relative or absolute environmental impact (28). One alternative that should be discussed is the no-action, or no-project, alternative (5). This alternative requires the proposing agency to predict what the future environment will be without the project, and it serves as the basis against which impacts of the proposed action can be compared. Concern regarding alternatives originally arose in conjunction with the question of the retroactivity of NEPA, and two well-known court cases that evolved from this point are the Gillham Reservoir in southwestern Arkansas and the Cross-Florida barge canal (17). One point to be considered in the alternatives section is an evaluation of the alternatives through a public participation program (5).

The fourth item is a description of "the relationship between local short-term uses of man's environment and the maintenance and enhancement of long-term productivity." This section is based on the principle that each generation should serve as trustee of the environment for succeeding generations; therefore, attention must be paid to the question of whether options for future use of the environment are being eliminated by the particular proposed action. In a pragmatic sense many impact statements have described the impacts associated with the construction and operational phases of a proposed action, considering the construction phase to be short term and the operational phase to be long term.

The last point is a discussion of "any irreversible and irretrievable commitments of resources which would be involved in the proposed action should it be implemented." Semantic difficulties are encountered with the terms irreversible and irretrievable. Again from a practical standpoint, most impact statements focus attention on possible changes in land usage as a result of a proposed action, loss of cultural features such as archeological or historical sites, preclusion of development of underground mineral resources, loss of habitat for plants and animals, loss or impact on rare and endangered plants and/or animals, material required for project construction, energy usage required during project utilization, and even the human and monetary expenditures involved.

COUNCIL ON ENVIRONMENTAL QUALITY GUIDELINES

Section II of NEPA established the CEQ, whose responsibilities in relation to the impact statement process include serving as a central repository for final EISs, preparing general guidelines applicable to all federal agencies in conjunction with their compliance with NEPA, reviewing draft EISs (particularly for controversial projects), and developing comparative analyses on the impact statement process. Federal agencies can request consultation and guidance from CEQ in conjunction with NEPA compliance or the preparation of agency procedures and guidelines.

The CEQ published guidelines for the preparation of EISs on April 23, 1971 (29) and August 1, 1973 (5). The guidelines issued in 1971 coordinated the impact statement process, particularly with regard to review of draft EISs (29). Two new items were added to the five basic points specified by NEPA to be included in an impact statement: a section describing the proposed action and a section oriented to a discussion of problems and objections raised by reviewers. The first new section precedes the basic five points, and the latter follows. A list of federal agencies with responsibility for various areas of environmental quality was included.

The CEQ guidelines of August 1, 1973 call for the addition of two more new sections in an impact statement, plus an expansion of a previously required section (5). The initial section of an impact statement became a description of the proposed action as well as a description of the existing environment. One new section pertains to the relationship of the proposed action to existing land-use plans, policies, and controls in the affected area, which requires a discussion of how the proposed action may conform or conflict with the objectives and specific terms of any federal, state, or local land-use matters, either approved or proposed. In addition, land-use plans developed in response to the requirements of the Clean Air Act (2) or the Federal Water Pollution Control Act Amendments of 1972 (1) should also be specified. The second new section calls for an indication of what other interests and considerations of federal policy are thought to offset the adverse environmental effects of the proposed action. This section is oriented to a discussion of other decision factors that the agency feels tend to counterbalance any adverse environmental effects. Agencies that prepare cost-benefit analyses of proposed actions should summarize these analyses in this section. Where nonenvironmental costs and benefits are part of the basis for decision, it is important that the agency specify the importance of these elements in the decision (30).

ROLE OF THE ENVIRONMENTAL PROTECTION AGENCY

The Environmental Protection Agency (EPA) was established in December 1970 as the environmental regulatory agency of the United States (31). It is not the chief administrative agency for EISs. The EPA reviews EISs prepared by others, particularly with regard to water pollution, air pollution, solid waste management, noise, radiation, and pesticides. Each statement reviewed is assigned a rating based on the proposed action and the EIS document itself (32). The EPA system of rating other agency actions is as follows:

1 Project rating (LO, ER, or EU).

LO (lack of objections). EPA has no objections to the proposed action as described in the draft impact statement or suggests only minor changes in the proposed action.

ER (environmental reservations). EPA has reservations concerning the environmental effects of certain aspects of the proposed action. EPA believes that further study of suggested alternatives or modifications is required and has asked the originating federal agency to reassess these aspects.

EU (environmentally unsatisfactory). EPA believes that the proposed action is unsatisfactory because of its potentially harmful effect on the environment. Furthermore, the agency believes that the potential safeguards that might be utilized may not adequately protect the environment from hazards arising from this action. The agency recommends that alternatives to the action be analyzed further (including the possibility of no action at all).

2 Adequacy of document (1, 2, or 3).

Category 1 (adequate). The draft impact statement adequately sets forth the environmental impact of the proposed project or action as well as alternatives reasonably available to the project or action.

Category 2 (insufficient information). EPA believes that the draft impact statement does not contain sufficient information to assess fully the environmental impact of the proposed project or action. However, from the information submitted, the agency is able to make a preliminary determination of the impact on the environment. EPA has requested that the originator provide the information that was not included in the draft statement.

Category 3 (inadequate). EPA believes that the draft impact statement does not adequately assess the environmental impact of the proposed project or action, or that the statement inadequately analyzes reasonably available alternatives. The agency has requested more information and analysis concerning the potential environmental hazards and has asked that substantial revision be made to the impact statement.

No rating of the proposed action is made if the EIS is assigned a category 3 rating because a basis does not generally exist on which to make a determination about the environmental impact of the proposed project. All 3 and EU ratings on draft impact statements must be cleared at the EPA headquarters in order to double check that all such ratings are consistent with policies and practices followed by EPA on a nationwide scale. The originating agency is notified of the assigned ratings at the time a comment letter is sent. The EPA also notifies the CEQ of its comments on all 3 or EU projects so that the CEQ can begin to follow up the project at an early stage in its development.

Although EPA does not have official responsibility with regard to the acceptance or rejection of impact statements, the review by this agency is critical, and the procurement of a satisfactory evaluation of the impact statement and the proposed action is very necessary.

The EPA has also developed guidelines and procedures for the preparation of impact statements associated with their proposed actions (33). The Federal Water Pollution Control Act Amendments of 1972 (1) exempted EPA from the NEPA requirement of preparing impact statements for activities leading to the setting of new source pollution standards, effluent limitations, and guidelines for water quality standards and in establishing "best practicable" and "best available" treatment standards (15, 34).

With regard to agency procedures, most federal agencies have developed guidelines for the EIS process. Moore (35) has indicated that each agency's NEPA-response regulations should

1 Identify those agency actions requiring an EIS.

2 Designate officials responsible for the statements.

3 Specify the general methods for obtaining information required in preparation of the statement.

4 Indicate the required content of an EIS.

5 Designate the appropriate time prior to decision to seek comments of other agencies.

6 Establish patterns for consulting with and taking into account comments of other agencies.

7 Describe the mechanism through which statements are to be made available to the public.

8 Provide for timely public announcement of plans and programs with environmental impact.

STATUS REPORT ON ENVIRONMENTAL IMPACT STATEMENTS

Many problems have arisen with regard to the EIS process. Three major categories of problems were identified in the third annual CEQ report (36), including procedural concerns, questions regarding the content of EISs, and the role of the CEQ. Procedural problems, at least in the period from 1970 through 1972, dealt largely with the following types of questions:

1 What actions require impact statements?

2 What are the impact statement requirements for actions initiated or authorized prior to NEPA enactment?

3 Can impact statements be prepared for programs rather than individual projects?

4 Which agency is responsible for preparing the impact statement for a proposed action involving multiple agencies?

5 What are the requirements for the review and comment process?

6 Do existing agencies with environmental regulatory activities have to prepare EISs?

7 Does an EIS have to be available for normal agency public hearings?

The required content of EISs has been clarified largely by the development and issuance of CEQ and agency guidelines and at present should cover the items included in the CEQ guidelines of August 1, 1973. It is the duty of the proposing agency to consider opposing views to the particular proposed action and to discuss alternatives to that action. If the agency proceeds in spite of the possible adverse environmental effects, the EIS should clearly identify the other interests that justify this action.

The role of CEQ includes serving in an advisory capacity to federal agencies and the president, issuing guidelines for implementing section 102 of NEPA, and identifying significant recurring substantive problems in the impact statement process.

Several comparative critiques of selected EISs were prepared in the initial years of the existence of NEPA. Ray (37) reported on major areas requiring improvement after reviewing about 250 impact statements from many agencies. The following major points were noted:

1 Agencies are not fully considering and applying the policies expressed in all applicable local, state, and federal laws.

2 Both primary and secondary impacts need to be delineated.

3 Alternatives should be discussed in sufficient detail so as not to foreclose choices other than the action proposed.

4 The cumulative effects of several time-phased segments of a major action on future choices should be presented.

5 Environmental considerations should not be limited to the framework of strict legislative mandates.

6 Proper attention and full consideration of all public and private reviews, comments, statements, and testimonies have not been adequately documented in all statements.

7 The preparation of environmental statements has not really been used as an effective vehicle for proper land-use planning.

In 1972 the Government Accounting Office (GAO) reported on a review of seven federal agencies' responses to NEPA requirements (38). The agencies included the Corps of Engineers (Civil Functions), Forest Service, Soil Conservation Service, Department of Housing and Urban Development, Bureau of Reclamation, Federal Aviation Administration, and Federal Highway Administration. The GAO indicated that improvements were needed in the following areas:

1 EISs as integral parts of decision-making processes.
2 Actions requiring EISs and the ranges of impacts that should be considered.
3 Public participation in environmental assessments.
4 Methods for obtaining views of federal, state, and local agencies.

A second study by GAO examined the adequacy of six impact statements (39). Three common problems associated with these statements were inadequate discussion of and support for the identified environmental impacts; inadequate treatment of reviewing agencies' comments on environmental impacts; and inadequate consideration of alternatives and their environmental impacts. The first and last of the three problems are also noted in an analysis of 395 EISs received in 1972 by region IV (Atlanta) of the EPA (40).

Ortolano and Hill (41) reviewed 234 Corps of Engineers environmental statements prepared through August 1971. They suggested that improvements could be made in the description of impacts by reducing levels of generality, dealing with uncertainty, and identifying the recipients of impacts. Attention was focused on educational needs with regard to basic biological concepts. It was noted that applicable water quality standards should be specified for the potentially impacted area. Finally, mention was made of the desirability of considering the impact of alternative project operating policies.

Pearson reported on a review of 50 EISs in 1973 (42). It was noted that inadequacies exist in describing the probable (or possible) impacts, especially secondary impacts; that qualitative statements instead of quantitative presentation are often used; that alternatives are not presented in detailed fashion; and that impact statements as now prepared represent myopic planning and fragmentation of projects. Impact statements can be a valuable aid in decision making, but they have often been written after the decisions have been made. A reason given for these inadequacies is that most statements are authored by persons with no technical environmental expertise.

In a survey of 26 EPA employees responsible for preparing draft and final impact statements at the regional level, Hudson (43) has reported an average length of federal service of 6.5 yr, with 1.4 yr of experience in impact statement work. The average grade level is GS-11. The average number of years of formal schooling per respondee is more than 17 but less than 18 yr. Of the 26 respondees 14 had a bachelor's degree in civil engineering.

The 1973 CEQ annual report (44), which identified several more recent trends in the EIS process, indicated that environmental considerations should be included early in the planning process; economic, technical, and other factors are to be included along with environmental factors in decision making; program statements are encouraged; secondary environmental impacts should receive greater emphasis; and public participation is encouraged in the environmental impact assessment process. Developments in the courts also were noted in that early cases dealt primarily with the response of agencies to the spirit and intent of the requirements of NEPA. These cases can be characterized as being basically administrative in scope. More recent litigation has dealt with the technical content of statement, and the cases are related to the substantive areas of impact assessment.

A survey of agency follow-up procedures during project implementation was conducted by Hudson during 1973 (45). The issue is who has the responsibility to ensure that a project is executed in accordance with the environmental analysis and mitigation measures discussed in the EIS. From a survey data base of 33 federal

Table 1-2 Environmental Impact Statements by Project Type Filed Annually to July 1, 1974

Project type	1970	1971	1972	1973	1974
Roads	41	1,123	543	305	150
Watershed, flood control	50	299	127	170	83
Energy-related projects	36	59	128	74	53
Airports	15	141	119	96	41
Navigation	47	93	83	93	66
Parks, wildlife refuges	2	24	84	111	36
Pesticides, herbicides	2	16	26	15	12
Timber management	5	1	26	58	54
All others	117	193	235	226	152

agencies, it was concluded that agencies do recognize that their responsibilities for protection extend beyond the mere preparation of an EIS. However, agencies have not developed adequate policies and procedures, including explicit monitoring and evaluation, to ensure that provisions or conditions contained in their EISs are subsequently carried out.

About 5,500 impact statements were prepared through July 1, 1974 (46). The total includes 2,100 draft statements and 3,400 final statements. Table 1-2 shows the number of environmental statements by project type filed annually to July 1, 1974. Figure 1-1 presents the same information in graphical form. Thc peak year for impact statement filing was 1971, representing the first major release of impact

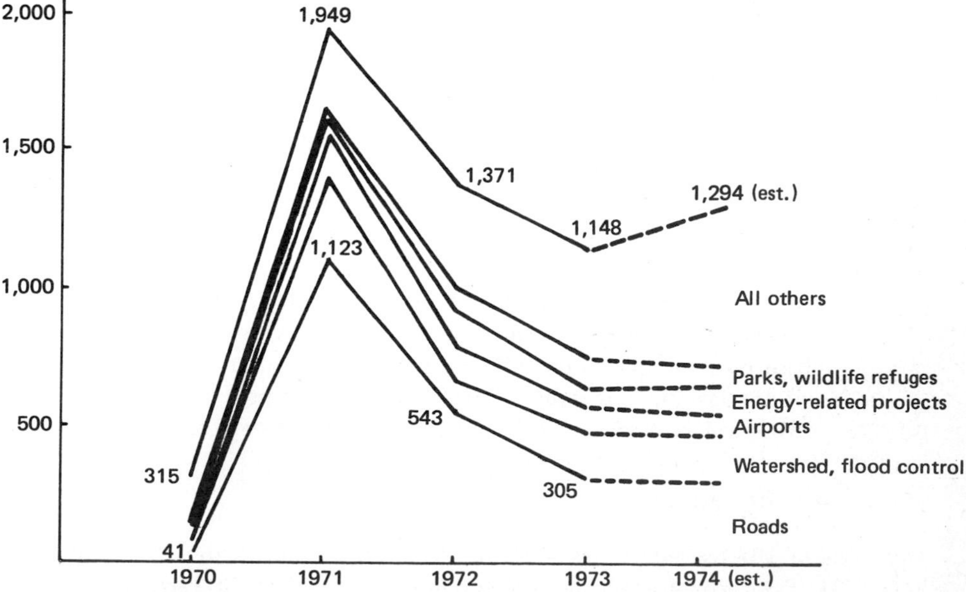

Figure 1-1 Environmental impact statements by project type filed annually to July 1, 1974.

statements following the passage of NEPA. The current pattern is for approximately 1,200 impact statements to be filed annually.

Concern is often expressed regarding the cost of impact statement preparation, the time required, and the effect of the impact statement process on agency decision making. Although data are minimal, some recent studies have provided information regarding cost and time for preparation.

The CEQ has estimated that the entire EIS process costs $65 million annually (47). Since about 1,300 statements are released per year, either as draft or final statements, the average cost per statement is $50,000. Hudson (48) has reported an average cost of $31,000 for 21 final statements prepared in the southeastern United States; however, this figure does not include expenses associated with agency reviews of impact statements prepared by a different agency. From August 1971 to July 1973 the average cost of preparation by the U.S. Navy of an EIS was $7,300 (49). There was considerable variance in preparation cost, ranging from a minimum of $200 to a maximum of almost $50,000.

The length of time required for preparation of an EIS is a function of project size and the point in the planning process when the impact statement is prepared. Some impact statements have been prepared in a matter of weeks, while others have required several years of effort on the part of a number of individuals. Hudson (50) has reported that on an average 8 workmonths are required per statement prepared by EPA regional offices. Other agencies require 18 workmonths per statement. About two-thirds of the time is spent on the draft statement and one-third on the final statement.

No cost-benefit analysis has ever been made for the EIS process. However, there are some examples available on the influence of NEPA. The fourth annual CEQ report (44) indicates that NEPA has influenced decision making in the Corps of Engineers, since 24 projects have been dropped due to the potential adverse impacts, 44 projects have been temporarily or indefinitely delayed, and 197 projects have been significantly modified. The Corps of Engineers has defined three phases in their EIS work since January 1, 1970 (51). These phases include a period of interpretation in 1970, a procedural phase in 1971 and 1972, and an integrated planning phase since 1972. The period of interpretation included considerable discussion on requirements, writing of impact statements on backlogged projects, and development of written procedures and requirements. During the procedural phase emphasis was placed on adhering to the letter of the law, and voluminous impact statements were prepared and many court cases evolved. The integrated planning phase places emphasis on fulfilling not just the letter of the law, but also the intent of the law. In this regard Corps procedures are now evolving that ensure the EIS process will be accomplished early in the planning stage for a given project.

STATE ENVIRONMENTAL POLICY ACTS

Beginning in 1970 several states adopted legislation equivalent to NEPA at the state level (52). The chronological order of development is shown in Table 1-3 (53). As of November 1973, 16 states plus Puerto Rico had passed state environmental

Table 1-3 Chronological Order of State Environmental Policy Act Passage

State or territory	Date
Puerto Rico	June 1970
California	1970
Montana	March 1971
New Mexico	April 1971
Washington	August 1971
Hawaii[a]	August 1971
Wisconsin	1971
North Carolina	1971
Michigan[a]	September 1971
Indiana	1972
Texas[b]	March 1972
Virginia	April 1972
Massachusetts	July 1972
Connecticut[a]	October 1972
Maryland	April 1973
Minnesota	May 1973
New Jersey[a]	October 1973

[a]Executive order or directive.
[b]Interagency Council on Natural Resources and Environment policy statement.

policy acts or their equivalent. Seven other states had attempted passage of such acts but had been unsuccessful. At least 20 other states are considering passage of environmental policy acts (54). State environmental policy acts are basically patterned after NEPA but are applicable to state funds for proposed actions. The requirements of the state acts do not replace the EIS process as required by the NEPA.

The private sector is obligated to provide EISs under rather broad circumstances in California, Montana, and Puerto Rico (55). The Indiana law precludes state agencies from requiring an environmental statement from a licensee prior to issuance of a permit or license.

FUTURE TRENDS IN THE ENVIRONMENTAL IMPACT ASSESSMENT PROCESS

The environmental impact assessment process is becoming an integral part of the planning process for federal agencies within the United States. The process is being accomplished at the initial stages of project planning as opposed to an after-the-fact statement prepared in accordance with the letter of the law of NEPA. Future trends indicate a focus on regional impacts, greater public involvement, more impact statements from the private sector, and more court cases dealing with the substantive issues of environmental impact assessment. It is possible that detailed requirements for separate impact statements will be altered in the future in response to the degree that environmental impact assessment becomes a part of project planning documentation (56).

SELECTED REFERENCES

1 *Federal Water Pollution Control Act Amendments of 1972*, PL 92-500, 92d Cong., S. 2770, Oct. 18, 1972.

2 *The Clean Air Act Amendments of 1970*, PL 91-604, 91st Cong., Dec. 31, 1974.

3 Kreith, Frank: Lack of Impact, *Environment*, vol. 15, no. 1, pp. 26–33, 1973.

4 *The National Environmental Policy Act of 1969*, PL 91-190, 91st Cong., S. 1075, Jan. 1, 1970.

5 Council on Environmental Quality: Preparation of Environmental Impact Statements: Guidelines, *Fed. Reg.*, vol. 38, no. 147, pp. 20550–20562, Aug. 1, 1973.

6 Hesketh, Howard E.: "Understanding and Controlling Air Pollution," chap. 3, Ann Arbor Science Publishers, Ann Arbor, Mich., 1972.

7 Nemec, Joseph, Jr.: The National Environmental Policy Act of 1969 and the Engineering Curriculum, *Civil Eng.*, pp. 64–65, Mar. 1973.

8 Ditton, Robert B.: National Environmental Policy Act of 1969 (PL 91-190): Bibliography on Impact Assessment Methods and Legal Considerations, *Counc. Planning Libr., Exchange Biblio.* 415, June 1973.

9 Viohl, Richard C., Jr., and Kenneth G. M. Mason: Environmental Impact Assessment Methodologies: An Annotated Bibliography, *Counc. Planning Libr., Exchange Biblio.* 691, Nov. 1974.

10 "A Handbook Approach to the Environmental Impact Report, 2d ed., Garing, Taylor and Associates, Arroyo Grande, Calif., 1974.

11 Caldwell, Lynton K.: The National Environmental Policy Act: Status and Accomplishments, *Proc. 38th N. Am. Wildlife Natural Resources Conf.*, Wildlife Management Institute, Washington, D.C., 1973.

12 Andrews, Richard N. L.: "Environmental Policy and Administrative Change: The National Environmental Policy Act of 1969, 1970–1971," pp. 76–109 and app. A, Ph.D. dissertation, University of North Carolina at Chapel Hill, 1972.

13 Yannacone, Victor J., Jr., and Bernard S. Cohen: "Environmental Rights and Remedies," chap. 5, The Lawyers Co-operative Publishing Co., Rochester, N.Y., 1972.

14 Andrews, Richard N. L.: Impact Statements and Impact Assessment, paper presented at Engineering Foundation Conference on Preparation of Environmental Impact Statements, Henniker, N.H., July 29–Aug. 4, 1973.

15 Barfield, Claude E.: Environment Report/Water Pollution Act Forces Showdown in 1973 Over Best Way to Protect Environment, *Nat. J.*, pp. 1871–1882, Dec. 9, 1972.

16 Johnson, William K.: Environmental Litigation: Lessons From the Courts, *Civil Eng.*, pp. 55–58, Jan. 1972.

17 Anderson, Frederick R.: "NEPA in the Courts," Resources for the Future, Washington, D.C., 1973.

18 Green, Harold P.: "The National Environmental Policy Act in the Courts (January 1, 1970–April 1, 1972)," The Conservation Foundation, Washington, D.C., May 1972.

19 Lynch, Robert S.: Complying with NEPA: The Tortuous Path to an Adequate Environmental Impact Statement, *Ariz. Law Rev.*, vol. 14, no. 4, pp. 717–745, 1973.

20 Wilmer, John W., Jr.: "Handbook of Judicial Decisions (through 20 October, 1972) Involving Environmental Impact Statements," Res. Contrib. 224, Institute of Naval Studies, Center for Naval Analyses, Arlington, Va., Nov. 1972.

21 Best, Judith A.: NEPA Impact Statements: Agency Efforts to Escape the Burden, *Natl. Tech. Info. Svc. Rept.* PB-227-807, June, 1972.

22 "Application of the National Environmental Policy Act to EPA's Environmental Regulatory Activities," Publ. PB-231-158, Environmental Protection Agency, Washington, D.C., Feb. 1973.

23 Baumgold, Marian S. and Gordon A. Enk (eds.): "Toward a Systematic Approach to Environmental Impact Review," The Institute on Man and Science, Rensselaerville, N.Y., June 1972.

24 "Environmental Impact and Related Statements," app. F, Policy and Procedure Memorandum 90-1, Federal Highway Administration, U.S. Department of Transportation, Washington, D.C., Sept. 1972.

25 Best, Judith, A.: The National Environmental Policy Act As A Full Disclosure Law, *Natl. Tech. Info. Svc. Rept.* PB-227-809, Dec. 1972.

26 Dickert, Thomas G., and Jens C. Sorensen: Some Suggestions on the Content and Organization of Environmental Impact Statements, in Thomas G. Dickert (ed.), "Environmental Impact Assessment: Guidelines and Commentary," p. 39, University Extension, University of California, Berkeley.

27 Kennedy, William V., and Bruce B. Hanshaw: The National Environmental Policy Act of 1969–Its Effectiveness and Limitations, unpublished paper for Western Pennsylvania Conservancy, Pittsburgh, Pa., 1973.

28 Hopkins, Lewis D.: "Environmental Impact Statements: A Handbook for Writers and Reviewers," Rept. IIEQ 73-8, Illinois Institute for Environmental Quality, Chicago, Ill., Aug. 1973.

29 Council on Environmental Quality: Guidelines for Statements on Proposed Federal Actions Affecting the Environment, *Fed. Reg.*, vol. 36, pp. 7724–7729, Apr. 23, 1971.

30 Jenny, Brian P.: CEQ–A View From the Top, in "How Effective are Environmental Impact Statements," pp. 11–14, Water Resources Research Institute, Oregon State University, Corvallis, July 1973.

31 "Reorganization Plan No. 3," The White House, Washington, D.C., July 9, 1970.

32 "Environmental Impact Statement Guidelines," rev. ed., p. 120, region X, Environmental Protection Agency, Seattle, Wash., Apr. 1973.

33 U.S. Environmental Protection Agency: Preparation of Environmental Impacts Statements: Interim Regulations, *Fed. Reg.*, vol. 38, pp. 1696–1712, Jan. 17, 1973.

34 Bowman, Wallace D.: The National Environmental Policy Act, in "Congress and the Nation's Environment–Environmental and Natural Resources Affairs of the 92nd Congress," chap. 38, pp. 1023–1046, U.S. Government Printing Office, Washington, D.C., 1973.

35 Moore, Sheppard N.: The Environmental Impact Statement Process, in Martin P. Wanielista and Waldron M. McLellon (eds.), "Proceedings of Workshop on

Environmental Impact Statement," pp. 7–20, Florida Technological University, Orlando, June 1973.

36 Council on Environmental Quality: "Environmental Quality, the Third Annual Report of the Council on Environmental Quality," chap. 7, U.S. Government Printing Office, Washington, D.C., Aug. 1972.

37 Ray, Hurlon C.: "Reviewing Environmental Impact Statements at the Regional Level," U.S. Environmental Protection Agency, region X, Seattle, Wash., Apr. 1972.

38 Comptroller General of the United States: Improvements Needed in Federal Efforts To Implement the National Environmental Policy Acts of 1969, report to the Subcommittee on Fisheries and Wildlife Conservation, Committee on Merchant Marine and Fisheries, House of Representatives, Washington, D.C., May 18, 1972.

39 Comptroller General of the United States: Adequacy of Selected Environmental Impact Statements Prepared Under the National Environmental Policy Act of 1969, report to the Subcommittee on Fisheries and Wildlife Conservation, Committee on Merchant Marine and Fisheries, House of Representatives, Washington, D.C., Nov. 27, 1972.

40 Hudson, Donald Ray: "Environmental Management and Public Policy: An Analysis of the Environmental Impact Statement Process with Emphasis on Procedures of the Environmental Protection Agency and Federal Agency Activities in the Southeastern United States," p. 267, Ph.D. dissertation, Georgia State University, Atlanta.

41 Ortolano, Leonard, and W. W. Hill: "An Analysis of Environmental Statements for Corps of Engineers' Water Projects," Rept. 72-3, U.S. Army Engineers' Institute for Water Resources, Alexandria, Va., 1972.

42 Pearson, James R.: Impact Statements–Present and Potential, *J. Prof. Activities, Proc. Am. Soc. Civil Eng.*, vol. 99, no. PP4, pp. 449–455, Oct. 1973.

43 Hudson, *op. cit.*, pp. 237–241.

44 Council on Environmental Quality: "Environmental Quality, the Fourth Annual Report of the Council on Environmental Quality," pp. 234–248, U.S. Government Printing Office, Washington, D.C., Sept. 1973.

45 Hudson, *op. cit.*, pp. 190– 192.

46 Council on Environmental Quality: *102 Monitor*, vol. 4, no. 7, pp. 116–117, Aug. 1974.

47 Anon.: Environmental Impact Statements Discussed, *Civil Eng.*, pp. 61–65, Feb. 1973.

48 Hudson, *op. cit.*, pp. 280–282.

49 "A Study of the Implementation of the National Environmental Policy Act by the United States Navy," p. 110, report prepared for Council on Environmental Quality by Presearch, Silver Spring, Md., Mar. 1974.

50 Hudson, *op. cit.*, pp. 242 and 279.

51 Ash, C. Grant: Three-Year Evolution, *Water Spectrum*, pp. 28–35, 1974.

52 Enk, Gordon: "Beyond NEPA–Criteria for Environmental Impact Review," The Institute on Man and Science, Rensselaerville, N.Y., May 1973.

53 Goidell, Lewis C.: "Analysis of State Environmental Policy Acts, M.S. thesis, University of Oklahoma, Dec. 1973.

54 Trzyna, Thaddeus C.: "Environmental Impact Requirements in the States: NEPA's Offspring," Rept. EPA-600/5-74-006, U.S. Environmental Protection Agency, Washington, D.C., Apr. 1974.

55 Hudson, *op. cit.*, pp. 127–128.

56 Curlin, James W., and H. Steve Hughes: "National Environmental Policy Act of 1969: Analysis of Proposed Legislative Modifications–First Session, 93rd Congress," U.S. Government Printing Office, Washington, D.C., June 1973.

Chapter 2

Framework for Environmental Assessment

Assessments of the environmental impacts of proposed actions require systematic, reproducible, and interdisciplinary approaches. Systematic denotes an all-inclusive, orderly, and scientific consideration of potential impacts on the physical, biological, cultural, and socioeconomic aspects of the environment. The results of an assessment should be reproducible by other groups of investigators. Finally, inputs are required from many disciplines to ensure that a complete analysis has been accomplished.

In this chapter a rational framework for conducting an environmental assessment is presented. Adherence to the steps in the framework yields a systematic, reproducible, and interdisciplinary assessment.

ENVIRONMENTAL ASSESSMENT PROCESS

Figure 2-1 shows five activities in the environmental assessment process. Certain basics, which are required to accomplish an environmental assessment, are related to the description of the environmental setting, impact prediction and assessment, and preparation of the EIS.

In order to be able to predict and assess the impacts associated with a proposed action, it is necessary to describe the environmental setting in which the proposed action is to take place. This gives the base-line information against which

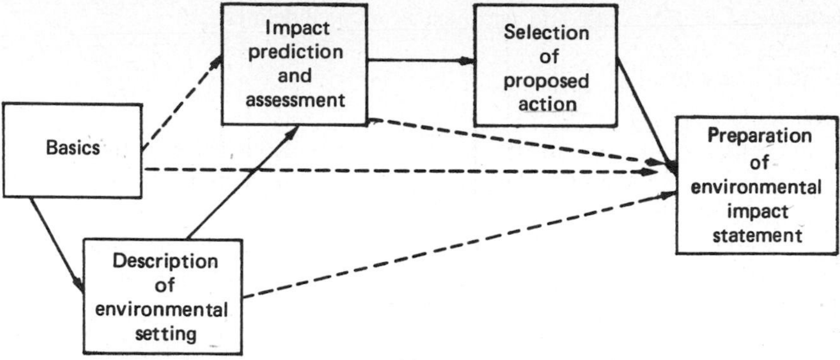

Figure 2-1 The environmental assessment process.

prediction and assessment can be made. The description of the environmental setting also provides input information for preparation of the EIS.

The major step in the environmental assessment process, and the step that requires the greatest degree of scientific application of technology, is impact prediction and assessment. This particular step involves projecting the environmental setting into the future without the proposed action and then performing the necessary calculations or studying the approaches for actually predicting the impact of the proposed action and assessing the consequences. Information from this step becomes a part of the next step involving aggregation of information and selection of the proposed action; also the impact and assessment information becomes a part of the EIS.

The next step is the aggregation of the impact information on each alternative. Based on this aggregated information, as well as on other decision factors such as technical and economic considerations, the proposed action is selected. Information about this step of the environmental assessment process also becomes a part of the EIS.

Finally, the last step involves the preparation of a draft EIS in accordance with specified guidelines of CEQ and the respective agency. The draft impact statement is circulated for review and comment, and following the incorporation of the review comments, a final EIS is prepared and filed with the CEQ. Thirty days following the filing, the proposed action may be initiated.

It is realized that this environmental assessment process is idealized and that it assumes the environmental assessment is conducted early in project planning so that environmental considerations can actually become a part of the decision among alternatives for solving a particular problem. Strict adherence to these steps is not absolutely necessary in all cases; however, this does represent an approach that can be used and is generally applicable.

BASICS TO THE PROCESS

There are several prerequisites that must be fulfilled for a proper environmental assessment to be accomplished. These prerequisites or basics are as follows:

Knowledge of NEPA, CEQ, and pertinent agency guidelines.
Knowledge of study.
Interdisciplinary team.

Table 2-1 (1) contains a list of current agency guidelines. Knowledge of the relevant guidelines enables more effective planning of the intermediate steps, which ultimately lead to the preparation of EIS.

Since different types of projects exert different impacts upon the environment and different alternatives for a given need exert different environmental impacts upon the environment, it is necessary to be familiar with the need for the project and the general types of possible solutions. This concept is shown in Fig. 2-2. After the identification of the need for a project or a study, it is necessary to identify certain possible alternative solutions for meeting the project need (2), including alternatives that might be outside the legislative responsibility of the proposing agency (3). Since one of the alternatives that must be considered in a proposed action is the no-project alternative, it is necessary to include that alternative along with other technical and managerial possibilities (3). Complete knowledge of the types of impacts associated with various alternatives is necessary. For example, alternatives to a multipurpose reservoir project include levees, channel straightening and deepening, single-purpose reservoirs, and development of alternative water supplies. The environmental impacts of the construction of a levee differ from those associated with the construction of a multipurpose reservoir. Therefore it is necessary to know the types of alternatives that are under consideration.

Based on the general knowledge of project need and possible alternative solutions, it is necessary to assemble an interdisciplinary team to work on the particular project (4). An interdisciplinary team should, at a minimum, consist of a physical scientist or engineer, a biologist, and an archeologist. Additional inputs are appropriate for many types of projects from economists, sociologists, geographers, planners, geologists, and many others. The most appropriate interdisciplinary team for a given project should be assembled after information is aggregated regarding the type of project to be considered.

DESCRIPTION OF ENVIRONMENTAL SETTING

A proper description of the environmental setting can provide base-line information necessary to assess the predicted impacts associated with the various alternatives under consideration. There is no single list of environmental items that is appropriate for every environmental assessment. A listing of environmental items should be developed on a project-by-project basis or at least a project-type-by-project-type basis. Four bases for the items to be included in the environmental setting are

1 Knowledge of impacts.
2 Guidelines.
3 Other EISs.
4 Methodologies for impact analysis.

Table 2-1 Agency NEPA Procedures as of August 1, 1974

Agency	Current procedures		Proposed revisions (if any)	
	Date	Citation[a]	Date	Citation[a]
Department of Agriculture				
Departmental	May 29, 1974	39 F.R. 18678		
Agriculture Stabilization and Conservation Service	May 29, 1974	39 F.R. 18678	May 31, 1974	39 F.R. 20490
Animal and Plant Health Inspection Service	Jan. 29, 1974[b]	39 F.R. 3696[b]		
Farmers Home Administration	Aug. 29, 1972	37 F.R. 17459		
Forest Service	May 3, 1973	38 F.R. 20919	Nov. 19, 1973	38 F.R. 31922
Rural Electrification Administration	May 20, 1974	39 F.R. 23240		
Soil Conservation Service	June 3, 1974	7 C.F.R. Part 650 39 F.R. 19646		
Appalachian Regional Commission	June 7, 1971	36 F.R. 23676		
Atomic Energy Commission				
Regulatory	July 18, 1974	10 C.F.R. Part 51 39 F.R. 26279		
Nonregulatory	Feb. 14, 1974	10 C.F.R. Part 11 39 F.R. 5620		
Canal Zone Government	Oct. 20, 1972	37 F.R. 22669		
Central Intelligence Agency	Jan. 28, 1974	39 F.R. 3579		
Civil Aeronautics Board	July 1, 1971	14 C.F.R. §399.110 36 F.R. 12513	May 24, 1974	39 F.R. 18288
Department of Commerce	Oct. 23, 1971	36 F.R. 21368	Dec. 6, 1973	38 F.R. 33625
Department of Defense	Apr. 26, 1974	32 C.F.R. Part 214 39 F.R. 14699		
Corps of Engineers	Apr. 8, 1974	33 C.F.R. §209.410 39 F.R. 12737		
Delaware River Basin Commission	July 11, 1974	18 C.F.R. Part 401 39 F.R. 25473		
Environmental Protection Agency	Jan. 17, 1973	40 C.F.R. Part 6 38 F.R. 1696	July 17, 1974	93 F.R. 26253
Federal Communications Commission	July 24, 1972	37 F.R. 15711		

Table 2-1 (*continued*) Agency NEPA Procedures as of August 1, 1974

Agency	Current procedures		Proposed revisions (if any)	
	Date	Citation[a]	Date	Citation[a]
Federal Power Commission	Dec. 18, 1972	Commission Order No. 415-C 37 F.R. 28412		
Federal Trade Commission	Nov. 19, 1971	16 C.F.R. §1.81-1.85 36 F.R. 22814		
General Services Administration			Apr. 16, 1974	GSA Order ADM 1095 39 F.R. 13722
Federal Supply Service	Dec. 11, 1971	FSS 1095.1A 36 F.R. 23702		
Transportation and Communications Service	June 30, 1971	TCS 1095.1		
Property Management and Disposal Service	Dec. 30, 1971	PMD Order 1095.1A 36 F.R. 23704		
Public Buildings Service	Mar. 2, 1973	PBS Order 1095.1B		
Department of Health, Education, and Welfare				
Departmental	Oct. 17, 1973	HEW General Administration Manual, chaps. 30-10 through 30-16		
Food and Drug Administration	Mar. 15, 1973	21 C.F.R. Parts 6,601 38 F.R. 7001	Apr. 16, 1974	39 F.R. 13741
Department of Housing and Urban Development	July 18, 1973	38 F.R. 19182	Feb. 22, 1974	39 F.R. 6815
Department of the Interior				
Departmental	Sept. 27, 1971	36 F.R. 19343		
Bonneville Power Administration	Jan. 19, 1972	37 F.R. 815		
Bureau of Indian Affairs	Sept. 17, 1970	Departmental Manual Release		
Bureau of Land Management	July 31, 1974	Departmental Manual Release		
Bureau of Mines	Feb. 9, 1972	37 F.R. 2895		
Bureau of Outdoor Recreation	Mar. 24, 1972	37 F.R. 6501		
Bureau of Reclamation	Jan. 18, 1972	37 F.R. 1126		
U.S. Fish and Wildlife Service	Dec. 1971	37 F.R. 207		
Geological Survey	Mar. 11, 1972	37 F.R. 5263		
National Park Service	July 29, 1974	Internal National Park Service Manual		

Interstate Commerce Commission	Mar. 28, 1972	49 C.F.R. §1100.250 37 F.R. 6318		
Department of Justice (Law Enforcement Assistance Administration)	Feb. 6, 1974	28 C.F.R. Part 19 39 F.R. 4736		
Department of Labor	Mar. 15, 1974	29 C.F.R. Part 1999 39 F.R. 9959		
National Aeronautics and Space Administration	Apr. 10, 1974	14 C.F.R. § 1204.11 39 F.R. 12999		
National Capital Planning Commission	Aug. 1972	37 F.R. 16039		
National Science Foundation	Jan. 28, 1974	45 C.F.R. Part 640 39 F.R. 3544		
Small Business Administration	Oct. 20, 1972	37 F.R. 22697		
Department of State				
Departmental	Aug. 31, 1972	37 F.R. 19167		
International Boundary and Water Commission	Mar. 14, 1974	39 F.R. 9868		
Tennessee Valley Authority	Feb. 14, 1974	39 F.R. 5671		
Department of Transportation				
Departmental	Nov. 1, 1973	38 F.R. 30215		
Federal Aviation Administration	June 19, 1973	FAA Order 1050.1A		
Federal Highway Administration	Sept. 7, 1972	Policy and Procedure Manual (PPM) 90-1 37 F.R. 21803	Nov. 1, 1973	38 F.R. 30192
United States Coast Guard	Dec. 11, 1973	Commandant Instruction 5922.10A Series 38 F.R. 34135		
Urban Mass Transportation Administration	Feb. 1, 1972	DOT Order 5610.1 37 F.R. 22692		
National Highway Traffic Safety Administration	Nov. 20, 1972	DOT Order 560-1 38 F.R. 30215	Dec. 21, 1973	38 F.R. 35018
Saint Lawrence Seaway Development Corporation	Nov. 1971	Procedure SLS 2-5610.1A	Nov. 21, 1973	38 F.R. 32179
Department of the Treasury	Apr. 26, 1974	39 F.R. 14796		
Internal Revenue Service	Aug. 12, 1971	36 F.R. 15061		
Veterans Administration	June 17, 1974	39 F.R. 21016		
Water Resources Council	Feb. 10, 1971	36 F.R. 23711		

[a]Citations are given to an agency's procedures where they have been published in the *Federal Register* or otherwise formally issued.
[b]These procedures, while issued in proposed form, are currently being followed on an interim basis.

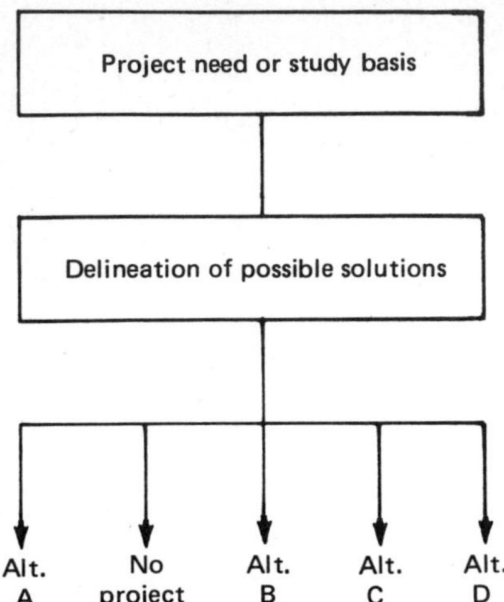

Figure 2-2 Knowledge of study.

One of the best approaches for assembling a list of items is to have knowledge of the types of impacts associated with given types of projects. For example, in the preparation of airport environmental impact statements it is necessary to consider the impact of airport operations on both air quality and noise; thus it is required to have a description of the air quality and the noise levels in the vicinity of the proposed airport prior to its development.

Many guidelines contain detailed descriptions of the items to be included in an environmental setting. Guidelines vary as to the detail suggested for these various environmental items. At a minimum it is necessary to cover those subjects specified in the guidelines associated with the agency proposing the action.

Great value can be derived from analyzing the EIS work by others. The approximately 6,000 EISs that have been prepared (5) represent an extensive pool of information, which, if properly analyzed, can be of value in preparing a description of the environmental setting. For example, if the project is to describe the environmental setting associated with a dredging operation, it would be of great value to examine the environmental setting in several dredging impact statements and to develop a master list of items to be included in the statement at hand.

Another approach that can serve as a basis for describing the environmental setting is the listing of environmental items included in several developed methodologies for impact analysis (6). Many of the methodologies contain from 50 to 100 different environmental items for which impact prediction and assessment is suggested. In order to be able to properly accomplish this assessment, environmental setting information must be accrued for each item in the methodology.

Using the four suggested bases for developing a list of environmental items, it is then desirable to compartmentalize the environment into several areas. One suggested arrangement is to divide the environment into physical-chemical parameters or items, biological parameters, cultural parameters, and socioeconomic parameters.

IMPACT PREDICTION AND ASSESSMENT

To properly predict and assess the impact of a proposed action, it is first necessary to describe the future environmental setting in the area without the project. Various techniques are available for projecting current conditions into the future based on historical trends. Less quantitative approaches are also available for predicting alternative futures based on management decision without the continuance of historical trends.

Prediction and assessment of the impact of each alternative on the physical-chemical, biological, cultural, and socioeconomic environment are required. There are many scientific approaches and models that can be used to predict impacts on the air environment, water environment, and noise environment. Less quantitative, although not entirely subjective, methods are available for predicting the impacts of alternatives on the biological environment. The prediction of impacts on the cultural environment is largely associated with determination of the geographic distribution of cultural resources in the project area. Since this determination requires on-site survey work, this impact prediction is perhaps the most precise one that can be accomplished. However, since many studies do not have the necessary funding for on-site archeological surveys, if scientific extrapolation has to be made from known history and known sites in the general vicinity, this impact prediction then becomes the least precise area of prediction. Finally, prediction of impacts on the socioeconomic environment does have technological bases. However, most of the projections for socioeconomic considerations, such as population and population distribution, are largely based on historical trends.

Assessment of the consequences of predicted impacts is approached in different ways depending upon the environmental compartment. For example, predicted changes in air and water quality can be assessed against environmental quality standards, whereas predicted changes in land use and the resultant consequences on floral and faunal species are more difficult to assess.

Most of the methodologies of impact analysis involve the concept of an impact scale and an impact importance (6). Examples of scaling include comparison with established environmental quality standards, use of straight-line interpolations, and use of environmental parameter functions. Professional judgment can also be used in the absence of any standards or formalized assessment methodologies. The subjective assignment of significance or importance to each of the environmental items is an important consideration. Perhaps the most-used technique involves considering each item as having nearly equal importance, with focus being given to the major items determined through professional judgment. More formalized approaches have been developed, including ranked pairwise comparison and importance grouping (6).

SELECTION OF PROPOSED ACTION

In an idealized environmental assessment, selection of the proposed action should be accomplished only after consideration of the environmental, technical, and economic factors associated with each alternative for meeting the project need (3). In many instances selection of the proposed action is made prior to the environmental assessment; however, current CEQ guidelines encourage the selection based on grounds other than technical and economic factors (3). The first element of this step is aggregation and display of the comparative environmental impacts of each alternative, including the no-project alternative. Various techniques are available for aggregating the information, including the multiplication of scaling and importance values in some techniques and the use of professional judgment in others (6).

In addition to the relative environmental impacts associated with alternatives, information should also be displayed on other decision factors for each alternative, including unique technical considerations or technical difficulties associated with an alternative; economic analyses, perhaps based on the benefit–cost ratio or excess benefits minus costs; and general public preference for an alternative if public meetings have been held in conjunction with the proposed action.

The selection of the alternative to become the proposed action for the project can be based on professional judgment or on some decision-making technique such as the use of weighted rankings (7).

PREPARATION OF ENVIRONMENTAL IMPACT STATEMENT

The last step in the environmental assessment process involves the preparation of a draft EIS, the subjection of this draft statement to review and comment by others, and the preparation of a final EIS and subsequent filing of this final statement with the CEQ (3). The general sequence of events in this step is as follows:

Prepare draft EIS

1 Description of proposed action, project purposes, and environmental setting.

2 Relationship of the proposed action to land-use plans, policies, and controls for the affected area.

3 Probable impact of the proposed action on the environment.

4 Alternatives to the proposed action.

5 Any probable adverse environmental effects that cannot be avoided.

6 Relationship between local short-term uses of the human environment and the maintenance and enhancement of long-term productivity.

7 Any irreversible and irretrievable commitments of resources that would be involved in the proposed action should it be implemented.

8 An indication of what other interests and considerations are thought to offset the adverse environmental effects of the proposed action.

Circulate draft EIS to others for review and comment

1 Federal agencies.

2 State and local agencies.

3 Public and private interest groups.

Prepare final EIS

1 Discuss and incorporate review comments of others.

2 File with CEQ.

The draft EIS should include either separate or combined sections that address each of the eight points shown above. Following preparation of a draft EIS and its internal review by an agency, it should be circulated to others for review and comment. The reviewers should include federal agencies as specified by CEQ guidelines, state and local agencies where the proposed action is to occur, and relevant local public and private interest groups such as the Sierra Club, the League of Women Voters, River Basin Development Commissions, and others.

Review comments from these groups should be incorporated into the draft EIS, and a separate section should be written on these review comments and other public participation associated with the proposed action. Then the final EIS should be filed with the CEQ. A filing period of at least 30 days is required prior to initiation of the proposed action.

SUMMARY

This chapter has presented a rational framework for the conduction of an environmental assessment. It represents in an idealized fashion the sequence of events that should be followed to prepare a final EIS.

SELECTED REFERENCES

1 Council on Environmental Quality: "Environmental Quality, the Fifth Annual Report of the Council on Environmental Quality," pp. 382–385, U.S. Government Printing Office, Washington, D.C., Dec. 1974.

2 Leopold, Luna B., et al.: A Procedure for Evaluating Environmental Impact, *Geolog. Surv. Circ.* 645, pp. 2–4, 1971.

3 Council on Environmental Quality: Preparation of Environmental Impact Statements: Guidelines, *Fed. Reg.*, vol. 38, no. 147, pp. 20550–20562, Aug. 1, 1973.

4 *The National Environmental Policy Act of 1969*, PL 91-190, 91st Cong., S. 1075, Jan. 1, 1970.

5 Council on Environmental Quality: *op. cit.*, pp. 388–391.

6 Warner, Maurice L., and Edward H. Preston: A Review of Environmental Impact Assessment Methodologies, Contr. 68-01-1871, U.S. Environmental Protection Agency, Washington, D.C., Oct. 1973.

7 Dean, Burton V., and Meir J. Nishry: Scoring and Profitability Models for Evaluating and Selecting Engineering Projects, *J. Oper. Res. Soc. Am.*, vol. 13, no. 4, pp. 550–569, July–Aug. 1965.

Chapter 3

Description of the Environmental Setting

One of the first steps in the environmental assessment process is to describe the environmental setting for the project study area. This description provides base-line data against which prediction and assessment of the impacts of the proposed action and alternatives can be compared. It is useful in describing the environmental setting to consider arranging the various factors into the following categories: physical-chemical, biological, cultural, and socioeconomic. This chapter is oriented toward defining the environmental setting, by examining the criteria for inclusion or exclusion of specific factors and the suggested approaches for developing a list of environmental parameters. In addition, reference will be made to sources for relevant information and data for numerous environmental factors.

PURPOSES FOR DEFINING THE ENVIRONMENTAL SETTING

There are several purposes for defining the environmental setting. One is to form a basis for assessment of the environmental impact of the proposed action and alternatives, including the no-action alternative. Another is to provide sufficient information so that decision makers and reviewers unfamiliar with the general location can develop an understanding of the project need as well as the environmental characteristics of the study area. Lack of a sufficient setting description is evidenced by the fact that many review comments on draft EISs are related to inadequate descriptions and data for one or more environmental parameters.

In defining the environmental setting for a project, it may be possible to establish the basis for the project need, whether the project involves construction of a highway, reservoir, or sewage-treatment plant; expansion and/or modification of an airport facility; or development of an industrial park. Even though there may be a section in the environmental assessment report that deals specifically with project need, the basis can be delineated in the description of the environmental setting.

One of the most important purposes for describing the environmental setting is to identify any environmentally significant items prior to initiation of the proposed action, as well as to enumerate any potentially critical environmental changes should the project be implemented. Critical environmental changes should be addressed in the prediction and assessment step in the environmental assessment process. Examples of environmentally significant items include stream segments with poor water quality, geographical areas with marginal air quality, rare and endangered plant or animal species, significant historical and/or archeological sites, high unemployment, and societal cohesion.

In a more general sense, it is necessary to prepare a description of the environmental setting in order to comply with the general intent of the NEPA and specific requirements in guidelines issued by the CEQ and various federal agencies. The general intent of the NEPA is that a systematic (all-inclusive) approach be utilized in the environmental assessment process. This systematic approach has been described with subsequent guidelines that are responsive to the NEPA. Basic to the systematic approach is the description of the environmental setting.

INCLUSION OR EXCLUSION OF ENVIRONMENTAL ITEMS

The basic structure of the description of the environmental setting is best determined relative to the project need and potential alternatives for meeting that need. For example, if a project is located in one area of a state, it would be appropriate to describe the setting in terms of the general categories of the physical-chemical, biological, cultural, and socioeconomic environment. Individual factors associated with each of these groups should be discussed under each category. On the other hand, for projects involving several states, such as a gas pipeline or transmission line, it would be appropriate to discuss the environmental setting on either a state-by-state basis or an environmental category basis. In either case it would be necessary to present a series of environmental characteristics in order to appropriately describe the environmental setting.

One of the key aspects in describing the environmental setting is to ensure that all environmental factors that need to be considered are included, while at the same time excluding those items that require extensive effort to procure and interpret but that have little relevance to the environmental impact of the proposed action or any of its alternatives. There are no general guidelines with regard to the number and type of environmental factors that must be included in a description of the environmental setting. In order to aid in the selection process, two questions can be used as criteria for the inclusion or exclusion of various environmental factors:

1 Will the proposed action or any of its alternatives have an impact, either beneficial or detrimental, on the environmental factor?

2 Will the environmental factor exert an influence on project construction scheduling or subsequent operation?

The application of these two questions to extensive lists of environmental factors as contained in various environmental assessment methodologies can be a useful tool for developing a composite of environmental factors to be included in a description of the environmental setting. Several impact assessment methodologies have been developed since 1970, and many of them define from 50 to 100 different environmental items for inclusion in an environmental impact assessment.

An additional area of concern is related to the extent of coverage for a particular environmental factor. There is no simple general rule that can be given in this regard. The 1973 CEQ guidelines suggest that the description of the environmental setting should include information–summary technical data and maps and diagrams where relevant–adequate to permit an assessment of the potential environmental impact of the proposed action and alternatives. Highly technical and specialized analyses and data should more appropriately be placed in appendices. Ample use should be made of footnotes and bibliographic references in order to provide proper documentation for the sources of information utilized (1).

As a practical suggestion in conjunction with describing the environmental setting, it is important that one or more site visits be made to the project study area by each person having responsible roles in the environmental assessment process.

SOME SUGGESTED APPROACHES FOR DEVELOPING A LIST OF ENVIRONMENTAL FACTORS

No single approach for developing a list of environmental factors should be used to the exclusion of other approaches. The most effective means of preparing a list of environmental factors is to employ a combination of approaches specifically tailored for the individual project need and alternatives under consideration. Four approaches that have been useful include using various agency guidelines, general knowledge regarding anticipated impacts of projects of a given type, listed environmental factors and impacts from other EISs for similar projects, and environmental factors that are components of environmental assessment methodologies. Many agencies have done an excellent job of identifying the environmental factors that should be included in a description of the environmental setting for a project. Environmental factors associated with describing the environmental setting for a nuclear power plant (2) include

1 Site location and topography.

2 Regional demography, land and water use–Do population distribution for 10-mi radius and 50-mi radius; do present and projected land use and zoning restrictions for 5-mi radius; indicate water use and surface and groundwater sources within 50-mi radius; note locations of any stack discharges in area.

3 Regional historical, scenic, cultural, and natural landmarks–Check national and state registries of historical places; do archeological survey.

4 Geology (topography, stratigraphy, soils, and rock types)–Relevant to earthquake potential and cooling ponds.

5 Hydrology–Describe physical, chemical, biological, and hydrological characteristics (and their seasonal variations) of surface waters and groundwaters of the site and immediate environs; note existing pollution sources; note applicable low flow; cite any applicable water quality standards.

6 Meteorology–Describe diurnal and monthly averages and extremes of temperature, dewpoint, and humidity, wind speed and direction, atmospheric stability, mixing heights, precipitation, and storms such as hurricanes and tornadoes; also include air quality data, sources of air pollution, and applicable air quality standards.

7 Ecology–Identify important flora and fauna of region and their habitats, distribution, and relationships to other species; note rare and endangered species; show distribution maps; discuss ecological succession and define any preexisting environmental stresses.

8 Background radiological characteristics.

A similar listing for a gas pipeline (3) is as follows:

1 Land features and uses–Identify present uses and describe the characteristics of the land area.

Land uses. Describe the extent of present uses, as in agriculture, business, industry, recreation, residence, wildlife, and other uses, including the potential for development; locate major nearby transportation corridors, including roads, highways, ship channels, and air traffic patterns; locate transmission facilities and their placement (underground, surface, or overhead); identify water resources.

Topography, physiography, and geology. Provide a detailed description of the topographical, physiographical, and geological features within the area of the proposed action. Include U.S. Geological Survey Topographic Maps, aerial photographs (if available), and other such graphic material.

Soils. Describe the physical characteristics and chemical composition of the soils, including the relationship of these factors to land slope.

Geological hazards. Indicate the potential occurrence of geological hazards in the area, such as earthquakes, slumping, landslides, subsidence, permafrost, and erosion.

2 Species and ecosystems–Identify those species and ecosystems that will be affected by the proposed action.

Species. List in general categories, by common and scientific names, the plant and wildlife species found in the area of the proposed action and indicate those having commercial and recreational importance.

Communities and associations. Describe the dominant plant and wildlife communities and associations located within the area of the proposed action. Provide an estimate of the population densities of major species. If data are not available for the immediate area of the proposed action, data from comparable areas may be used.

Unique and other biotic resources. Describe unique ecosystems or rare or endangered species and other biotic resources that may have special importance in the area of the proposed action.

3 Socioeconomic considerations–If the proposed action could have a significant socioeconomic effect on the local area, discuss the socioeconomic future of the area without the implementation of the proposed action; describe the economic development in the vicinity of the proposed action, particularly the local tax base and per capita income; and identify trends in economic development and/or land use of the area, from both an historical and a prospective viewpoint. Describe the population densities of both the immediate and generalized area. Include distances from the site of the proposed action to nearby residences, cities, and urban areas and list the populations of these areas. Indicate the number and type of residences, businesses, and industries that will be directly affected and those requiring relocation if the proposed action occurs.

4 Air and water environment–Describe the prevailing climate and the quality and quantity of the air and water resources of the area.

Climate. Describe the climatic conditions that have prevailed in the vicinity of the proposed action: extremes and means of monthly temperatures, precipitation, and wind speed and direction. In addition, indicate the frequency of temperature inversions, fog, smog, and destructive storms such as hurricanes and tornadoes.

Hydrology and hydrography. Describe surface waters, fresh, brackish, or saline, in the vicinity of the proposed action and discuss drainage basins, physical and chemical characteristics, water use, water supplies, and circulation. Describe the groundwater situation, water uses and sources, aquifer systems, and flow characteristics.

Air, noise, and water quality. Provide data on the existing quality of the air and water [indicate the distance(s) from the proposed action site to monitoring stations] and the mean and maximum noise levels at the site boundaries.

5 Unique features–Identify unique or unusual features of the area, including historical, archeological, and scenic sites and values.

These two lists represent a broad, all-inclusive approach for describing the environmental setting for a given proposed action. Familiarity with the guidelines from several agencies is useful for developing lists of environmental factors.

One of the better ways for identifying environmental factors is to utilize knowledge related to the anticipated impacts of types of projects. Potential environmental impacts of a transportation facility, arranged by phase of project development (4), are

Planning and design phase

1 Impact on land use through speculation in anticipation of development.
2 Impact of uncertainty on economic and social attributes of nearby areas.
3 Impact on other planning and provision of public services.
4 Acquisition and condemnation of property for project, with subsequent dislocation of families and businesses.

Construction phase

1 Displacement of people.
2 Noise.

3 Soil erosion and disturbance of natural drainage.
4 Interference with water table.
5 Water pollution.
6 Air pollution (including dust and dirt and burning of debris).
7 Destruction of or damage to wildlife habitat.
8 Destruction of parks, recreation areas, and historic sites.
9 Aesthetic impact of construction activity and destruction of or interference with scenic values.
10 Impact of ancillary activities (e.g., disposal of earth, acquisition of gravel and fill).
11 Commitment of resources to construction.
12 Safety hazards.

Operation of facility—Direct impacts

1 Noise.
2 Air pollution.
3 Water pollution.
4 Socioeconomic.
5 Aesthetic.
6 Effects on animal and plant life (ecology).
7 Demand for energy resources.

Operation of facility—Indirect impacts

1 Contiguous land use.
2 Regional development patterns.
3 Demand for housing and public facilities.
4 Impact on use of nearby environmental amenities (e.g., parks, woodlands, and recreation areas).
5 Differential usefulness for different economic and ethnic groups (resulting problems and solutions).
6 Impact on life-styles of increased mobility and other factors.
7 Impact of improved facility on transportation and related technological development (and consequent impacts).

A description of the characteristics for each of these potential impact factors would represent a good approach for describing the environmental setting for a project. Another example is related to the general impacts associated with dredging projects, as shown in Fig. 3-1 (5). This figure depicts a network analysis of dredging and focuses on those aspects of dredging that have an impact on various environmental factors. An additional feature of the network analysis is that it can be used to identify the interrelationships of various environmental changes.

Another example of knowledge of anticipated impacts is related to the general effects of impoundments on water quality (6). Considering water quality only, it is known that impoundment of water will lead to the following beneficial effects:

Turbidity reduction.
Hardness reduction.

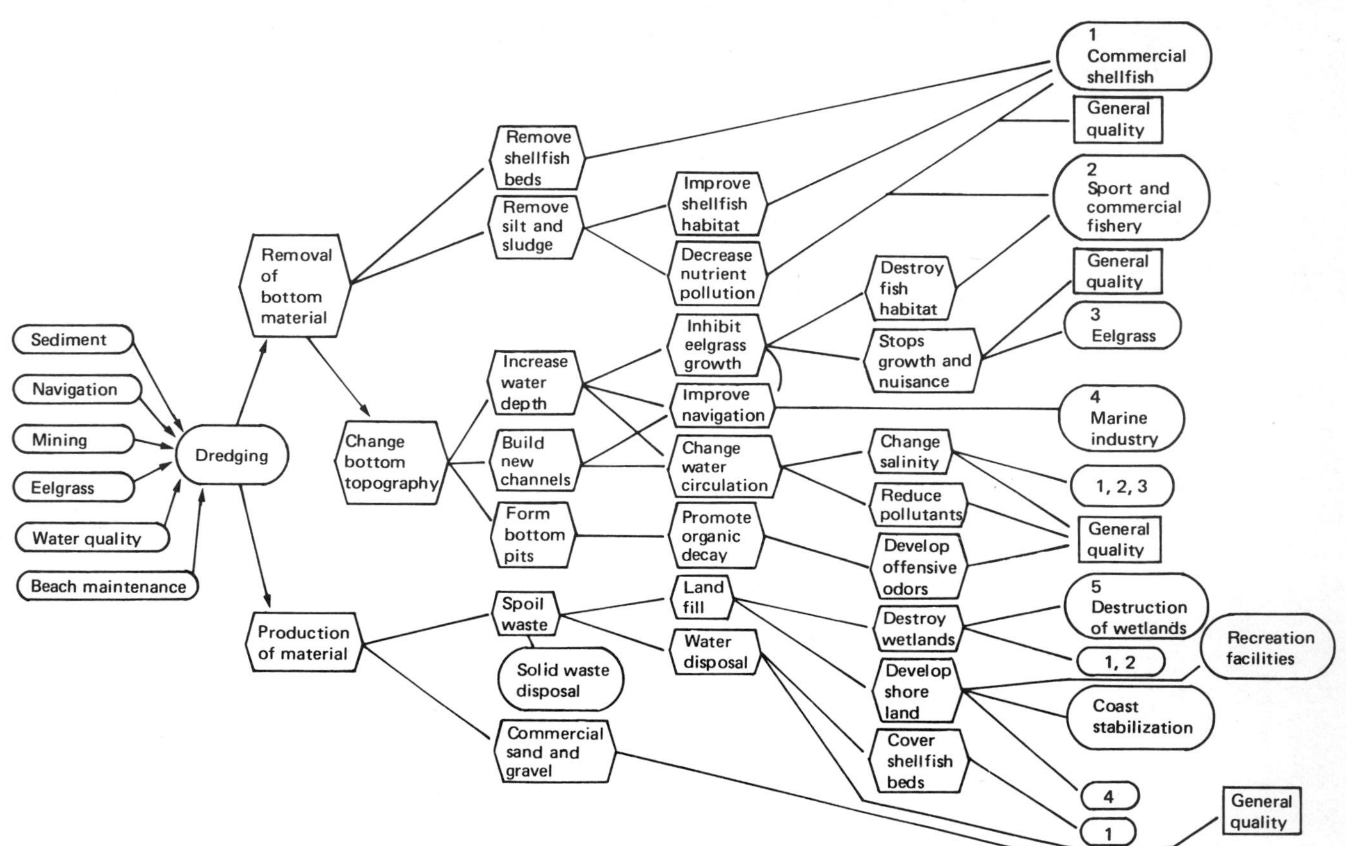

Figure 3-1 A network analysis of dredging.

Oxidation of organic material.
Coliform reduction.
Flow equalization.

The detrimental effects of impoundment include

Lower reaeration.
Buildup of inorganics.
Algae blooms.
Stratified flow.
Thermal stratification.

Perhaps the most significant impact on water quality is due to thermal stratification with the following additional changes in water quality:

Decreased dissolved oxygen in hypolimnion.
Anaerobic conditions in hypolimnion.
Dissolution of iron and manganese from bottom deposits.

In addition to changes in water quality resulting from thermal stratification, changes in mixing patterns also occur. A stratified or density current is the movement, without loss of identity by turbulent mixing at the bounding surfaces, of a stream of fluid under, through, or over a body of fluid with which it is miscible, the density difference being a function of the differences of temperature, salt content, and/or the silt content of the two fluids. In the case of thermal stratification, the difference is a function of differences in temperature. Thermal stratification can result in overflow (warmer water flowing over the surface of colder water), interflow (cool water flowing between upper layers of warmer water and lower layers of colder water), or underflow (cooler water flowing underneath warmer surface water). These conditions of thermal stratification prevent complete mixing from occurring in normal time frames. An additional concern of water impoundment is the reduction in waste assimilative capacity of the body of water being impounded. Waste assimilative capacity is the capacity of a body of water to receive organic wastes and purify itself through natural reaeration. In general, water impoundment decreases the reaeration ability of a body of water, thus reducing the waste loading that the body of water can receive without having the dissolved oxygen concentration decreased below a prescribed water quality standard.

Another example of knowledge of anticipated impacts is related to construction activities. Most proposed actions involve a construction phase, with the activities being similar for many types of projects. Table 3-1 contains an example of potential environmental impacts resulting from construction activities associated with nuclear power plant site preparation and plant and transmission facilities construction (7). Although this list was developed for the nuclear industry, many of the construction activities are relevant to numerous projects. The assemblage of data and information to describe the environmental setting for each of the areas listed

Table 3-1 Potential Environmental Impacts Resulting from Construction Practices

Construction phase	Construction practice	Potential environmental impacts
Preconstruction	Site inventory	Short term and nominal
	Vehicular traffic	Dust, sediment, and tree injury
	Test pits	Tree root injury, sediment
	Environmental monitoring	Negligible if properly done
	Temporary controls	Short term and nominal
	Storm water	Vegetation, water quality
	Erosion and sediment	Vegetation, water quality
	Vegetative	Fertilizers in excess
	Dust	Negligible if properly done
Site work	Clearing and demolition	Short term
	Clearing	Decrease in the area of protective tree, shrub, and ground covers, stripping of topsoil; increased soil erosion, sedimentation, and storm-water runoff; increased stream water temperatures; modification of stream banks and channels, water quality
	Demolition	Increased dust, noise, solid wastes
	Temporary facilities	
	Shops and storage sheds	Increased surface areas impervious to water infiltration, increased water runoff, petroleum products
	Access roads and parking lots	Increased surface areas impervious to water infiltration, increased water runoff, generation of dust on unpaved areas
	Utility trenches and backfills	Increased visual impacts, soil erosion, and sedimentation for short periods
	Sanitary facilities	Increased visual impacts, solid wastes
	Fences	Barriers to animal migration
	Lay-down areas	Visual impacts, increased runoff
	Concrete batch plant	Increased visual impacts; disposal of wastewater, increased dust and noise
	Temporary and permanent pest control (termites, weeds, insects)	Nondegradable or slowly degradable pesticides accumulated by plants and animals, then passed up the food chain to humans; degradable pesticides having short biological half-lives preferred for use
	Earthwork	Long term
	Excavation Grading Trenching Soil treatment	Stripping, soil stockpiling, and site grading; increased erosion, sedimentation, and runoff, soil compaction; increased soil levels of potentially hazardous materials; side effects on living plants and animals, and the incorporation of decomposition products into food chains; water quality
	Site drainage	Long term
	Foundation drainage Dewatering Well points Stream channel relocation	Decrease in the volume of underground water for short and long time periods, increased stream flow volumes and velocities, downstream damages, water quality

Table 3-1 (*continued*) **Potential Environmental Impacts Resulting from Construction Practices**

Construction phase	Construction practice	Potential environmental impacts
	Landscaping	
	Temporary seeding Permanent seeding and sodding	Decreased soil erosion and overland flow of storm water, stabilization of exposed cut and fill slopes, increased water infiltration and underground storage of water, visual impacts
Permanent facilities	Transmission lines and heavy traffic areas	Long term
	Parking lots	Storm-water runoff, petroleum products
	Switchyard	Visual impacts, sediment, runoff
	Railroad spur line	Storm-water runoff
	Buildings	Long term
	Warehouses	Impervious surfaces, storm-water runoff, solid wastes, spillages
	Sanitary waste treatment	Odors, discharges, bacteria, viruses
	Cooling towers	Visual impacts
	Related facilities	Long term
	Reactor intake and discharge channel	Shoreline changes, bottom topography changes, fish migration, benthic fauna changes
	Water supply and treatment	Waste discharges, water quality
	Storm-water drainage	Sediment, water quality
	Wastewater treatment	Sediment, water quality, trace elements
	Dams and impoundments	Dredging, shoreline erosion
	Breakwaters, jetties, etc.	Circulation patterns in the waterway
	Fuel-handling equipment	Spillages, fire, and visual impacts
	Oil-storage tanks, controls, and piping	Visual impacts
	Conveying systems (cranes, hoists, chutes)	Visual impacts
	Waste-handling equipment (incinerators, wood chippers, trash compactors)	Noise, visual impacts
	Security fencing	Long term
	Access road	Increased runoff
	Fencing	Barriers to animal movements
Project closeout	Removal of temporary offices and shops	Short term
	Demolition	Noise, solid waste, dust
	Relocation	Storm water, runoff, traffic blockages, soil compaction

Table 3-1 (*continued*) **Potential Environmental Impacts Resulting from Construction Practices**

Construction phase	Construction practice	Potential environmental impacts
	Site restoration	Short term
	Finish grading	Sediment, dust, soil compaction
	Topsoiling	Erosion, sediment
	Fertilizing	Nutrient runoff, water quality
	Sediment controls	Vegetation
	Preliminary start-up	Short term
	Cleaning	Water quality, oils, phosphate and other nutrients
	Flushing	

under potential environmental impacts in Table 3-1 would be another way of developing a list of environmental factors for consideration.

Over 6,000 draft and final EISs have been prepared since the passage of the NEPA. For certain project types, there may be several hundred impact statements that could be examined for the environmental factors included in the description of the environmental setting, as well as the environmental impacts discussed in the impact statement proper. An example of reported impacts for 55 EISs dealing with Corps of Engineers dams and reservoirs (8) is as follows:

- **I** Direct loss of land and/or productivity.
  - **A** Specified land uses.
    - **1** Agricultural or grazing land.
    - **2** Forests or timberland.
    - **3** Wetlands or marshes.
  - **B** Commercial productivity.
    - **1** Mineral resources (gravel, limestone, oil, gas, etc.).
    - **2** Commercial fisheries.
- **II** Loss or relocation of manufactured structures and archeological or historical sites.
  - **A** Archeological or historical sites.
  - **B** Homes or villages.
  - **C** Highways, railroads, and other transportation facilities.
  - **D** Cemeteries.
  - **E** Recreation facilities.
- **III** Loss of wildlife habitat.
  - **A** Specification of habitat type.
  - **B** Loss of hunting opportunities.
- **IV** Change in aesthetic quality.
  - **A** Decreased aesthetics.
  - **B** Increased aesthetics.
- **V** Loss or inundation of the natural stream.
  - **A** Loss of the stream fishery.
  - **B** Loss of recreation potential.

VI Environmental impacts due to the reservoir per se.
- **A** Substitution of a lake environment for a stream environment.
- **B** Creation of a warm-water fishery.
- **C** Creation of wildlife habitat.

VII Alterations in water quality due to impoundment.
- **A** Thermal stratification.
- **B** Growth of algae.
- **C** Impoundment of nutrients and wastes.
  - 1 Decrease in water quality.
  - 2 Increase in rate of eutrophication.

VIII Impacts resulting from the dam as a barrier.
- **A** Increased sediment deposition.
- **B** Loss of anadromous fish runs.

IX Impacts due to spillways.

X Downstream effects.
- **A** Decreased silt or sediment in downstream channel.
  - 1 Increased erosion downstream.
  - 2 Increased water quality downstream.
- **B** Improvement or enhancement of downstream fishery.
- **C** Flow regulation or low flow augmentation.
  - 1 Improvement of water quality downstream.
  - 2 Improvement of downstream aesthetics.
  - 3 Improvement of recreation downstream.
  - 4 Reduction of mosquito problems downstream.

XI Effects on groundwater recharge.

XII Effects of periodic inundation or a fluctuating shoreline.
- **A** Adverse effects on wildlife.
- **B** Adverse effects on vegetation.
- **C** Decreased aesthetics at low reservoir stages.

Again, the use of these reported impacts would be beneficial in preparing a list of environmental factors for the description of the environmental setting. A list of questions devised from a review of eight airport EISs relative to impacts on artificial and natural environments (9) includes

I Impacts upon human environment.
- **A** Real estate impacts.
  - 1 Any public lands involved within boundaries of proposed project?
  - 2 Number of acres involved in the project?
  - 3 Number of residences within boundaries of project?
  - 4 Number of relocated people as a result of the proposed project?
  - 5 Numbers of replacement housing as a result of the proposed project?
  - 6 Any land-use studies of areas adjacent to the proposed project included?
  - 7 Any roads/highways to be closed due to the project?
  - 8 Any need for road/highway relocation due to the project?
  - 9 Vicinity maps included in statement?
  - 10 Maps showing proposed facility layout?
  - 11 Soil description in statement?
  - 12 Any mention to future expandability of the proposed project?

13 Any mention of methods of disposing of debris from clearing (i.e., burning, burial, solid waste treatment)?

B Aesthetics and visual impacts.

1 Positive or negative effect disclosed from statement regarding land clearing?
2 What physical features will be retained or removed by the project?
3 Any unique interest areas in the boundaries of the project mentioned?
4 Any scenic beauty areas affected by the project?
5 Any recreational areas affected by the project?
6 Any forest areas directly affected by the project (i.e., within the boundaries)?
7 Any landscaping required?
8 Any residential developments directly within project boundaries?
9 Any industrial developments directly within project boundaries?
10 Any hospitals within project boundaries?
11 Any churches within project boundaries?
12 Any schools within project boundaries?

C Community impacts.

1 Any division/disruption of an existing community?
2 Surface transportation due to the airport discussed in relation to traffic congestion?

D Public services impact.

1 Water supply discussed for project facility?
2 Sewage treatment for facility arranged?
3 Solid waste treatment for facility outlined?
4 Liquid waste and accidental spill precautions discussed?
5 Storm drainage (runoff) precautions discussed?

E Displaced persons discussed?

F Noise impact.

1 Any plans or considerations given to compensate for vehicular noise at the facility?
2 Overlay map showing Composite Noise Rating (CNR) zones of project included?
3 Detailed noise pollution analysis included?
4 Survey opinions of similar situations in other locations included in statement?
5 Air traffic pattern in the area changed due to the project?
6 Areas under the new air traffic patterns described thoroughly?
7 Projected air traffic volumes described for affected areas?
8 Argument of "... noise due to takeoff and climbout will be decreased due to runway extension or realignment" used?
9 Any indication given of the land-use compatibility studies for areas adjacent to the project?

G Airport configuration, airspace patterns, and safety problems discussed in the statement?

H Projected level of use for the project included?

1 Expected future operation increases anticipated?
2 Expected future use of facility by larger aircraft discussed in statement?
3 Number of passengers expected to increase as the population area increases?

I Employment impact due to jobs created by project discussed?
J Stimulated population growth in a particular direction discussed?
K Social-psychological impact discussed?

II Natural environmental impacts.
A Wildlife in the project area.
1 Any rare/endangered species indicated in the project area or boundaries?
2 Bird sanctuary or wildlife reservation involved in or near the project?
3 Existing inhabitants seek more suitable habitats in adjacent areas?
4 Any wildlife relocation necessary?
5 Clearing procedures coincide with periods of least biological activity to ease hardships of relocating?
6 Detailed listing of residing wildlife in the area given?
7 Expected change in the wildlife population in the project area?
8 Fewer hunting opportunities due to the project in the area?
9 Fewer fishing opportunities due to the project in the area?
B Water pollution impacts.
1 Any streams/rivers within the project boundaries?
2 Any ponds/lakes within the project boundaries?
3 Any marine life or fish in the project area?
4 Any water relocation necessary due to the project?
5 Any relocation of marine life or fish necessary?
6 Mention made of the existing water-table characteristics?
7 Mention made of adherence to *FAA Circ.* 150/5370-7, Airport Construction Controls to Prevent Air and Water Pollution?
8 Water pollutant statements qualified with specific data?
9 Methods discussed for handling runoff associated with runways, etc.?
10 Mention made of the possibility of water-table pollution?
11 Mention made of associated turbidity due to runoff?
C Timberland impacts.
1 Any timber within the project boundaries?
2 Detailed listing given in statement?
3 Any timber marketable?
4 Timber utilized in landscaping?
5 Mention made of disposal means of cleared timber and brush?
D Any existing physical features (buildings, cemeteries, etc.) in the project boundaries? Any requiring relocation?
E Air pollution impacts.
1 Any mention of the effects of the facility upon the O_2-CO_2 exchange process?
2 Mention made of dust generation during construction?
3 Means for dust alleviation discussed?
4 Predicted pollutant emissions a significant additive over existing pollutant levels?
5 Air pollutant statements qualified with data?
6 Any assessment given to visual impact of particulate smoke trails from operations of aircraft in area?
7 Meteorological effects upon air pollutants discussed (i.e., wind currents relative to adjacent cities)?
8 Meteorological effects upon neighboring cities discussed (carrying added pollutants)?

F Erosion control impact.
  1 Grading of erosion-resistant slopes discussed?
  2 Turfing discussed?
  3 Area topography given?
G General ecological implications (suggestions).
  1 Effect (increase/decrease) on the full productivity of the area?
  2 Influence on migratory and wintering waterfowl in the area?

Another approach that could be used for preparing a list of environmental factors is to select factors utilized in current environmental assessment methods. Only two methods will be cited at this point, although over 50 have been developed since the passage of the NEPA. The following list of 88 environmental items incorporated in the Leopold interaction matrix was prepared for use by the U.S. Geological Survey (10):

I Physical and chemical characteristics.
  A Earth.
    1 Mineral resources.
    2 Construction material.
    3 Soils.
    4 Land form.
    5 Force fields and background radiation.
    6 Unique physical features.
  B Water.
    1 Surface.
    2 Ocean.
    3 Underground.
    4 Quality.
    5 Temperature.
    6 Recharge.
    7 Snow, ice, and permafrost.
  C Atmosphere.
    1 Quality (gases, particulates).
    2 Climate (micro, macro).
    3 Temperature.
  D Processes.
    1 Floods.
    2 Erosion.
    3 Deposition (sedimentation, precipitation).
    4 Solution.
    5 Sorption (ion exchange, complexing).
    6 Compaction and setting.
    7 Stability (slides, slumps).
    8 Stress-strain (earthquake).
    9 Air movements.
II Biological conditions.
  A Flora.

1 Trees.
2 Shrubs.
3 Grass.
4 Crops.
5 Microflora.
6 Aquatic plants.
7 Endangered species.
8 Barriers.
9 Corridors.

B Fauna.
1 Birds.
2 Land animals including reptiles.
3 Fish and shellfish.
4 Benthic organisms.
5 Insects.
6 Microfauna.
7 Endangered species.
8 Barriers.
9 Corridors.

III Cultural factors.

A Land use.
1 Wilderness and open spaces.
2 Wetlands.
3 Forestry.
4 Grazing.
5 Agriculture.
6 Residential.
7 Commercial.
8 Industrial.
9 Mining and quarrying.

B Recreation.
1 Hunting.
2 Fishing.
3 Boating.
4 Swimming.
5 Camping and hiking.
6 Picnicking.
7 Resorts.

C Aesthetics and human interest.
1 Scenic views and vistas.
2 Wilderness qualities.
3 Open space qualities.
4 Landscape design.
5 Unique physical features.
6 Parks and reserves.
7 Monuments.
8 Rare and unique species or ecosystems.
9 Historical or archeological sites and objects.
10 Presence of misfits.

- D Cultural status.
  - 1 Cultural patterns (life-style).
  - 2 Health and safety.
  - 3 Employment.
  - 4 Population density.
- E Constructed facilities and activities.
  - 1 Structures.
  - 2 Transportation network (movement, access).
  - 3 Utility networks.
  - 4 Waste disposal.
  - 5 Barriers.
  - 6 Corridors.

IV Ecological relationships.
- A Salinization of water resources.
- B Eutrophication.
- C Disease-insect vectors.
- D Food chains.
- E Salinization of surficial material.
- F Brush encroachment.
- G Other.

The environment is divided into the following major categories: physical and chemical characteristics, biological conditions, cultural factors, and ecological relationships. Figure 3-2 itemizes 78 environmental factors used in the environmental evaluation system prepared for water resource projects by Battelle-Columbus (11). In this method the environment is divided into ecology, environmental pollution, aesthetics, and human interest. It should be noted that these two methods discussed here have very little emphasis on the socioeconomic environment.

INFORMATIONAL SOURCES FOR ENVIRONMENTAL FACTORS

Appendix C contains a list of informational sources for environmental factors. This list of factors is from the Leopold interaction matrix prepared for the U.S. Geological Survey (12). The sources cited in Appendix C are not intended to be all inclusive, but rather suggestive of agencies where information and data on various environmental factors can be procured. Examination of this appendix indicates that some sources can be employed for many environmental factors, while other sources can provide summary data and information that could be obtained from many individual sources. In other words, it is not necessary to contact all the sources listed for information for each of the 88 items included in Appendix C. Depending upon the type of project, the number of informational sources can be reduced considerably.

SUMMARY

The preparation of a proper description of the environmental setting is one of the key steps in the environmental assessment process. Proper delineation of a list of

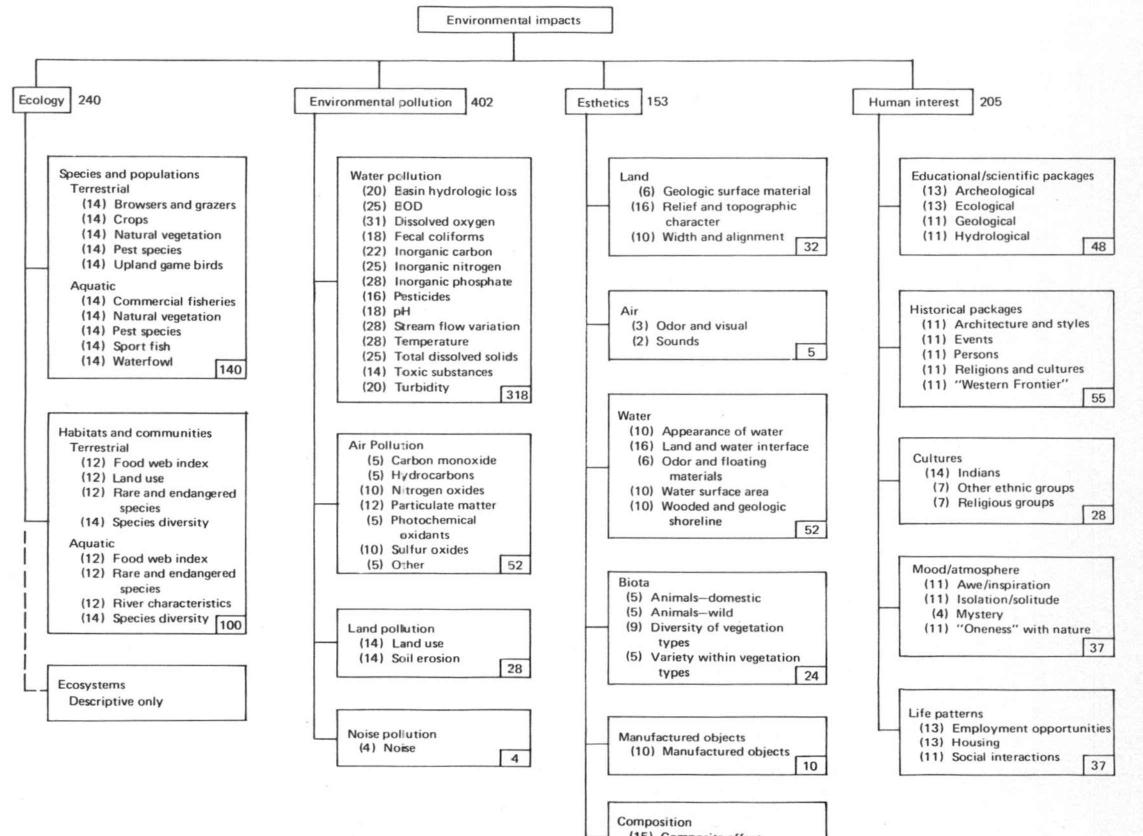

Figure 3-2 Battelle environmental evaluation system for water resource projects.

environmental factors to be addressed, as well as documentation of information and data utilized, is a necessary component of this phase. A summary of detailed technical information and data should be provided, with the referenced information appropriately contained in tables, maps, footnotes, and appendices. Several approaches can be used for developing a list of pertinent environmental factors for a given project. No single approach is universal, and utilization of a combination of approaches leads to the best results. The informational sources for environmental factors are manifold; however, these can generally be narrowed to a minimal number of key sources for obtaining relevant information and data.

SELECTED REFERENCES

1 Council on Environmental Quality: Guidelines for Preparation of Environmental Impact Statements, *Fed. Reg.*, vol. 38, no. 147, pp. 20550–20562, Aug. 1, 1973.

2 Atomic Energy Commission: "Preparation of Environmental Reports for Nuclear Power Plants," Regulatory Guide 4.2, Mar. 1973.

3 Federal Power Commission: Implementation of the National Environmental Policy Act of 1969, Order 485, Order Amending Part 2 of the General Rules to Provide Guidelines for the Preparation of Applicants' Environmental Reports Pursuant to Order 415-C, Washington, D.C., June 7, 1973.

4 A. D. Little, Inc.: "Transportation and Environment: Synthesis for Action: Impact of National Environmental Policy Act of 1969 on the Department of Transportation," 3 vols., prepared for Office of the Secretary, Department of Transportation, July 1971.

5 Sorensen, J. C.: "A Framework for Identification and Control of Resource Degradation and Conflict in the Multiple Use of the Coastal Zone," University of California, Berkeley, June 1971.

6 Krenkel, P. A.: Thermal Pollution, unpublished notes, Vanderbilt University, Nashville, Tenn., 1971.

7 Hittman Associates, Inc.: "General Environmental Guidelines for Evaluating and Reporting the Effects of Nuclear Power Plant Site Preparation, Plant and Transmission Facilities Constructions," pp. B-13–B-18, prepared for Atomic Industrial Forum, Washington, D.C., Feb. 1974.

8 Ortolano, Leonard, and W. W. Hill: "An Analysis of Environmental Statements for Corps of Engineers' Water Projects," Rept. 72-3, U.S. Army Engineers' Institute for Water Resources, Alexandria, Va., 1972.

9 Gilliam, J. D., and L. W. Canter: Comparative Analysis of Eight Airport Environmental Impact Statements, unpublished report, University of Oklahoma, Norman, 1973.

10 Leopold, Luna B., et al.: A Procedure for Evaluating Environmental Impact, *Geolog. Surv. Circ.* 645, 1971.

11 Dee, Norbert, et al.: Environmental Evaluation System for Water Resource Planning, final report prepared by Battelle-Columbus for Bureau of Reclamation, Jan. 31, 1972.

12 Vandertulip, J. A.: Informational Sources for Environmental Factors, unpublished report, International Boundary and Water Commission, El Paso, Tex., 1972.

Chapter 4

Prediction and Assessment of Impacts on the Air Environment

One of the major impacts of many actions is on the air quality in the vicinity of the project area. Construction of highways, airports, dams, waterways, power plants, industrial parks, apartment houses, and pipelines generates construction dusts and exhaust emissions from construction equipment. The operation of airports and power plants and the use of highways and industrial parks also cause emission of gaseous and particulate air pollutants. This chapter is addressed to data needs and associated technology for predicting and assessing the impact of a proposed action on air quality.

BASIC STEPS FOR PREDICTION AND ASSESSMENT

The basic steps associated with prediction of changes in air quality and assessment of the impact of these changes are as follows:

1 Identify air pollutants emitted from the alternatives under consideration for meeting a given need.
2 Describe or determine the existing air quality levels in the area. If possible, present historical trends in air quality. Examine the frequency distribution and the median and mean concentrations for each gaseous or particulate air pollutant that has an ambient air quality standard.

3 Determine the air pollution dispersion potential for the area; this can be accomplished by aggregating information on seasonal or monthly variations of mean mixing depth, inversion heights, wind speeds, high air pollution potential, and episode-days. Historical records of air pollution episodes in the area should be described.
4 Summarize the basic meteorological data for the area; this should include monthly summaries of precipitation, temperature, wind speed and direction, solar radiation, and other parameters if deemed appropriate. Included in this information should be the monthly, seasonal, and annual wind roses. Any unique meteorological phenomena that have occurred in the area should be identified.
5 Procure ambient air quality standards and emissions standards, if they are relevant. Consideration should be given to the time schedule required for meeting these standards.
6 Summarize emission inventory data for the smallest applicable scale of the region and include the regional emission inventory. Identify the major point sources of air pollution in the area and indicate the quantities of pollutants emitted as well as the specific location of these point sources relative to the sites of the alternatives under study.
7 Determine the *mesoscale* impact due to construction and operation of each alternative. This can be accomplished by calculating the estimated annual quantity of air pollutants from each alternative and determining the percentage increase in the regional and local emission inventory for each pollutant emitted. Particular attention should be addressed to increases in priority I or priority II pollutants (these priorities will be defined later).
8 Calculate the ground level concentrations of air pollutants from the alternatives under varied meteorological conditions. Develop isopleths of concentration in the vicinity of the sources of emission. In order to determine the *microscale* impact, compare the calculated air quality levels with the applicable ambient air standards. If emissions standards are applicable to the alternatives, then the emission standards for the action should be considered in relationship to the anticipated emission levels.
9 If ambient air or emission standards are exceeded by the proposed action, consider mitigation or control measures in order to minimize the air quality impact.

These nine steps are directed toward determining the air quality impacts of alternatives and a proposed action on the mesoscale and microscale levels. The mesoscale level assessment measures the contribution of the proposed action to area and regional emission inventories. The microscale level assessment is a comparison of calculated concentration levels of air pollutants at specific locations to applicable ambient air quality standards. Both levels of impact assessment are necessary in order to adequately address the air quality impacts associated with proposed actions. Steps 2–6 should be summarized in the environmental setting and the remainder in the environmental impact section.

The organization of this chapter is primarily oriented to the nine steps identified above. However, prior to the step-by-step discussion and analysis, some brief information on basic air pollution considerations is presented.

BASIC INFORMATION ON AIR POLLUTION

Air pollution may be defined as

> the presence in the outdoor atmosphere of one or more contaminants such as dust, fumes, gas, mist, odor, smoke, or vapor in quantities, of characteristics, and of duration, such as to be injurious to human, plant or animal life or to property, or which unreasonably interferes with the comfortable enjoyment of life and property (1).

Several key ideas are embodied in the above definition of air pollution: (a) the focus is on the outdoor atmosphere and does not include the industrial working environment, (b) air pollution may be caused by single contaminant gases or particulates or combinations of these contaminants, (c) the concentration or quantity of material is a basic determinant in causing air pollutant effects, (d) the time of exposure, or the persistence of a given concentration level of a pollutant, is also a basic determinant in the effects of air pollution, and (e) the effects of air pollutants can occur on living things, inanimate objects, and the aesthetic features of an area.

There are a number of significant dates in the history of air pollution occurrences (2). One of the first recorded events was about 1300, when King Edward I issued a proclamation prohibiting the use of sea coal during sessions of the English Parliament. In 1866 the first paper on the health effects of air pollution was presented, and in 1875 cattle deaths in London were shown to be due to an air pollution episode. In the early 1900s the term "smoke-fog" began to be used, with this term being shortened to "smog," which has become a synonym for air pollution. Some major air pollution episodes occurred in the Meuse Valley in Belgium in 1930, in Donora, Pennsylvania in 1948, and in London, England in 1952. In 1955 the first U.S. legislation dealing with air pollution was enacted, and in 1963 the Clean Air Act was passed. This act was amended in 1967, 1970, and 1974 (3).

The sources of air pollution can be categorized according to type, that is, whether natural or artificial, by number and spatial distribution, or by type of emissions such as gases and particulates. The number and spatial distribution category includes single or point sources, area or non-point sources, and line sources.

The two major classes of gaseous air pollutants are inorganic gases and organic vapors. Examples of widely occurring inorganic gases include sulfur dioxide, oxides of nitrogen, carbon monoxide, and hydrogen sulfide; and organic vapors include hydrocarbons, mercaptans, alcohols, ketones, and esters. Organic vapors are generally localized pollutants. Secondary gaseous air pollutants resulting from photochemical reactions include oxidants, with the primary component being ozone. Sulfur-, nitrogen-, and carbon-containing inorganic gases can be oxidized in the atmosphere to their most oxidized form and then be combined with water vapor to create acidic mists such as sulfuric acid, nitric acid, and carbonic acid.

Particulate air pollutants are any dispersed matter, solid or liquid, in which the individual aggregates are larger than single small molecules (about 0.0002 μm in diameter) but smaller than about 500 μm. Particles persist in the air from a few seconds to several months (4). Particulate matter is basically divided into two broad categories depending upon the sampling technique. Total suspended particulates are those materials that can be filtered from the atmosphere through the use of a large-volume air sampler. Settleable solids ("fallout") or dust fall refers to those materials that are deposited by gravity into a dust-fall sampler over a period of 1 month. The most-used particulate measurement for air quality control is the total suspended particulates.

One of the primary concerns is the effect of air pollutants on aesthetics, economic viability, safety, personal discomfort, and health (5). The aesthetic effects include loss of clarity of the atmosphere due to the presence of particulates and/or photochemical smog and the presence of objectionable odors primarily associated with gases such as ammonia and sulfur-containing mercaptans. Economic losses attributable to air pollution include the soiling effect of particulates; damage to vegetation and crops resulting from exposure to excessive concentrations of gases such as sulfur dioxide, oxides of nitrogen, and ozone; damage to livestock associated with exposure to fluorine; and deterioration of exposed materials by a variety of air pollutants. Materials deterioration includes corrosion of metals by sulfur dioxide, weathering of stone by acidic mists, darkening of lead-based white paint by hydrogen sulfide, accelerated cracking of rubber by ozone, and deterioration of fabrics such as nylons by sulfur dioxide. Safety hazards associated with air pollution result primarily from decreased visibility, and they can become of major concern in conjunction with airport operations as well as ground transportation. Personal discomfort is associated with eye irritation from photochemical oxidants and irritation to individuals with respiratory difficulties from a variety of pollutant forms. Actual health hazards may result from air pollution, with an example being the short-term effects of carbon monoxide in urban areas with heavy traffic. There is some evidence regarding acute illness and even death resulting from air pollution, but substantive data regarding long-term effects of exposures to lower concentrations of air pollutants are minimal. Tables 4-1 and 4-2 are a summary of the results of a 1968 survey of air pollution damages (6).

Table 4-1 National Costs[a] of Pollution Damage by Pollutants in 1968

Effects (loss category)	SO_x	Particulate	Oxidant	NO_x	Total
Residential property	2.808	2.392	–	–	5.200
Materials	2.202	0.691	1.127	0.732	4.752
Health	3.272	2.788	–	–	6.060
Vegetation	0.013	0.007	0.060	0.040	0.120
Total	8.295	5.878	1.187	0.772	16.132

[a]In $ billion.

Table 4-2 National Costs[a] of Pollution Damage by Source and Effect in 1968

Effects	Stationary source fuel combustion	Transportation	Industrial processes	Solid waste	Miscellaneous	Total
Residential property	2.802	0.156	1.248	0.104	0.884	5.200
Materials	1.853	1.093	0.808	0.143	0.855	4.752
Health	3.281	0.197	1.458	0.119	1.005	6.060
Vegetation	0.047	0.028	0.020	0.004	0.021	0.120
Total	7.983	1.474	3.534	0.370	2.765	16.132

[a] In $ billion.

STEP 1: IDENTIFICATION OF AIR POLLUTANTS

The first step associated with prediction and assessment of air quality impacts involves identification of the type and quantities of air pollutants emitted from the construction and operation of each alternative under consideration for a proposed action. One approach for identifying the anticipated air pollutants is to review other environmental statements prepared for projects of a similar type. Perhaps the best approach is to use emission factors arranged according to the various activities of people. An emission factor is the statistical average of the rate at which a pollutant is released into the atmosphere as a result of some activity, such as production by industry or combustion, divided by the activity (7). Examples of construction phase emission factors are found in Table 4-3 (8) for asphalt plants and Table 4-4 (9) for concrete batching plants. To give an idea of the accuracy of the factors presented

Table 4-3 Particulate Emission Factors for Asphaltic Concrete Plants, Emission Factor Rating—A

Type of control	Emissions	
	lb/ton	kg/MT
Uncontrolled[a]	45.0	22.5
Precleaner	15.0	7.5
High-efficiency cyclone	1.7	0.85
Spray tower	0.4	0.20
Multiple centrifugal scrubber	0.3	0.15
Baffle spray tower	0.3	0.15
Orifice-type scrubber	0.04	0.02
Baghouse[b]	0.1	0.05

[a] Almost all plants have at least a precleaner following the rotary dryer.

[b] Emissions for a properly designed, installed, operated, and maintained collector can be as low as 0.005–0.020 lb/ton (0.0025–0.010 kg/MT).

Table 4-4 Particulate Emission Factors for Concrete Batching,[a] Emission Factor Rating–C

	Emissions	
Concrete batching	lb/yd^3 concrete	kg/m^3 concrete
Uncontrolled	0.2	0.12
Good control	0.02	0.012

[a] 1 yd^3 concrete weighs 4,000 lb (1 m^3 = 2,400 kg). The cement content varies with the type of concrete mixed, but 735 lb cement/yd^3 (436 kg/m^3) may be used as a typical value.

for a specific process, each process is ranked as A, B, C, D, or E. For a process with an A ranking, the emission factor is considered excellent, based on field measurements of a large number of sources. A process ranked B is considered above average, based on a limited number of field measurements. A ranking of C is considered average; D, below average; and E, poor (7). Examples of operational phase emission factors are shown in Table 4-5 (10) for incinerators, Table 4-6 (11) for zinc smelters, and Table 4-7 (12) for aircraft. The primary thing to note from Tables 4-3–4-7 is that information is provided, at various levels of reliability, to enable the calculation of the total quantity of air pollution anticipated from the given activity. This information is basic to the prediction of the mesoscale air quality impact of the alternatives for a proposed action.

STEP 2: DESCRIPTION OF EXISTING AIR QUALITY LEVELS

The second step in the process is to assemble information on the existing air quality levels in the area of the project, particularly for those air pollutants anticipated to be emitted from the construction and operational phases of the project. Sources of information include the relevant county and state air pollution control agency and private industries in the area that might be maintaining an air quality monitoring program for their particular interests. One of the best sources of information is the SAROAD system of the EPA, with SAROAD meaning Storage And Retrieval Of Aerometric Data (13). Typical information from the SAROAD system is presented in Table 4-8 (14), with the data based on a frequency distribution as well as on various maximum and minimum values and statistical parameters. If possible, it is desirable to examine the complete history of air quality for the sampling stations in the particular locale. To utilize this information, one must carefully describe the characteristics of the sampling site, including any unique factors about the site. Some information relevant to this step could possibly be procured from the Air Quality Implementation Plan encompassing the area or from special studies that have been conducted, such as transportation control strategy studies for various air quality regions (3). Graphical presentation of information of this type is useful, particularly if there appear to be trends either upward or downward in the air quality levels of any of the air pollutants.

Table 4-5 Emission Factors for Refuse Incinerators without Controls,[a] Emission Factor Rating—A

Incinerator type	Particulates		Sulfur oxides[b]		Carbon monoxide		Hydrocarbons[c]		Nitrogen oxides[d]	
	lb/ton	kg/MT	lb/ton	kg/MT	lb/ton	kg/MT	lb/ton	kg/MT	lb/ton	kg/MT
Municipal										
Multiple chamber, uncontrolled	30	15	2.5	1.25	35	17.5	1.5	0.75	3	1.5
With settling chamber and water spray system	14	7	2.5	1.25	35	17.5	1.5	0.75	3	1.5
Industrial/commercial										
Multiple chamber	7	3.5	2.5	1.25	10	5	3	1.5	3	1.5
Single chamber	15	7.5	2.5	1.25	20	10	15	7.5	2	1
Trench										
Wood	13	6.5	0.1	0.05	NA[e]	NA	NA	NA	4	2
Rubber tires	138	69	NA	NA	NA	NA	NA	NA	NA	NA
Municipal refuse	37	18.5	2.5	1.25	NA	NA	NA	NA	NA	NA
Controlled air	1.4	0.7	1.5	0.75	Neg[f]	Neg	Neg	Neg	10	5
Flue-fed single chamber	30	15	0.5	0.25	20	10	15	7.5	3	1.5
Flue-fed, modified	6	3	0.5	0.25	10	5	3	1.5	10	5
Domestic single chamber										
Without primary burner	35	17.5	0.5	0.25	300	150	100	50	1	0.5
With primary burner	7	3.5	0.5	0.25	Neg	Neg	2	1	2	1
Pathological	8	4	Neg	Neg	Neg	Neg	Neg	Neg	3	1.5

[a] Average factors given based on EPA procedures for incinerator stack testing.
[b] Expressed as sulfur dioxide.
[c] Expressed as methane.
[d] Expressed as nitrogen dioxide.
[e] NA, not available.
[f] Neg, negligible.

Table 4-6 Emission Factors for Primary Zinc Smelting without Controls,[a] Emission Factor Rating—B

Type of operation	Particulates		Sulfur oxides	
	lb/ton	kg/MT	lb/ton	kg/MT
Roasting (multiple-hearth)	120	60	1100	550
Sintering	90	45	*b*	*b*
Horizontal retorts	8	4	—	—
Vertical retorts	100	50	—	—
Electrolytic process	3	1.5	—	—

[a]Approximately 2 unit weights of concentrated ore are required to produce 1 unit weight of zinc metal. Emission factors expressed as units per unit weight of concentrated ore produced.

[b]Included in SO_2 losses from roasting.

Information on existing air quality levels in the area will be useful as a base line for establishing the impact of the particular proposed action.

STEP 3: DETERMINATION OF AIR POLLUTION DISPERSION POTENTIAL

The next step is to gather information on the general characteristics of the area with regard to air pollution dispersion. This information will be useful in evaluating the mesoscale impact of the proposed action. There are several useful parameters for this step, including mixing height, inversion height, annual wind speeds, high air pollution potential advisories, and episode-days.

The first general indicator of air pollution dispersion potential is mixing height, which can be defined as the vertical distance available above the earth's surface at a given location and at a given time period for the mixing of pollutants. The mixing heights vary daily, seasonally, and topographically (15). Figures 4-1 (16) and 4-2 (17) indicate the mean annual morning and afternoon mixing heights for the entire United States, respectively. Information on this parameter, as well as essentially all of the parameters in this step, can be obtained from climatological offices in states or from the National Oceanographic and Atmospheric Administration.

Inversions occur when temperature increases with height above the earth's surface (18). Inversions typically are present during the night or early morning hours due to the heating and cooling pattern at the earth's surface. In general, inversions are more frequent during the fall season of the year than during any other season. One of the characteristics of inversions is that they are often accompanied by wind speeds less than 7 mi/hr; thus they often represent time periods when there is limited horizontal and vertical dispersion. The maximum inversion height in most areas of the country is limited to about 500 m above the earth's surface. Figure 4-3 (19) shows seasonal maps of the percentage of total hours of the occurrence of inversions or isothermal conditions below 500 ft during the winter and the summer.

Another indicator of air pollution dispersion potential is the high air pollution

Table 4-7 Emission Factors[a] per Aircraft Landing-Takeoff Cycle, Emission Factor Rating—B

Aircraft	Solid particulates		Sulfur oxides[b]		Carbon monoxide		Hydrocarbons		Nitrogen oxides[c]	
	lb	kg	lb	kg	lb	kg	lb	kg	lb	kg
Jumbo jet	1.30	0.59	1.82	0.83	46.8	21.2	12.2	5.5	31.4	14.2
Long-range jet	1.21	0.55	1.56	0.71	47.4	21.5	41.2	18.7	7.9	3.6
Medium-range jet	0.41	0.19	1.01	0.46	17.0	7.71	4.9	2.2	10.2	4.6
Air carrier turboprop	1.1	0.49	0.40	0.18	6.6	3.0	2.9	1.3	2.5	1.1
Business jet	0.11	0.05	0.37	0.17	15.8	7.17	3.6	1.6	1.6	0.73
General aviation turboprop	0.20	0.09	0.18	0.08	3.1	1.4	1.1	0.5	1.2	0.54
General aviation piston	0.02	0.01	0.014	0.006	12.2	5.5	0.40	0.18	0.047	0.021
Piston transport	0.56	0.25	0.28	0.13	304.0	138.0	40.7	18.5	0.40	0.18
Helicopter	0.25	0.11	0.18	0.08	5.7	2.6	0.52	0.24	0.57	0.26
Military transport	1.1	0.49	0.41	0.19	5.7	2.6	2.7	1.2	2.2	1.0
Military jet	0.31	0.14	0.76	0.35	15.1	6.85	9.93	4.5	3.29	1.49
Military piston[d]	0.28	0.13	0.14	0.04	152.0	69.0	20.4	9.3	0.20	0.09

[a]Expressed as lb/engine and kg/engine.
[b]Based on 0.05% sulfur content fuel.
[c]Expressed as nitrogen dioxide.
[d]Engine emissions based on Pratt & Whitney R-2800 engine scaled down 2 times.

Table 4-8 1973 Suspended Particulate Concentrations[a]

Region-city	Site number	Year	Minimum value	Frequency distribution, % (% of values less than the stated one) 10	30	50	70	90	95	99	Highest value reached/exceeded more than one time
Region 184		73	11	79	106	124	149	183	196	233	233
Okla. City	022	72	43	92	118	150	172	245	321	487	
		73	24	38	58	67	81	99	126	140	140
Kingfisher	030	72	7	24	46	60	77	122	165	207	
		73	8	20	32	41	51	69	77	96	96
Norman	040	72	1	20	33	42	52	65	73	94	
		73	2	30	39	46	64	81	85	113	108
Moore	044	72	6	27	44	49	62	106	117	130	
		73	11	25	32	39	46	61	75	85	85
Shawnee	052	72	1	26	32	45	55	69	82	110	
		73	7	12	18	25	39	53	99	226	99
Alex	063	72	10	15	25	38	59	84	95	141	
		73	9	31	42	54	63	88	92	212	103
Purcell	072	72	26	32	45	50	57	106	116	125	
		73	11	15	31	47	58	77	79	113	113
Chandler	080	72	11	27	43	58	70	83	101	140	
		73	8	23	43	59	71	84	97	122	107
Guthrie	090	72	17	34	50	60	68	95	117	235	
		73	14	21	43	55	73	100	118	172	172
El Reno	100	72	10	22	48	62	86	124	137	166	

[a] In $\mu g/m^3$.

potential advisory (HAPPA). A HAPPA is issued following the occurrence of limiting dispersion conditions over a 36-hr period and covering an area of approximately 75,000 mi^2 (20). The conditions are as follows: the morning urban mixing height is $\leqslant 500$ m, the morning wind speed is $\leqslant 4$ m/sec, the afternoon wind speed is $\leqslant 4$ m/sec, and the afternoon ventilation rate (which is wind speed times mixing height) is $\leqslant 6{,}000\ m^2/sec$. Figure 4-4 (21) is a map of forecast-days of high air pollution potential in the United States.

Another air pollution advisory is episode-days, with the criteria required for defining an episode-day varying for mixing height, average wind speed in the mixing layer, degree of precipitation, and time period of persistence. Figure 4-5 (22) indicates isopleths of the total number of episode-days in the United States in a 5-yr period for certain criteria.

Another general indicator of air pollution dispersion potential is the mean annual wind speed at a given location. Figures 4-6 (23) and 4-7 (24) represent isopleths of the mean annual wind speed averaged through the morning and afternoon mixing layers, respectively, in the United States. Some areas are characterized by mean annual wind speeds as low as 3 m/sec in the morning, whereas others are as high as 9 m/sec in the afternoon.

One other thing that should be considered is the historical record of air pollution episodes in the area. Information of this type has been compiled, and if previous air pollution episodes have occurred, these need to be documented in the EIS (25).

Maximum value	Total no. observed	Arithmetic SD	Arithmetic mean	Geometric SD	Geometric mean	Change in geometric mean (1972-1973)
329	53	48.5	132.2	1.6	122.1	–24.8
487	74	77.1	162.3	1.6	146.9	
171	110	25.4	71.9	1.4	67.5	12.0
207	54	41.2	67.7	2.0	55.5	
147	112	21.0	44.0	1.6	39.5	1.5
105	109	18.2	42.9	1.8	38.0	
113	50	22.2	52.2	1.8	46.1	–4.1
130	57	28.5	57.1	1.7	50.2	
119	81	16.8	42.0	1.5	39.2	–1.9
110	64	19.0	46.0	1.8	41.1	
226	40	38.1	37.8	2.1	28.2	–8.3
141	60	30.6	45.5	2.0	36.5	
212	35	33.1	58.7	1.7	52.1	–0.3
125	47	25.7	57.0	1.5	52.4	
654	53	86.5	57.5	2.0	42.4	–10.3
140	54	25.2	58.4	1.6	52.7	
122	47	25.1	57.7	1.8	50.9	–6.4
235	75	30.6	62.8	1.5	57.3	
233	92	36.7	62.8	1.8	53.6	–2.2
166	46	37.6	67.4	2.0	51.8	

STEP 4: ASSEMBLAGE OF BASIC METEOROLOGICAL DATA

Information that is associated with general air pollution dispersion potential, but is more relevant to specific calculations of microscale impact, includes monthly records of precipitation, temperature, wind speed and direction, solar radiation, relative humidity, and other items. Basic weather data for a given area can be obtained from several sources, including state climatology offices, the National Oceanographic and Atmospheric Administration, and the Federal Aviation Administration. A typical record of weather data is shown in Table 4-9 (26). This information is generally presented for the most recent 30-yr period for those stations that have been collecting information for that length of time. Graphical presentations of monthly, seasonal, or annual patterns of various parameters should be considered.

Wind roses should be presented for the particular area or the nearest weather station. A wind rose is a diagram designed to show the distribution of wind direction in a given location over a considerable period of time (27). It is a pictorial graph showing the prevailing wind direction and speed, with the wind direction being the direction from which the wind is blowing. In Figure 4-8 (28) two wind roses representing the averages of 10 yr of data are shown. Information of this type is necessary for microscale calculations of air quality impact.

Any unique meteorological phenomena that occur in the area should be noted, particularly as related to the occurrence of tornadoes or characteristics such as fog

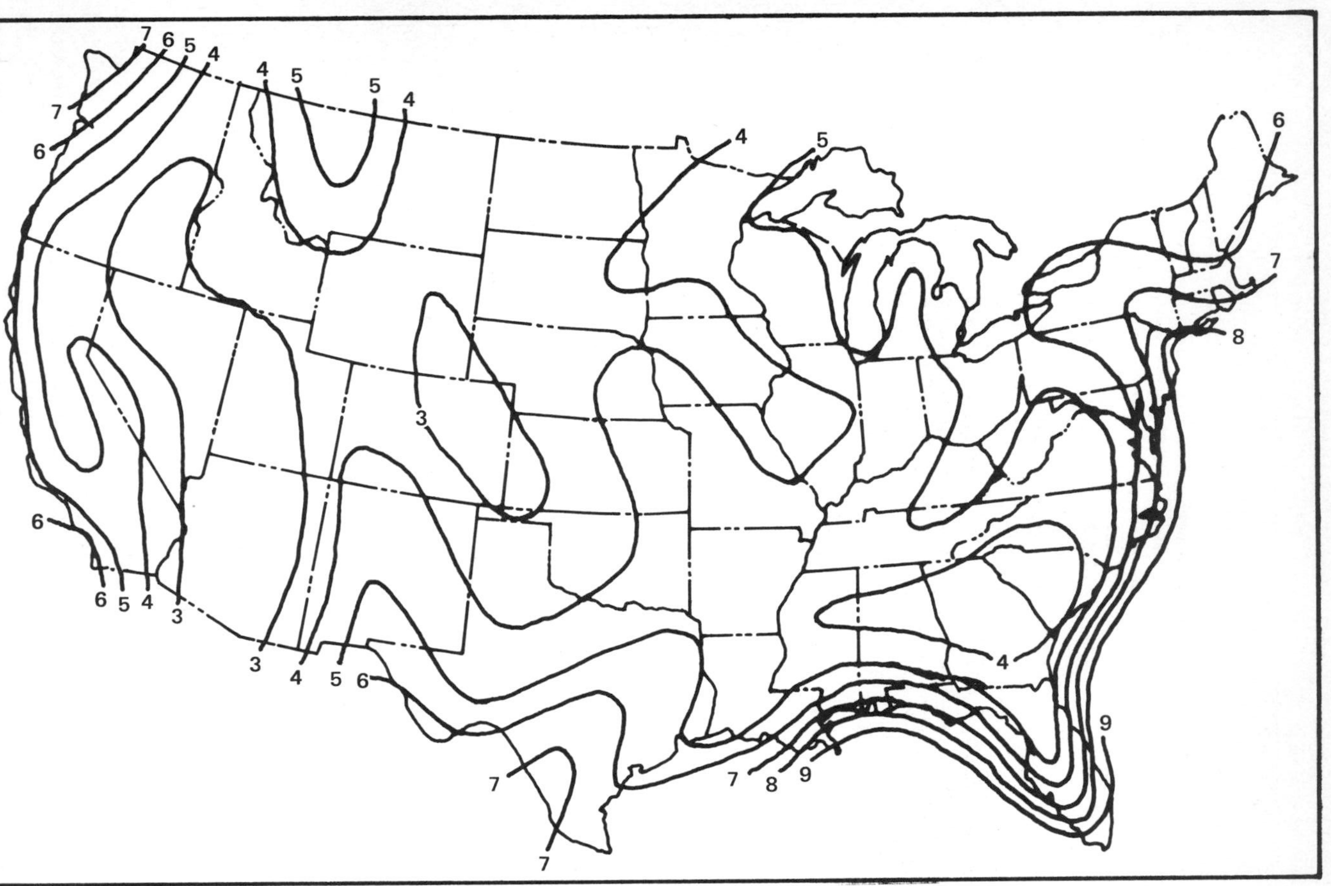

Figure 4-1 Isopleths (m × 10^2) of mean annual morning mixing heights.

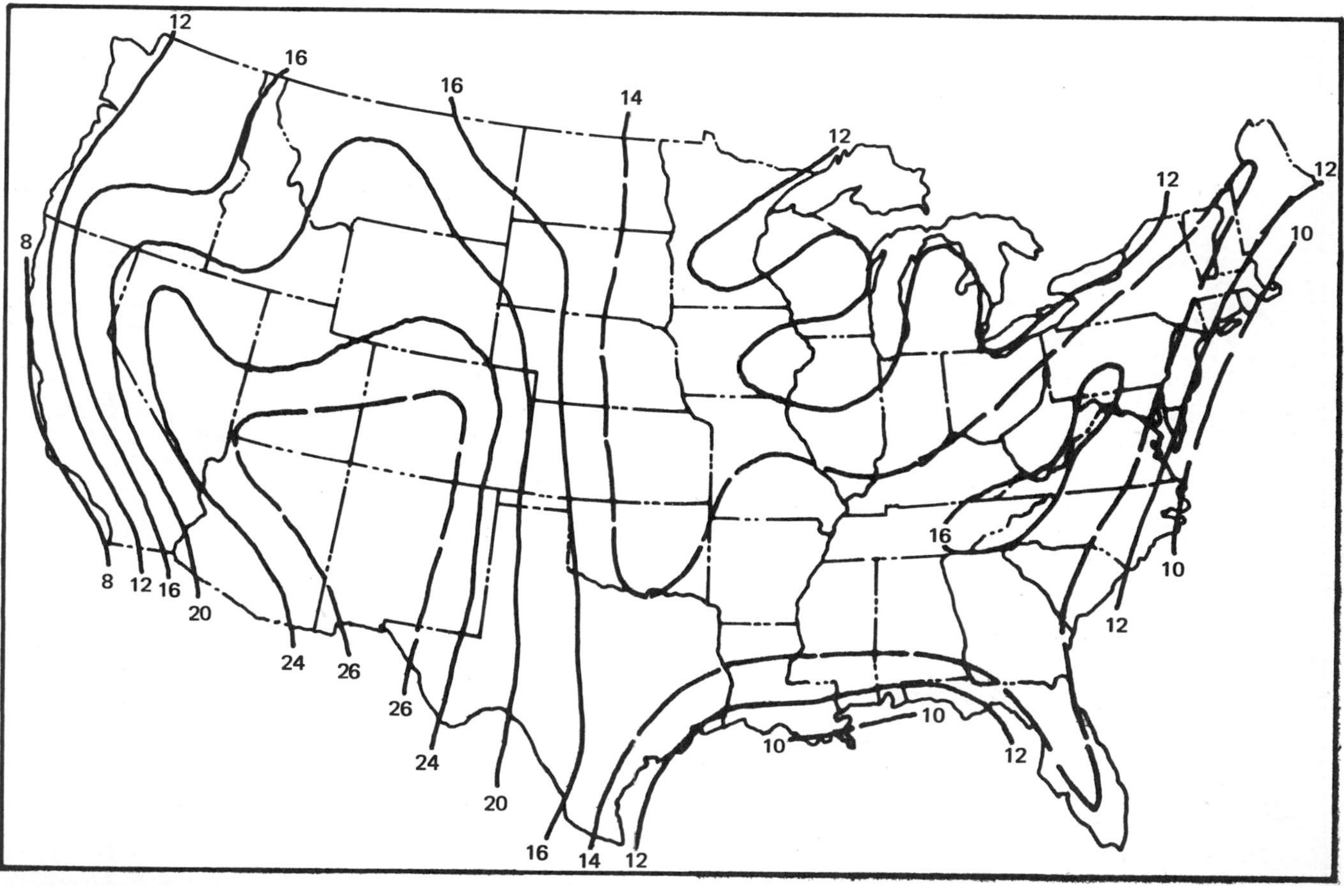

Figure 4-2 Isopleths (m × 10²) of mean annual afternoon mixing heights.

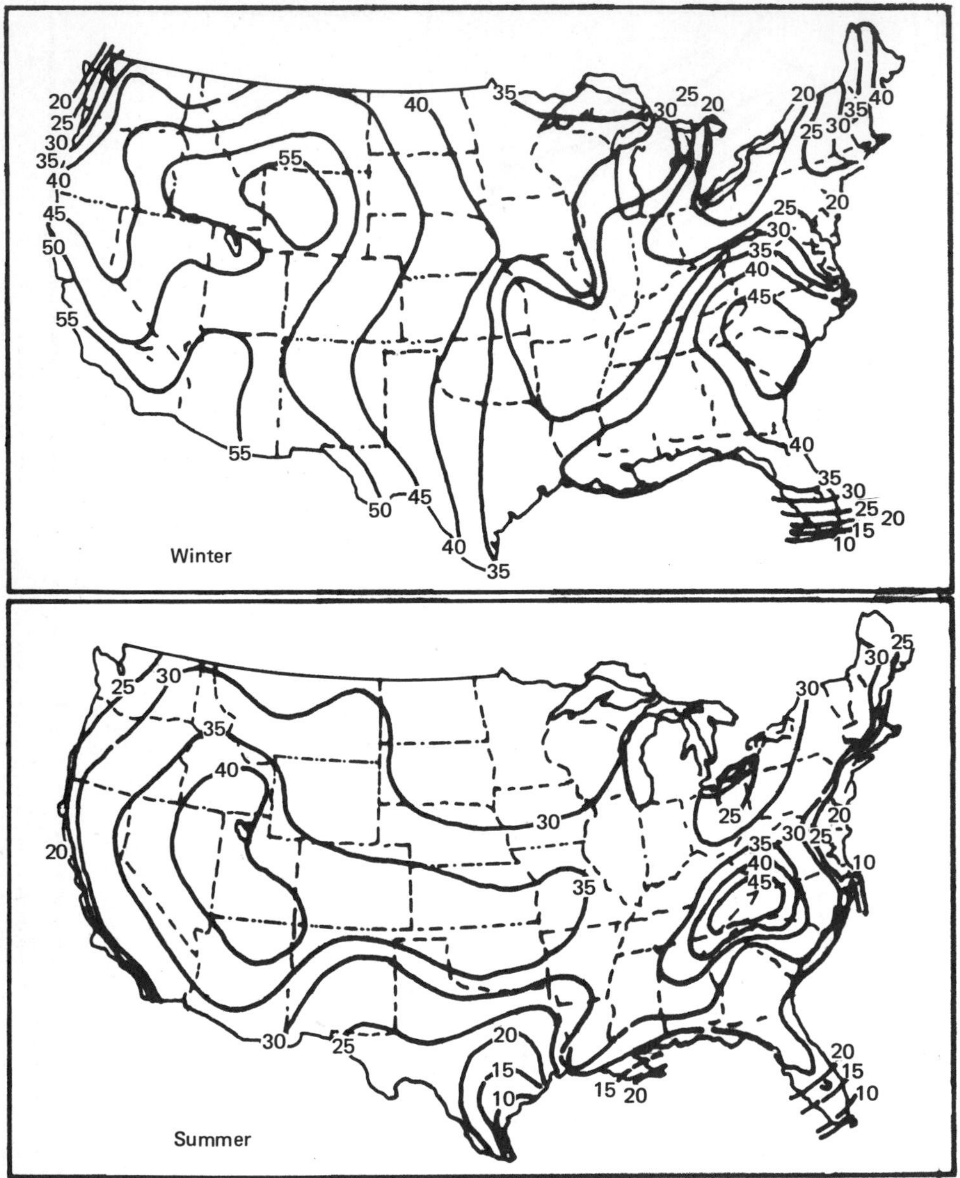

Figure 4-3 Percentage frequency (percent of total hours) of the occurrence of inversions or isothermal conditions based below 500 ft during the winter and summer.

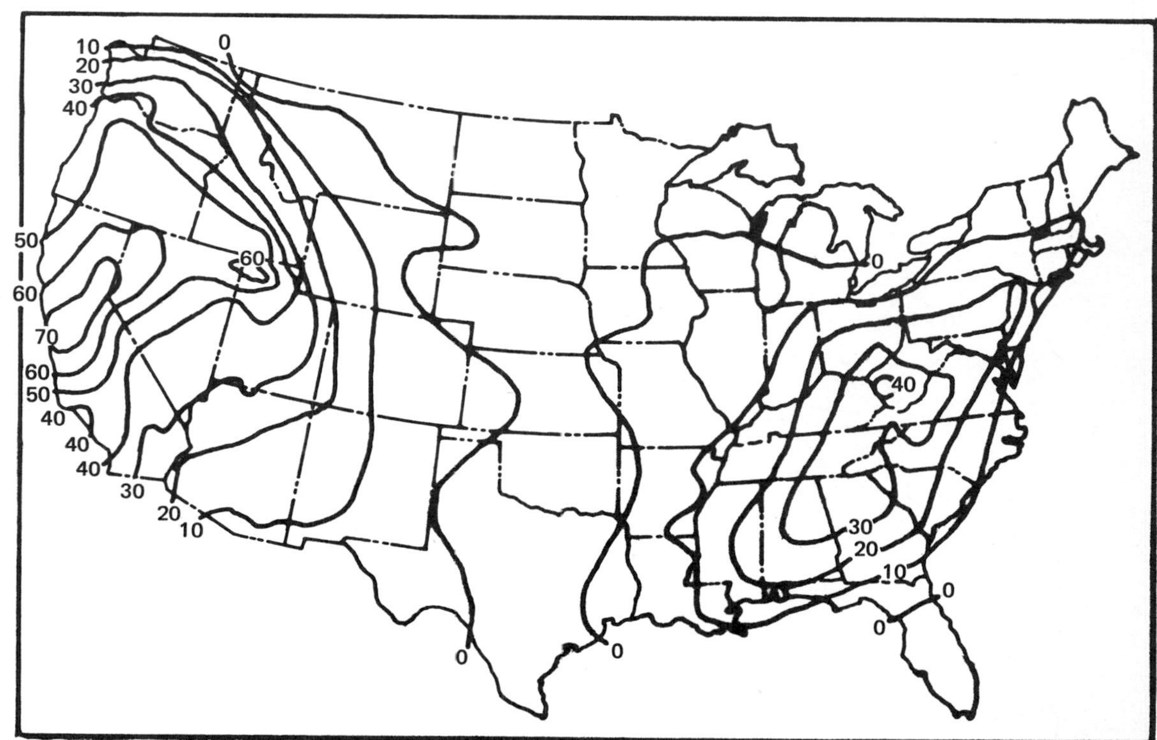

Figure 4-4 Isopleths of total number of forecast-days of high meteorological potential for air pollution in a 5-yr period. Data are based on forecasts issued since the program began, August 1, 1960 and October 1, 1963 for eastern and western parts of the United States, respectively, through April 3, 1970.

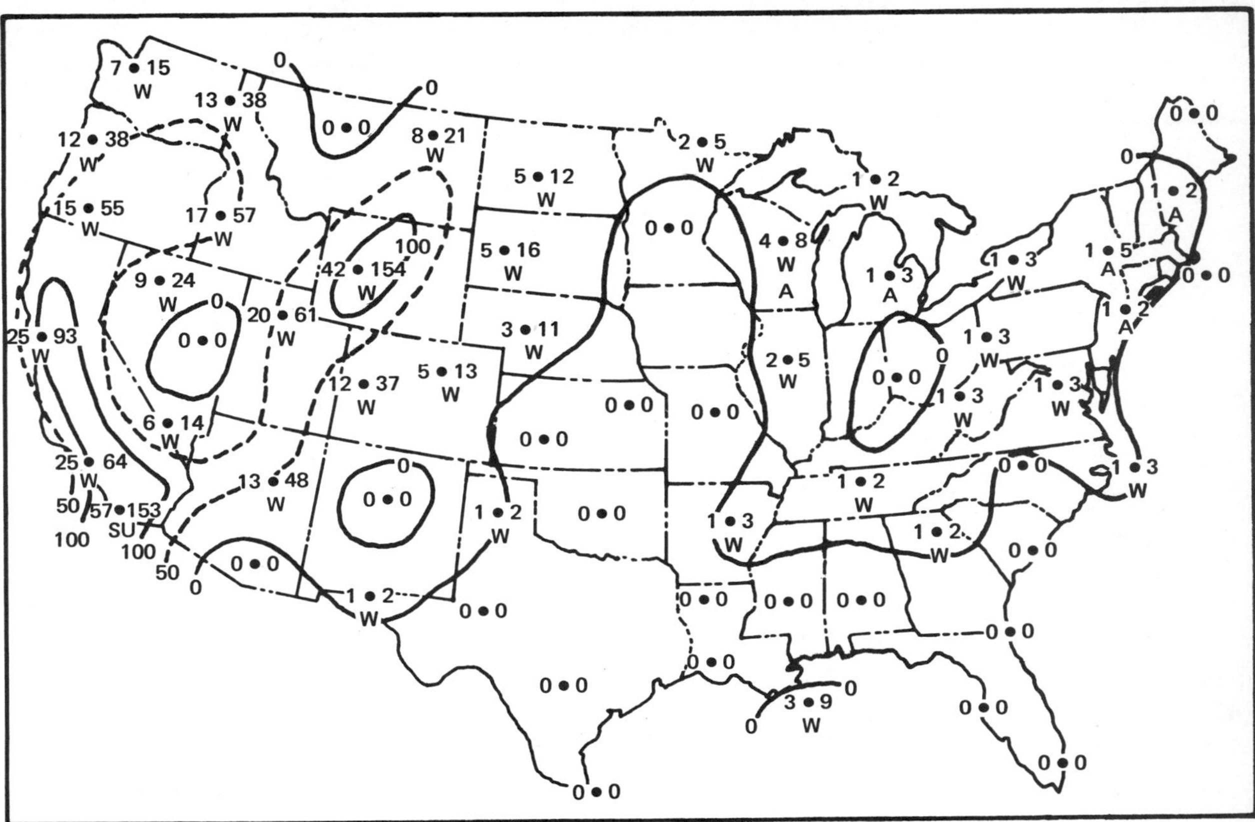

Figure 4-5 Isopleths of total number of episode-days in 5 yr with mixing heights ⩽ 500 m, wind speeds ⩽ 4.0 m/sec, and no significant precipitation for episodes lasting at least 2 days. Numerals on left and right give total number of episodes and episode-days, respectively. Season with greatest number of episode-days indicated as W (winter), SP (spring), SU (summer), or A (autumn).

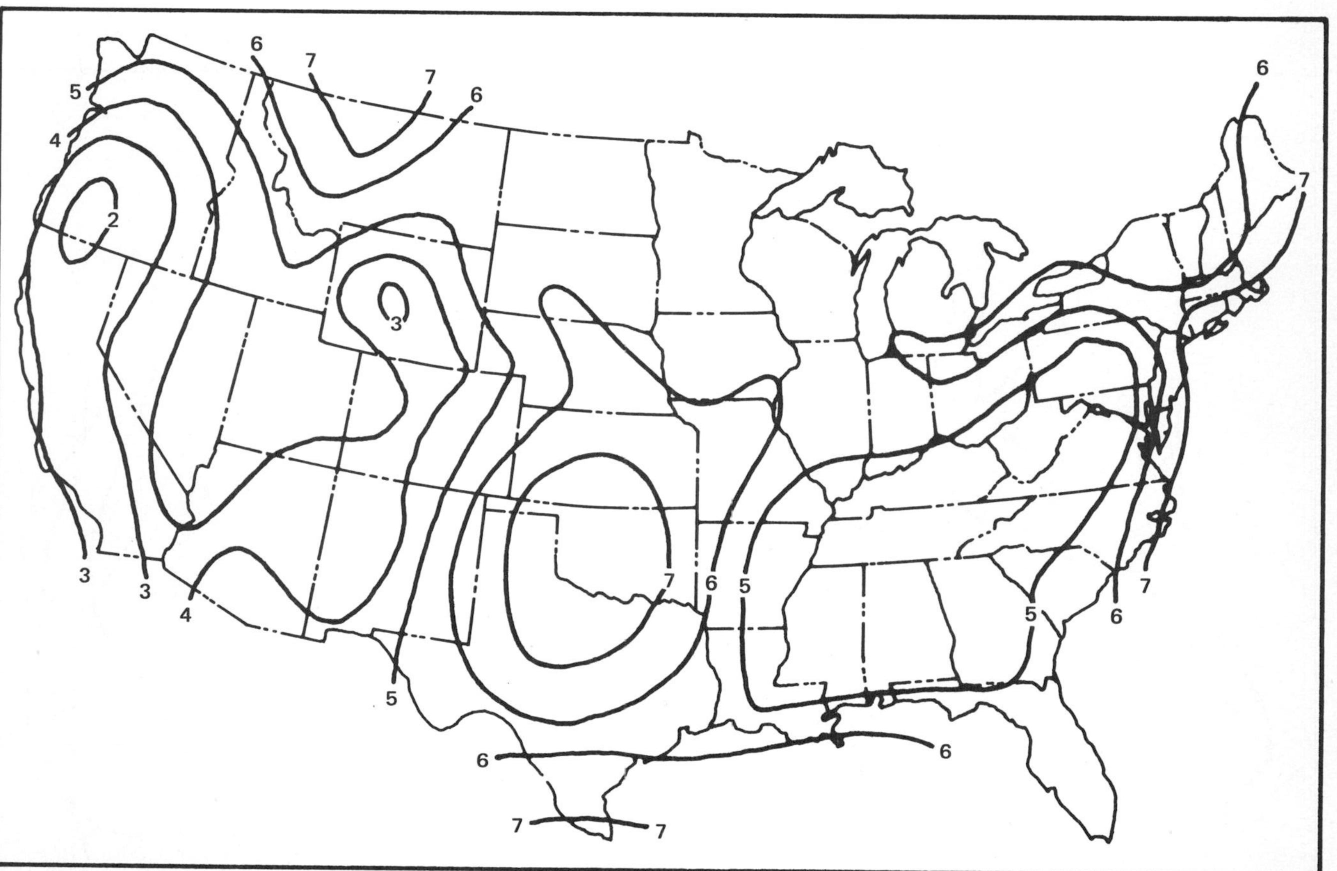

Figure 4-6 Isopleths (m/sec) of mean annual wind speed averaged through the morning mixing layer.

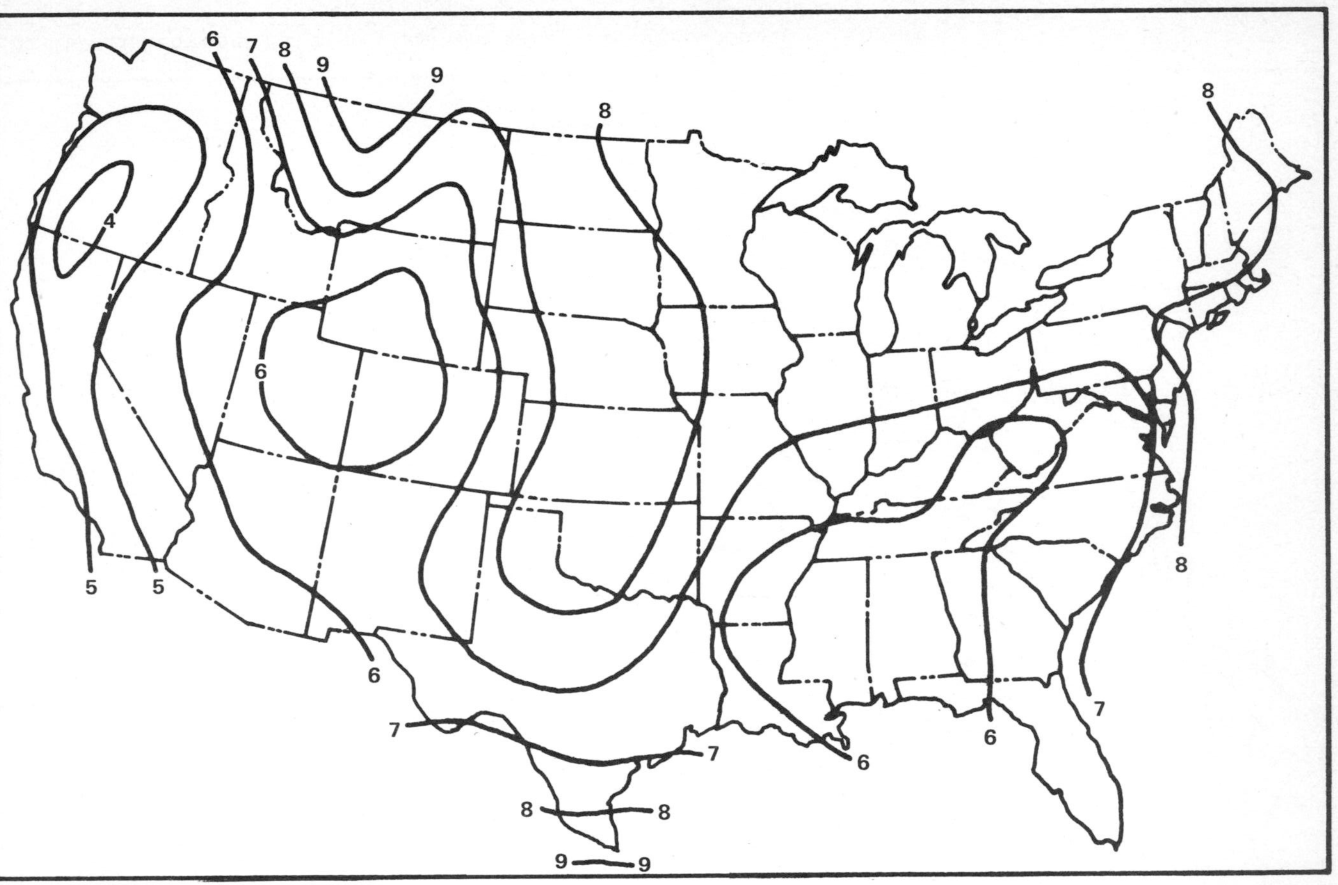

Figure 4-7 Isopleths (m/sec) of mean annual wind speed averaged through the afternoon mixing layer (see Fig. 4-6).

Table 4-9 Summary of Climatological Data in New Orleans[a,b]

	Length of record, yr	Month												
		J	F	M	A	M	J	J	A	S	O	N	D	Yr
Temp.														
Normal														
Daily max.	*e*	64.4	66.7	71.2	77.7	64.4	89.6	90.6	90.7	67.2	60.3	70.3	65.3	78.2
Daily min.	*e*	44.8	47.5	51.6	58.1	64.4	70.5	72.6	73.0	69.3	60.5	49.6	45.5	59.0
Monthly	*e*	54.6	57.1	61.4	67.9	74.4	80.1	81.6	81.9	78.3	70.4	60.0	55.4	68.6
Extremes														
Record high	22	83	84	87	91	96	100	99	100	97	92	86	83	100
Year		1957	1948	1955	1948	1953[f]	1954	1951	1951	1954[f]	1962[f]	1951	1951	June 1954[f]
Record low	22	14	19	26	38	41	55	60	60	42	35	28	17	14
Year		1963[f]	1951	1968	1962	1960	1966	1967	1968	1967	1968	1968[f]	1962	Jan. 1963[f]
Normal degree-days	*e*	363	258	192	39	0	0	0	0	0	19	192	322	1,385
Precipitation														
Normal total	*e*	3.84	3.99	5.34	4.55	4.38	4.43	6.72	5.34	5.03	2.84	3.34	4.10	53.90
Max. monthly	22	12.62	10.56	19.09	8.78	14.33	8.87	11.46	11.77	13.53	6.45	14.58	10.77	19.09
Year		1966	1959	1948	1949	1959	1962	1954	1955	1948	1959	1947	1967	Mar. 1948
Min. monthly	22	0.54	1.02	0.24	0.33	0.99	1.12	3.45	2.00	0.24	0.00	0.21	1.46	0.00
Year		1968	1962	1955	1965	1949	1952	1951	1952	1953	1952	1949	1958	Oct. 1952
Max. in 24 hr	22	4.77	5.60	7.87	4.35	9.86	4.19	4.30	2.48	5.46	2.58	6.38	3.94	9.86
Year		1955	1961	1948	1953	1959	1953	1966	1957	1957	1960	1953	1952	May 1959
Snow, sleet														
Mean total	22	T[g]	0.1	T	0.0	0.0	0.0	0.0	0.0	0.0	0.0	T	0.1	0.2
Max. monthly	22	T	2.0	T	0.0	0.0	0.0	0.0	0.0	0.0	0.0	T	2.7	2.7

Table 4-9 (*continued*) Summary of Climatological Data in New Orleans[a,b]

	Length of record, yr	Month												Yr
		J	F	M	A	M	J	J	A	S	O	N	D	
Year		1963[f]	1958	1959								1950	1963	Dec. 1963
Max. in 24 hr	22	T	2.0	T	0.0	0.0	0.0	0.0	0.0	0.0	0.0	T	2.7	2.7
Year		1963[f]	1958	1959								1950	1963	Dec. 1963
Relative humidity[c]														
Midnight	20	84	83	83	85	87	87	89	89	87 ·	84	83	83	85
6 A.M.	20	86	86	85	88	89	90	90	91	89	87	86	86	88
Noon	20	67	64	60	60	60	63	65	66	65	58	59	67	63
6 P.M.	20	73	69	65	66	65	68	73	73	74	72	73	75	71
Wind[d]														
Mean hourly speed	20	9.6	10.3	10.2	9.7	8.4	7.0	6.4	6.2	7.6	7.7	9.0	9.3	8.5
Fastest mile														
Speed	9	33	39	33	35	31	38	30	35	69	40	29	31	69
Direction	9	28	15	15	07	23	27	25	20	09	17	07	32	09
Year		1966	1966	1964	1960	1962	1962	1967[f]	1962	1965	1964	1968	1960	Sep. 1965
Mean sky cover sunrise to sunset	20	6.6	6.3	6.1	5.7	5.1	5.0	6.1	5.4	5.2	4.0	5.0	6.1	5.6
Mean number of days														
Sunrise to sunset														
Clear	20	7	8	9	8	11	10	5	9	11	16	11	8	113
Partly cloudy	20	8	6	8	11	11	13	16	14	10	8	9	8	122
Cloudy	20	16	14	14	11	9	7	10	8	9	7	10	15	130
Precipitation, 0.01 in. or more	20	10	10	9	7	8	10	15	13	9	6	6	10	112
Snow, sleet, 1.0 in. or more	20	0	<1/2	0	0	0	0	0	0	0	0	0	<1/2	<1/2

Thunderstorms	20	2	2	3	5	6	10	16	13	7	2	1	2	69
Heavy fog	20	7	5	5	2	1	<1/2	<1/2	<1/2	<1/2	2	5	5	32
Temp.														
Max.														
90 and above	22	0	0	0	<1/2	4	17	20	19	8	1	0	0	69
32 and below	22	<1/2	0	0	0	0	0	0	0	0	0	0	<1/2	<1/2
Min.														
32 and below	22	5	3	1	0	0	0	0	0	0	0	1	3	12
0 and below	22	0	0	0	0	0	0	0	0	0	0	0	0	0

[a]New Orleans International Airport. Latitude, 29° 59.2′N, longitude, 90° 15.3′W, ground elevation 3 ft.

[b]Means and extremes are from the existing or comparable location(s). Annual extremes have been exceeded at other locationas as follows: Highest temperature 102 in June 1954 and earlier; lowest temperature 7 in February 1899; maximum monthly precipitation 25.11 in October 1937; maximum precipitation in 24 hr 14.01 in April 1927; maximum monthly snowfall 8.2 in February 1895; maximum snowfall in 24 hr 8.2 in February 1895.

Unless otherwise indicated, dimensional units used are temperature, °F; precipitation, including snowfall, in.; wind movement, mph; and relative humidity, %. Degree-day totals are the sums of the negative departures of average daily temperatures from 65°F. Sleet was included in snowfall totals beginning with July 1948. Heavy fog reduces visibility to 1/4 mi or less.

Sky cover is expressed as a range of 0 for no clouds or obscuring phenomena to 10 for complete sky cover. The number of clear days is based on average cloudiness 0–3; partly cloudy days 4–7; and cloudy days 8–10 tenths.

[c]Central standard time.

[d]Figures instead of letters in the direction column indicate direction in tens of degrees from true north; i.e., 09—east, 18—south, 27—west, 36—north, and 00—calm. Resulting wind is the vector sum of wind directions and speeds divided by the number of observations. If figures appear in the direction column, the corresponding speeds are fastest observed 1-min values.

[e]Climatological standard normals, 1931–1960.

[f]Also on earlier dates, months or years.

[g]T, trace, an amount too small to measure.

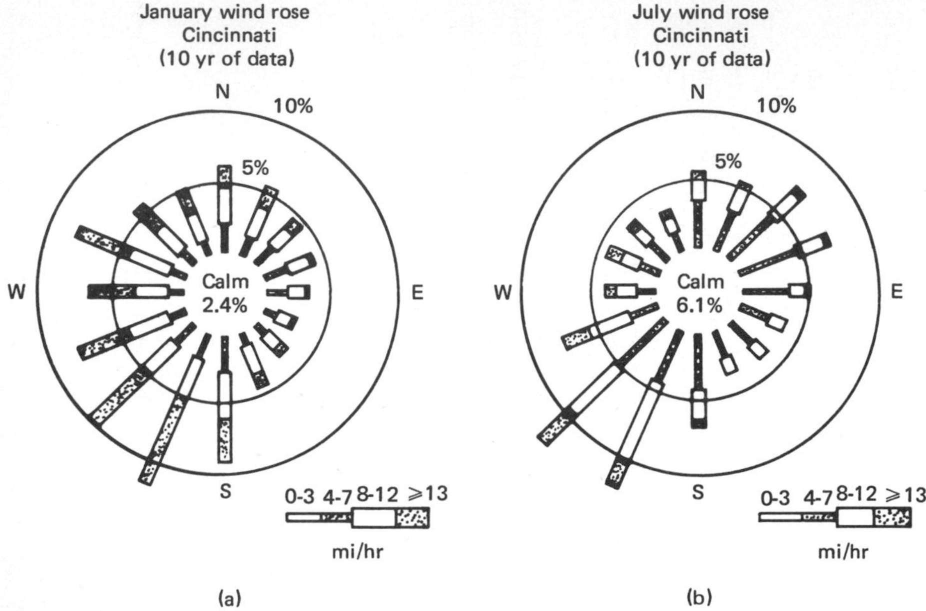

Figure 4-8 January and July wind roses, Cincinnati. The monthly distributions of wind direction and wind speed are summarized on polar diagrams. The positions of the spokes show the direction from which the wind was blowing; the length of the segments indicates the percentage of the speeds in various groups.

formation and persistence. Some agencies require a discussion of the probability of a tornado's occurring in an area, and this probability can be calculated by using the approach suggested by Thom (29).

STEP 5: PRESENTATION OF AIR QUALITY STANDARDS

One of the major points of concern in assessment of air quality impacts is the question of whether air quality standards will be exceeded. Two types of standards are relevant: ambient air standards and emission standards. Ambient air quality standards apply to the general ambient atmosphere, whereas emission standards refer to pollutant materials that can be emitted from a source into the ambient atmosphere.

The Federal Clean Air Act of 1970 established a program for the creation of air quality standards (3). A summary of the ambient air standards is contained in Table 4-10 (30). Primary standards are oriented to the protection of public health, whereas secondary standards are geared to the projection of public welfare. The criteria for the primary standards are shown in Table 4-11 (31). Some new source performance standards, which are emission standards for new point sources constructed following the passage of the Clean Air Act of 1970, are shown in Table 4-12 (32).

Every state has air quality standards at least as stringent as those at the national level, and some states have standards that are stricter. If the state or local standards do not match the national standards, this should be noted and both levels of standards presented. Where there are time schedules required for meeting standards, these schedules should be discussed.

STEP 6: EMISSION INVENTORY

An emission inventory is the compilation of the quantities of air pollutants from all sources in a defined area entering the air in a given time period, which is typically 1 yr. The areas are usually associated with a county or perhaps with a multicounty region (33). Typical emission inventory data contained in a state air quality implementation plan are shown in Table 4-13 (34).

It should be noted that emission inventory information is useful for general air quality management and trend analysis. It does not include consideration of atmospheric reactions and the damage associated with a given weight of pollutant.

Another use to be made of emission inventory information is the identification of major point sources of air pollution in the area. A point source emits 25 tons/yr or more of air pollutants. Figure 4-9 (35) shows the major point sources of air pollution in southwestern Oklahoma. For each point source that might be located close to the area of a proposed action, information should be obtained on the nature of the source, the pollutants, and the quantities of the emissions.

Table 4-10 National Ambient Air Quality Standards[a]

Pollutant	Primary	Secondary
Particulate matter		
Annual geometric mean	75	60
Max. 24-hr conc.[b]	260	150
Sulfur oxides		
Annual arithmetic mean	80 (0.03)	60 (0.02)
Max. 24-hr conc.[b]	365 (0.14)	260 (0.1)
Max. 3-hr conc.[b]		1,300 (0.5)
Carbon monoxide		
Max. 8-hr conc.[b]	10 (9)	
Max. 1-hr conc.[b]	40 (35)	Same as primary
Photochemical oxidants		
Max. 1-hr conc.[b]	160 (0.08)	Same as primary
Hydrocarbons		
Max. 3-hr (6–9 A.M.) conc.[b]	160 (0.24)	Same as primary
Nitrogen oxides		
Annual arithmetic mean	100 (0.05)	Same as primary

[a]All measurements are expressed in $\mu g/m^3$ except for carbon monoxide, which is expressed in mg/m^3. Equivalent measurements in ppm are given in parentheses for the gaseous pollutants.

[b]Not to be exceeded more than once a year.

Table 4-11 National Air Quality Criteria and National Air Quality Standards in the United States, 1971

Pollutant	Adverse health effects observed at these concentrations	Air quality standard, protective for health
Particulate matter	80 μg/m^3, annual mean	75 μg/m^3, annual geometric mean 260 μg/m^3, max. 24-hr value may occur once each year
Sulfur dioxide	115 μg/m^3, annual mean 300 μg/m^3, 24-hr avg. for 3–4 days	80 μg/m^3, annual arithmetic mean 365 μg/m^3, max. 24-hr value may occur once each year
Carbon monoxide	12–17 mg/m^3, for 8 hr produces conc. of 2–2.5% carboxyhemoglobin; 35 mg/m^3 for 8 hr produces conc. of 5% carboxyhemoglobin	10 mg/m^3, max. 8-hr value may occur once each year 40 mg/m^3, max. 1-hr value may occur once each year
Photochemical oxidants	130 μg/m^3 hourly avg. impaired performance of student athletes; 200 μg/m^3 instantaneous level increased eye irritation; 490 μg/m^3 peaks with 300 μg/m^3 hourly avg. increased asthma attacks	160 μg/m^3, max. 1-hr value may occur once each year
Hydrocarbons	With nonmethane hydrocarbon, 200 μg/m^3 in 3 hr (6–9 A.M.) produced (2–4 hr later) photochemical oxidant of up to 200 μg/m^3 that lasted 1 hr; by extrapolation downward, conc. of HC of 100 μg/m^3 can produce lowest injurious level of photochemical oxidant	160 μg/m^3, max. level that may occur 6–9 A.M. once each year
Nitrogen oxides	118–156 μg/m^3, 24-hr mean over 6 months, produced increase in acute bronchitis in infants and schoolchildren; this avg. associated with a 24-hr max. of 284 μg/m^3; 117–205 μg/m^3, 24-hr mean over 6 months, and mean suspended nitrate level of 3.8 μg/m^3 or more produced increased respiratory disease in family groups	100 μg/m^3, annual arithmetic mean

STEP 7: CALCULATION OF MESOSCALE IMPACT

This step involves calculation of the estimated annual quantity of air pollutants that would be emitted from the construction and operational phases of each alternative for the proposed action. This calculation can be done based on emission factors and information regarding the size or type of activity. The calculated quantities of air pollutants should then be compared with regional or local emission inventory information, and percentage increases should be calculated based on the proposed action.

Attention should be focused on any priority I or priority II air pollutants. Priority I pollutant levels in the ambient air violate primary standards, priority II levels are at or near primary standards, and priority III levels are better than primary standards. This particular approach is based on the concept that a given percentage increase in a pollutant is more important for a priority I or priority II pollutant than for a priority III pollutant. Table 4-14 (36) shows the priority classifications in Oklahoma (37).

STEP 8: MICROSCALE IMPACT DETERMINATION

The next step is to calculate ground level concentrations of air pollutants anticipated from each alternative for the proposed action during both the

Table 4-12 New Source Performance Standards

Source category	Emission
Fossil fuel-fired steam generators (250 million Btu/hr heat input or greater)	
Particulates	0.1 lb/10^6 Btu (max. 2-hr avg.)
Sulfur dioxide	
Oil fired	0.80 lb/10^6 Btu
Coal fired	1.2 lb/10^6 Btu (max. 2-hr avg.)
Nitrogen oxides	
Gas fired	0.20 lb/10^6 Btu
Oil fired	0.30 lb/10^6 Btu
Coal fired	0.70 lb/10^6 Btu (max. 2-hr avg. expressed as NO_2)
Visible emissions	Not to exceed 20% opacity, except that for 2 min in any 1 hr, emissions may be as great as 40% opacity
Incinerators	
Particulates	0.08 grains/standard cubic foot
Nitric acid plants	
Nitrogen oxides	3 lb/ton acid
Visible emissions	<10% opacity
Sulfuric acid plants	
Sulfur dioxide	4 lb/ton acid
Acid mist	0.15 lb/ton acid
Visible emissions	<10% opacity
Portland cement plants	
Particulates	
Kilns	0.3 lb/ton feed
Clinker coolers	0.1 lb/ton feed
Visible emissions	
Kilns	10% opacity
Others	<10% opacity

Table 4-13 Emissions Inventory Summary for the State of Oklahoma, Southwestern Oklahoma Intrastate, Air Quality Control Region 189, Data Representative of Calendar Year 1970

	Pollutant, tons/yr					Fuel, etc.	
Source category	Particulate	SO_2	CO	HC	NO_X	Quantity	Units
Fuel combustion							
Residential fuel—Area source							
Distillate oil	8	60	4	2	10	1,597	10^3 gal
Natural gas	82	3	86	35	324	8,641	10^6 ft^3
Total	90	63	90	37	334		
Commercial—Institutional and industrial							
Bituminous coal—Point source	0	4	0	0	3	0	tons
Distillate oil—Area source	64	309	0	0	279	8,591	10^3 gal
Distillate oil—Point source	0	0	0	13	0	0	—
Residual oil—Area source	15	217	0	2	41	1,363	10^3 gal
Residual oil—Point source	0	0	0	0	0	0	—
Natural gas—Area source	56	2	30	74	379	6,002	10^6 ft^3
Natural gas—Point source	161	6	4	374	11,695	18,695	10^6 ft^3
Other (specify)—Area source	72	53	75	29	235	23,483	10^3 gal
Other (specify)—Point source	8	0	0	15	6,474		
Total	376	591	109	507	19,106		
Steam—Electric power plant							
Bituminous coal	0	0	0	0	0	0	—
Distillate oil	0	0	0	0	0	0	—
Residual oil	0	0	0	0	0	0	—
Natural gas	276	11	7	735	7,165	36,744	10^6 ft^3
Total	276	11	7	735	7,165		
Total fuel combustion	752	665	206	1,279	26,605		
Process losses							
Area sources	0	0	0	30	0		
Point sources	47,079	2,579	2,323	2,323	309		

Solid waste disposal							
Incineration							
Municipal, etc.—Point source	0	0	0	0	0	0	—
Open burning							
On site—Area source	2,238	140	11,889	4,196	839	179,749	tons
Total solid waste disposal	2,238	140	11,889	4,196	839		
Transportation—Area source							
Motor vehicles—Gasoline	335	203	72,553	13,584	11,475	1,014,335	10^3 gal
Motor vehicles—Diesel	122	219	1,584	317	1,658	9,750	10^3 gal
Off-highway fuel usage	78	327	59	27	169	20,087	10^3 gal
Aircraft	555	202	7,528	1,875	226		
Railroad	43	112	121	86	129	3,444	10^3 gal
Vessels	0	0	0	0	0	0	—
Gasoline handling evaporation losses	0	0	0	1,669	0		
Other (specify) petroleum storage loss	0	0	0	25,088	0		
Total transportation	1,133	1,063	81,845	42,646	13,657		
Miscellaneous—Area sources							
Other (specify)	0	0	0	294	0	294	tons
Total miscellaneous	0	0	0	294	0	294	tons
Grand total							
Area source	3,668	1,847	93,929	47,308	14,764		
Point source	47,524	2,600	2,334	3,460	25,646		
Total	51,192	4,447	96,263	50,768	40,440		

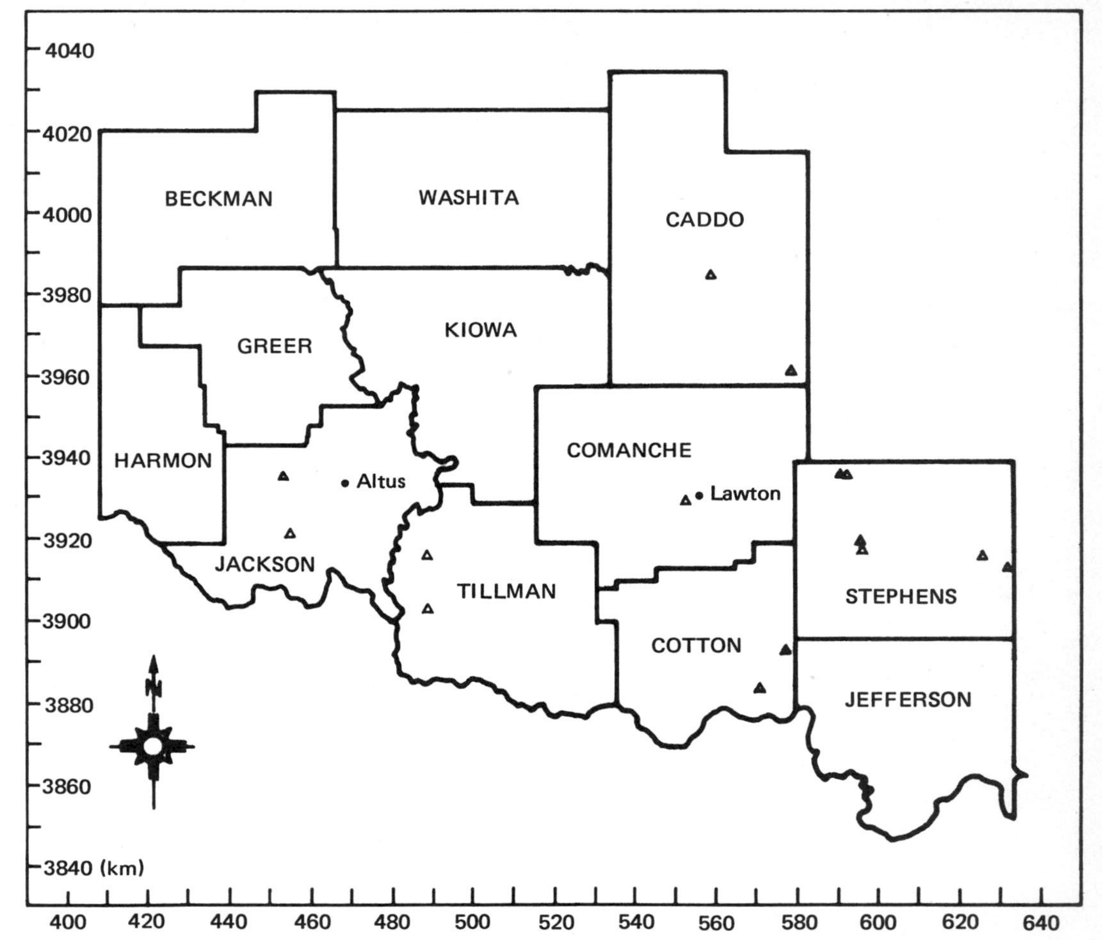

Figure 4-9 Major point sources of air pollution, southwestern Oklahoma, Air Quality Control Region 189. △, point sources emitting 25 tons/yr or more.

Table 4-14 State of Oklahoma Air Quality Control Region (AQCR) Priority Classifications

AQCR (no.)	1970 population	Priority				
		Particulates	SO_2	CO	NO_2	Oxidants (HC)
Central Oklahoma (184)	779,518	I	III	III	III	I
Northeastern Oklahoma (186)	771,412	I	III	III	III	I
Southeastern Oklahoma (188)	305,750	I	III	III	III	III
North Central Oklahoma (185)	171,970	III	III	III	III	III
Southwestern Oklahoma (189)	284,279	III	III	III	III	III
Northwestern Oklahoma (187)	123,876	III	III	III	III	III
Fort Smith Interstate (017)	93,822	II	III	III	III	III
Shreveport–Texarkana–Tyler Interstate (022)	28,642	II	III	III	III	III

construction and operational phases. It is beyond the scope of this presentation to derive the various mathematical models for gaseous and particulate dispersion. This section will be oriented to the simple presentation of several mathematical models for use in microscale impact calculations and the identification of reference sources that will lead to a more detailed study for an individual project.

The two basic factors that influence movement of pollutants from their point of origin to some other location are horizontal wind speed and direction and the vertical temperature structure of the atmosphere. These two parameters influence the vertical and horizontal motion of pollutants released into the atmosphere. The influence of these two parameters can be combined and is called "atmospheric stability," with representative values shown in Table 4-15 (38). Class A indicates the greatest amount of spreading in the most unstable atmospheric conditions, whereas class F indicates the least amount of spreading in the most stable atmospheric conditions. Each of the mathematical models for prediction of microscale impact involves the use of a stability classification.

The first model can be used to calculate ground-level concentrations from an elevated point source. The mathematical model is as follows (38):

$$C_{x,y,0} = \frac{Q}{\pi \sigma_y \sigma_z \bar{u}} \exp -\left(\frac{H^2}{2\sigma_z{}^2} + \frac{y^2}{2\sigma_y{}^2}\right)$$

where C = ground level concentration, $\mu g/m^3$
Q = release rate from stack, $\mu g/sec$
σ_y = crosswind standard deviation, m (function of stability classification and X, location of maximum ground concentration)
σ_z = vertical standard deviation, m (function of stability classification and X)
$\bar{u}$ = mean wind speed, m/sec

Table 4-15 Key to Stability Categories

Surface wind speed at 10 m height, m/sec	Insolation stability classes[a]				
	Day			Night	
	Strong[b]	Moderate[c]	Slight[d]	Thinly overcast or >1/2 cloud[e]	Clear to <1/2 cloud
>2 (4.5 mi/hr)	A[f]	A–B	B	—	—
2–3 (4.5–6.7)	A–B	B	C	E	F
3–5 (6.7–11)	B	B–C	C	D	E
5–6 (11–13.5)	C	C–D	D	D	D
>6 (>13.5 mi/hr)	C	D	D	D	D

[a]Insolation, amount of sunshine.
[b]Sun >60° above horizontal; sunny summer afternoon; very convective.
[c]Summer day with few broken clouds.
[d]Sunny fall afternoon; summer day with broken low clouds; or summer day with sun from 15 to 35° with clear sky.
[e]Winter day.
[f]Class A indicates greatest amount of spreading and most unstable atmospheric conditions, and class F indicates least spreading and most stable atmospheric conditions.

H = effective stack height, m
x,y = downwind and crosswind distances, respectively, m

The vertical standard deviation and crosswind standard deviation, also called dispersion coefficients, are a function of atmospheric stability and the distance downwind for which the calculation is being made. Figures 4-10 and 4-11 summarize these dispersion coefficients in the vertical and horizontal directions, respectively (39). An example problem utilizing this model is as follows:

$Q = 10^6\ \mu g/sec$
$\bar{u} = 1.0$ m/sec
$H = 30$ m

Property line is at 1000 m; find the ground-level concentration at the property line under the most stable conditions (class F).

From Figs. 4-10 and 4-11,

$$\sigma_y = 35 \text{ m}$$
$$\sigma_z = 14 \text{ m}$$

$$C_{1{,}000,\,0,\,0} = \frac{10^6}{(3.14)(35)(14)(1)} \exp - \left[\frac{(30)^2}{2\,(14)^2}\right]$$

$$= 65\ \mu g/m^3$$

It is possible to calculate directly the maximum gaseous ground-level concentration from an elevated point source (38). The location X of the maximum ground concentration will occur approximately where $\sigma_z = H/\sqrt{2}$ for a given stability condition. The maximum concentration can be calculated from

$$C_{x,0,0\ \max} = \frac{0.117\ Q}{\bar{u}\sigma_y\sigma_z} \qquad \sigma_y \text{ and } \sigma_z \text{ for given stability condition and distance } x$$

Effective stack height is equal to the actual height plus any rise of the plume that occurs as it leaves the stack. There are two basic reasons for plume rise: the momentum effect due to the vertical velocity of the gas leaving the stack and the buoyancy effect, which is related to warm stack gases tending to rise in a cooler surrounding atmosphere. The Holland equation can be utilized to calculate plume

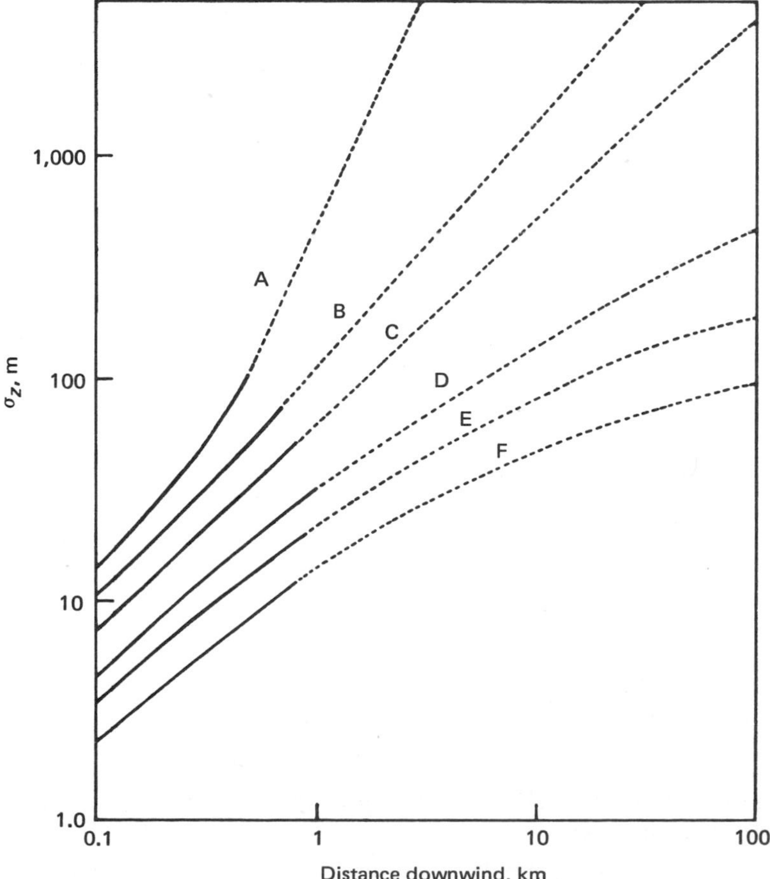

Figure 4-10 Vertical dispersion coefficient as a function of downwind distance from the source.

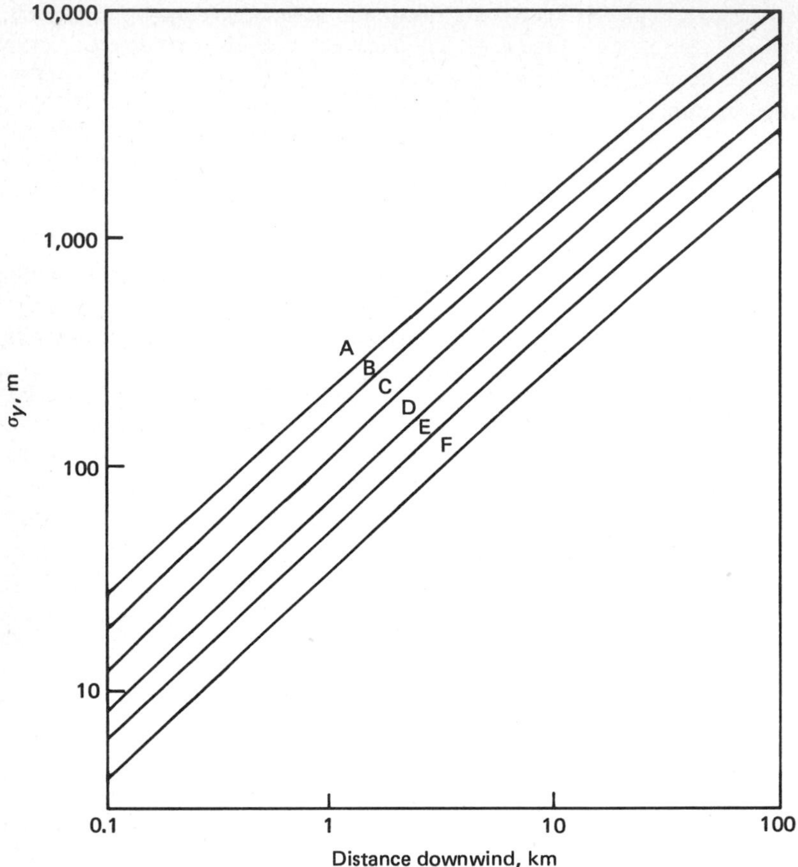

Figure 4-11 Horizontal dispersion coefficient as a function of downwind distance from the source.

rise when the vertical temperature gradient is equal to the adiabatic lapse rate (38):

$$\Delta h = \frac{V_e D}{\bar{u}}\left(1.5 + 2.68 \times 10^{-3}\, P \frac{\Delta T}{T_s} D\right)$$

where Δh = plume rise above stack, m
V_e = stack exit velocity, m/sec
D = stack internal diameter, m
$\bar{u}$ = mean wind speed, m/sec
P = atmospheric pressure, mbar (1 atm = 1,013 mbars)
T_s = stack gas temperature, K
$\Delta T = T_s$ − ambient temperature, K

The adiabatic lapse rate describes the rate of cooling with lifting (or heating upon descent) of a parcel of air with no heat exchange. The adiabatic lapse rate is

−5.4°F/1,000 ft. If the actual or environmental lapse rate is greater than the adiabatic rate, say −8°F/1,000 ft, then the Δh should be multiplied by 1.2; if the environmental lapse rate is less than the adiabatic value, then the Δh should be multiplied by 0.8.

The concentration of gases from ground-level point sources can be calculated on the basis of the following equation (38):

$$C_{x,y,z} = \frac{Q}{\pi \bar{u} \sigma_y \sigma_z} \exp - \left(\frac{y^2}{2\sigma_y{}^2} + \frac{z^2}{2\sigma_z{}^2} \right)$$

The concentrations of gases from ground-level area sources or line sources can be calculated utilizing the following mathematical model (38):

$$C_{x,0,0} = \frac{Q}{\pi \bar{u} (\sigma_y{}^2 + \sigma_{y_0}{}^2)^{1/2} \sigma_z}$$

where σ_{y_0} = 1/4 of emission width, m

The ground-level, center line concentrations of particulates from an elevated source can be calculated using the following model (38):

$$C_{x,0,0} = \frac{PQ}{\bar{u} \sigma_y \sigma_z} \exp - \left(\frac{1}{2} \frac{B^2}{\sigma^2} \right)$$

where P = weight fraction of effluent in a particular size range

$B = H - V_s X/100\,\bar{u}$

V_s = Stokes terminal settling velocity for particles in size range, cm/sec, determined from Stokes law as follows:

$$V_s = \frac{2 r_p{}^2 g \rho_p}{9 \mu_a}$$

where r_p = particle radius, cm

g = gravitational constant, 980 cm/sec^2

ρ_p = particle density, g/cm^3

μ_a = air viscosity, g/cm-sec^2

Utilization of the basic ground-level concentration model requires the selection of several P values and calculation of the resultant concentration. Summation of these concentrations over the range of particle sizes anticipated from the point source will yield the total anticipated ground-level concentrations.

The mathematical models shown above are basically oriented to point sources for gases and particulates and area sources for gases (40). Refinements of these models for use in highway impact statements have been made, and they are summarized elsewhere (41). Numerous other mathematical models have been developed for calculation of the dispersion of air pollution from aircraft, for urban transport, and for reactions of pollutants (42–45).

Utilization of mathematical models allows the development of isopleths of equal concentration around the source of emission for various types of air pollutants. These isopleths should be calculated bearing in mind the frequency of occurrence based on the meteorological data assembled earlier. The calculated ground-level concentrations at various positions in the project area should then be compared to the applicable ambient air standards. If the calculated ground-level concentrations are less than the standards, then fractional proportions should be indicated to present the degree of safety. If the calculated ground-level concentrations are greater than the ambient air quality standards, then control measures or abatement strategies must be developed. If there are other sources of air pollution in the area of the proposed action that would significantly contribute to the anticipated ground-level concentrations from the proposed action, these should also be included in the calculations so as to examine the possible additive or even potentiating effect of various air pollutant sources.

In addition to these determinations, it is necessary to examine the emissions from each alternative for the proposed action in light of applicable emission standards. It is presumed that the proposed action will be in compliance with the emission standards, and the extent of compliance involved in the proposed action should be discussed.

STEP 9: ABATEMENT STRATEGIES

If it is determined what ambient air standards are exceeded by the proposed action, then abatement strategies or control measures should be presented. Excellent references are available describing various control technologies that can be utilized for various gaseous and particulate emissions (46–54). It is beyond the scope of this presentation to give details on control methodologies.

SUMMARY

This chapter is oriented to the types of information and practical steps that can be employed in predicting the impacts of a given proposed action in terms of both the mesoscale viewpoint and the microscale viewpoint. Suggestions are provided for assessment or interpretation of the predicted impacts.

SELECTED REFERENCES

1 Perkins, Henry C.: "Air Pollution," p. 3, McGraw-Hill Book Company, New York, 1974.

2 *Ibid.*, p. 7.
3 U.S. Environmental Protection Agency: "Air Program Policy Statement," Washington, D.C., Aug. 1974.
4 National Air Pollution Control Administration: "Air Quality Criteria for Particulate Matter," p. 5, Publ. AP-49, Washington, D.C., Jan. 1969.
5 Rossano, August T.: "Air Pollution Control Guidebook for Management," pp. 87-91, Environmental Science Service Division, E.R.A., Stamford, Conn., 1969.
6 Barrett, Larry B., and Thomas E. Waddell: "Cost of Air Pollution Damage: A Status Report," pp. 59–60, Publ. AP-85, Environmental Protection Agency, Research Triangle Park, N.C., Feb. 1973.
7 Environmental Protection Agency: "Compilation of Air Pollutant Emission Factors," 2d ed., p. 1, Publ. AP-42, Research Triangle Park, N.C., Apr. 1973.
8 *Ibid.*, p. 8.1-4.
9 *Ibid.*, p. 8.10-1.
10 *Ibid.*, p. 2.1-3.
11 *Ibid.*, p. 7.7-1.
12 *Ibid.*, p. 3.2.1-4.
13 Chamblee, Carolyn P., and Gerald J. Nehls: "SAROAD Terminal User's Manual," pp. 1.0.1–1.0.2, Publ. EPA-450/2-73-004, Environmental Protection Agency, Research Triangle Park, N.C., Oct. 1973.
14 Oklahoma State Department of Health: "Oklahoma 1973 Annual Ambient Air Quality Report," p. 38, Oklahoma City, June 1974.
15 Holzworth, George C.: "Mixing Heights, Wind Speeds, and Potential for Urban Air Pollution Throughout the Contiguous United States," pp. 3–6, Publ. AP-101, Environmental Protection Agency, Research Triangle Park, N.C., Jan. 1972.
16 *Ibid.*, p. 26.
17 *Ibid.*, p. 31.
18 Hosler, Charles R.: Low-Level Inversion Frequency in the Contiguous United States, *Monthly Weather Rev.*, vol. 89, pp. 319–339, Sept. 1961.
19 *Ibid.*, p. 322.
20 Holzworth, *op. cit.*, p. 22.
21 Holzworth, *op. cit.*, p. 96.
22 Holzworth, *op. cit.*, p. 77.
23 Holzworth, *op. cit.*, p. 36.
24 Holzworth, *op. cit.*, p. 41.
25 Environmental Protection Agency: "Guide for Air Pollution Episode Avoidance," pp. 123–136, Publ. AP-76, Research Triangle Park, N.C., June 1971.
26 Environmental Science Service Administration: "Local Climatological Data, Annual Summary with Comparative Data, 1968, New Orleans, Louisiana," p. 2, U.S. Department of Commerce, Asheville, N.C., 1968.
27 Hesketh, Howard E.: "Understanding and Controlling Air Pollution," p. 50, Ann Arbor Science Publishers, Ann Arbor, Mich., 1972.
28 Smith, Maynard (ed.): "Recommended Guide for the Prediction of the Dispersion of Airborne Effluents," p. 2, The American Society of Mechanical Engineers, New York, May 1968.
29 Thom, Herbert C. S.: Tornado Probabilities, *Monthly Weather Rev.*, vol. 91, no. 10–12, pp. 730–736, 1963.

30 Seinfeld, John H.: "Air Pollution, Physical and Chemical Fundamentals, p. 28, McGraw-Hill Book Company, New York, 1975.
31 Chanlett, Emil T.: "Environmental Protection," pp. 270–271, McGraw-Hill Book Company, New York, 1973.
32 Perkins, *op. cit.*, p. 374.
33 Guide for Compiling a Comprehensive Emission Inventory, *Natl. Tech. Info. Svc., Rept.* PB-212-231, pp. 1-1-1-3, June 1972.
34 Oklahoma State Department of Health: State of Oklahoma Air Quality Control Implementation Plan, p. A-9, draft copy, Oklahoma City, Nov. 1971.
35 *Ibid.*, p. J-9.
36 Oklahoma State Department of Health: "Oklahoma 1973 Annual Ambient Air Quality Report," p. 7.
37 Environmental Protection Agency: "Federal Air Quality Control Regions," pp. 144–146, Publ. AP-102, Rockville, Md., Jan. 1972.
38 Hesketh, *op. cit.*, pp. 52–72.
39 Turner, D. Bruce: "Workbook of Atmospheric Dispersion Estimates," rev. ed., pp. 8–9, Publ. AP-26, Environmental Protection Agency, Research Triangle Park, N.C., 1970.
40 Mathis, Joe J., Jr., and William L. Grose: "A Review of Methods for Predicting Air Pollution Dispersion," Publ. NASA SP-322, National Aeronautics and Space Administration, Washington, D.C., 1973.
41 National Technical Information Service: "Air Quality Manual," 8 vols., Publs. PB-219811 to PB-219818, U.S. Department of Commerce, Dec. 1972.
42 Milford, S. N., et al.: Dispersion Modeling of Airport Pollution, *Proc. Aircraft Environ. Conf., Washington, D.C.*, pp. 147–162, Feb. 6–11, 1971.
43 Gillford, F. A., Jr.: Atmospheric Transport and Dispersion Over Cities, *Nucl. Safety*, vol. 13, no. 5, pp. 391–402, Oct. 1972.
44 Johnson, Warren B.: The Status of Air Quality Simulation Modeling, *Proc. Interagency Conf. Environ., EPA AEC, Livermore, Calif.*, pp. 114–127, 143–144, 1972.
45 Roth, Philip M.: Mathematical Modelling of Photochemical Air Pollution, vol. II, A Model and Inventory of Pollutant Emissions, *Atmos. Environ.*, vol. 8, no. 2, pp. 97–130, Feb. 1974.
46 Cross, Frank L., Jr.: "Handbook on Air Pollution Control," Technomic Publishing Co., Westport, Conn., 1973.
47 Danielson, John A. (ed.): "Air Pollution Engineering Manual," 2d ed., Environmental Protection Agency, Research Triangle Park, N.C., May 1973.
48 Ross, Richard D. (ed.): "Air Pollution and Industry," Van Nostrand Reinhold Company, New York, 1972.
49 National Air Pollution Control Administration: "Control Techniques for Particulate Air Pollutants," Publ. AP-51, Washington, D.C., Jan. 1969.
50 National Air Pollution Control Administration: "Control Techniques for Sulfur Oxide Air Pollutants," Publ. AP-52, Washington, D.C., Jan. 1969.
51 National Air Pollution Control Administration: "Control Techniques for Carbon Monoxide Emissions from Stationary Sources," Publ. AP-65, Washington, D.C., Mar. 1970.
52 National Air Pollution Control Administration: "Control Techniques for Carbon Monoxide, Nitrogen Oxide, and Hydrocarbon Emissions from Mobile Sources," Publ. AP-66, Washington, D.C., Mar. 1970.

53 National Air Pollution Control Administration: "Control Techniques for Hydrocarbon and Organic Solvent Emissions from Stationary Sources," Publ. AP-68, Washington, D.C., Mar. 1970.

54 National Air Pollution Control Administration: "Control Techniques for Nitrogen Oxide Emissions from Stationary Sources," Publ. AP-67, Washington, D.C., Mar. 1970.

Chapter 5

Prediction and Assessment of Impacts on the Water Environment

One of the major impacts from many actions is evidenced by changes in water quality both in the vicinity and downstream from project areas. Construction of reservoirs, power plants, industrial parks, and pipelines will cause short-term impacts on the water environment; and operation of these same facilities will result in longer-term impacts. General impacts on the water environment are related to hydraulic and hydrologic cycle changes as well as to the introduction of suspended and dissolved materials into receiving waters. This chapter addresses the data needs and associated technology for predicting and assessing impacts of proposed actions on water quality.

BASIC STEPS FOR PREDICTION AND ASSESSMENT

Basic steps associated with prediction of changes in the water environment and assessment of the impact of these changes are as follows:

1 Determine types and quantities of water pollutants emitted from all alternatives for meeting a given need during both construction and operational phases.
2 Determine the existing water quantity and quality levels for the surface watercourses in the area. Examine the frequency distributions and the median

and mean data for both water quantity and quality. If possible, consider historical trends of water quality. Note particularly the low flow utilized by the local regulatory agency for maintenance of water quality standards.

3 Document unique pollution problems that have occurred or are existing in local surface watercourses.

4 If relevant for the project alternatives, describe groundwater quantity and quality in the area, noting the depth of the groundwater table and direction of groundwater flow. Identify major local uses of groundwater, and delineate historical trends for groundwater depletion and pollution.

5 Assemble summary of key meteorological parameters for the area, noting particularly the monthly averages of precipitation, evaporation, and temperature.

6 Procure the applicable water quality standards for local surface watercourses and groundwater supplies if relevant. Specify applicability of effluent standards and required treatment technology and state whether the receiving stream is water-quality limited or effluent limited. Consider time schedules required for attaining applicable water quality standards.

7 Summarize the organic waste load allocation study for the area. Also procure extant information on inorganic, thermal, sediment, and bacterial waste loads. Identify known point sources of pollution, focusing specifically on unique discharges or wastewater constituents. Also enumerate the types of water uses in the area and summarize the quantities involved.

8 Determine the mesoscale impacts by calculating estimated daily quantities of water pollutants from the alternatives during both construction and operational phases and comparing these to existing waste loads in the drain area. Determine the percentage increase in these waste loads. Note existing water quality parameters that are good or poor relative to current or potential standards.

9 Consider construction phase impacts in terms of the following factors:

   a Time period of construction and the resultant time period of decreased water quality. Specify stream discharges and quality variations that would be anticipated during the construction phase.

   b Anticipated distance downstream of decreased water quality.

   c Implications of decreased water quality relative to downstream water users. If there are users that require certain water qualities, identify the required raw water quality characteristics and discuss the effects of decreased quality during the construction phase.

   d Specific construction specifications directed toward pollutant minimization.

10 Determine the microscale impacts by calculating specific downstream concentrations resulting from conservative pollutants, dissolved oxygen concentrations resulting from nonconservative (organic) pollutants, and temperatures resulting from thermal discharges. Consider these microscale impacts for both construction and operational phases. Compare calculated downstream concentrations with applicable water quality standards. Check if applicable effluent standards are met for existing facilities, or consider how they will be brought into compliance. In the case of new sources, identify necessary technology for compliance with new source performance standards.

11 If water quality or effluent standards are exceeded, consider mitigation or control measures.

12 Consider operational impacts of alternatives in terms of the following factors:

a Frequency distribution of decreased quality and quantity.
b Effects of sedimentation on the stream bottom ecosystem.
c Fate of nutrients by incorporation into biomass.
d Reconcentration of metals, pesticides, or radionuclides into the food web.
e Chemical precipitation or oxidation/reduction of inorganic chemicals.
f Anticipated distance downstream of decreased water quality and the implications for water users and related raw water quality requirements.
g General effects of any water quality changes on the stream ecosystem.
h Unique water quality changes that occur as a result of water impoundment and thermal stratification.

These twelve steps are directed toward determining the water impacts of alternatives on the mesoscale and microscale levels. Mesoscale assessment considers the contribution of alternatives to area water pollutant sources, both point and non-point. Microscale assessment involves comparisons of calculated concentrations of water pollutants to applicable water quality standards. Both levels of impact assessment are necessary in order to adequately address water quality impacts associated with proposed actions.

Steps 9 and 12 are directed toward a general discussion of the features and ramifications of construction phase and operational phase impacts. Step 12 is suggestive of the types of factors that should be considered under operational phase impacts, but this list is not intended to be all inclusive of items that should be addressed. Detailed coverage responsive to step 12 should be related to the type of project and the environmental setting. Steps 2-7 should be summarized in the environmental setting section of the impact statement, with the remaining steps included in the environmental impact section.

The organization of this chapter is primarily based on the twelve steps identified above. However, prior to a step-by-step discussion and analysis, brief information on basic water pollution considerations is presented.

BASIC INFORMATION ON WATER POLLUTION

Water pollution can be defined in a number of ways; however, the basic elements of most definitions are the concentrations of particular pollutants in water for sufficient periods of time to cause certain effects. If the effects are health related, such as those caused by pathogenic bacterial intrusion, the term "contamination" is appropriate. Effects that have to do with limitations on water availability due to certain water quality requirements related to usage can serve as a basis for defining a condition of water pollution. "Nuisance" refers to aesthetically displeasing effects created by oils, grease, or other floating materials.

Potential water quality impacts must be considered based on a clear delineation of various water quality characteristics. It is necessary to utilize a manifold evaluation of water quality characteristics in order to develop a total evaluation of existing water quality as well as microscale changes that might result from project alternatives. Water quality can be described in terms of physical, chemical, and bacteriological parameters.

Physical parameters include color, odor, temperature, solids (residues), oils, and grease. Color can be defined relative to type and density, the type being related to whether it is true color (dissolved) or apparent color (filterable). Odor is described by type and threshold odor number, which is related to the odor-free water required for diluting an odorous water sample to a nonodorous level. Figure 5-1 shows the relationships between various solids parameters used in water quality characterization. Total solids are comprised of suspended and dissolved solids, and each of these fractions can be further divided into organic (volatile) and inorganic (fixed) components. Turbidity is another measure of the solids content, and it is related to light transmittance through water. Settleable solids describe the materials present in solution that will settle by gravity in a 1-hr period. Specific conductance (conductivity) is a measure of the inorganic dissolved solids present in ionic form. In surface watercourses oil and grease is of interest relative to nuisance considerations.

Chemical parameters can be subdivided into organic and inorganic constituents. Several tests can be employed to describe the organic characteristics of water. The most-used test is the biochemical oxygen demand (BOD), which is defined as the amount of oxygen required by bacteria in decomposing organic material in a sample under aerobic conditions at 20°C over a 5-day incubation period. A typical BOD response curve for domestic sewage is shown in Fig. 5-2. The first-stage BOD represents the carbonaceous demand plus the nitrogeneous oxygen demand (NOD). Other tests that describe the organic content of water include the chemical oxygen demand, total organic carbon, and total oxygen demand.

Inorganic parameters of potential interest in water quality characterization include salinity, hardness, pH, acidity, alkalinity, and the content of iron, manganese, chlorides, sulfates, sulfides, heavy metals (Hg, Pb, Cr, Cu, Zn), nitrogen (organic, ammonia, nitrite, nitrate), and phosphorus. Salinity and chloride content are a measure of the salt in water. Hardness is caused primarily by divalent metallic

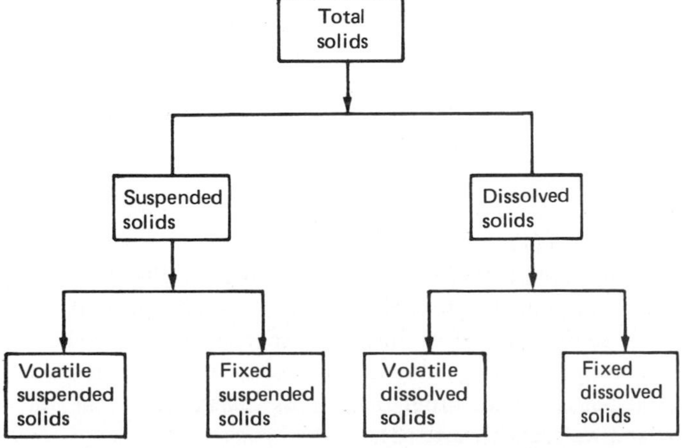

Figure 5-1 Relationships of various solids tests used for water quality characterization.

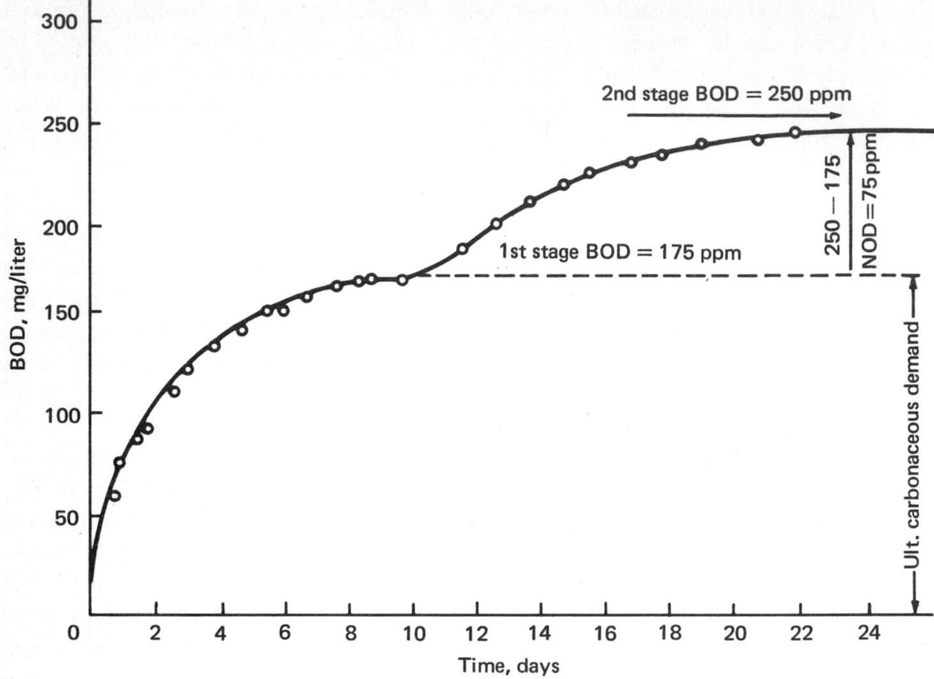

Figure 5-2 BOD response curve.

cations that have soap-consuming potential, the major ones being calcium and magnesium. Nitrogen and phosphorus contents are of interest due to their nutrient characteristics.

Bacteriological parameters include coliforms, fecal coliforms, specific pathogens, and viruses. Total coliform and fecal coliform organisms are used as indicators of the presence of pathogens. Specific pathogens such as salmonella organisms may be relevant for certain environmental impact studies. Minimal information is available on virus types and concentrations in surface waters and pollutant discharges therein.

There are two main sources of water pollutants in surface watercourses, namely, point sources and non-point sources. The total waste load in a stream is represented by the sum of all point and non-point pollutant sources. Figure 5-3 illustrates the relationship between land usage and these sources relative to the resulting quality of receiving waters (1).

Individual sources of water pollutants in aqueous systems are manifold, and, as shown in Fig. 5-3, resultant water quality probably will not be attributable to single types of pollutants or single sources. Some of the major sources of water pollution in the United States are municipal wastes, industrial wastes, agricultural wastes, soil erosion, accidental spillage of oil and other hazardous substances, acid mine drainage, mine sediments, and watercraft waste (2).

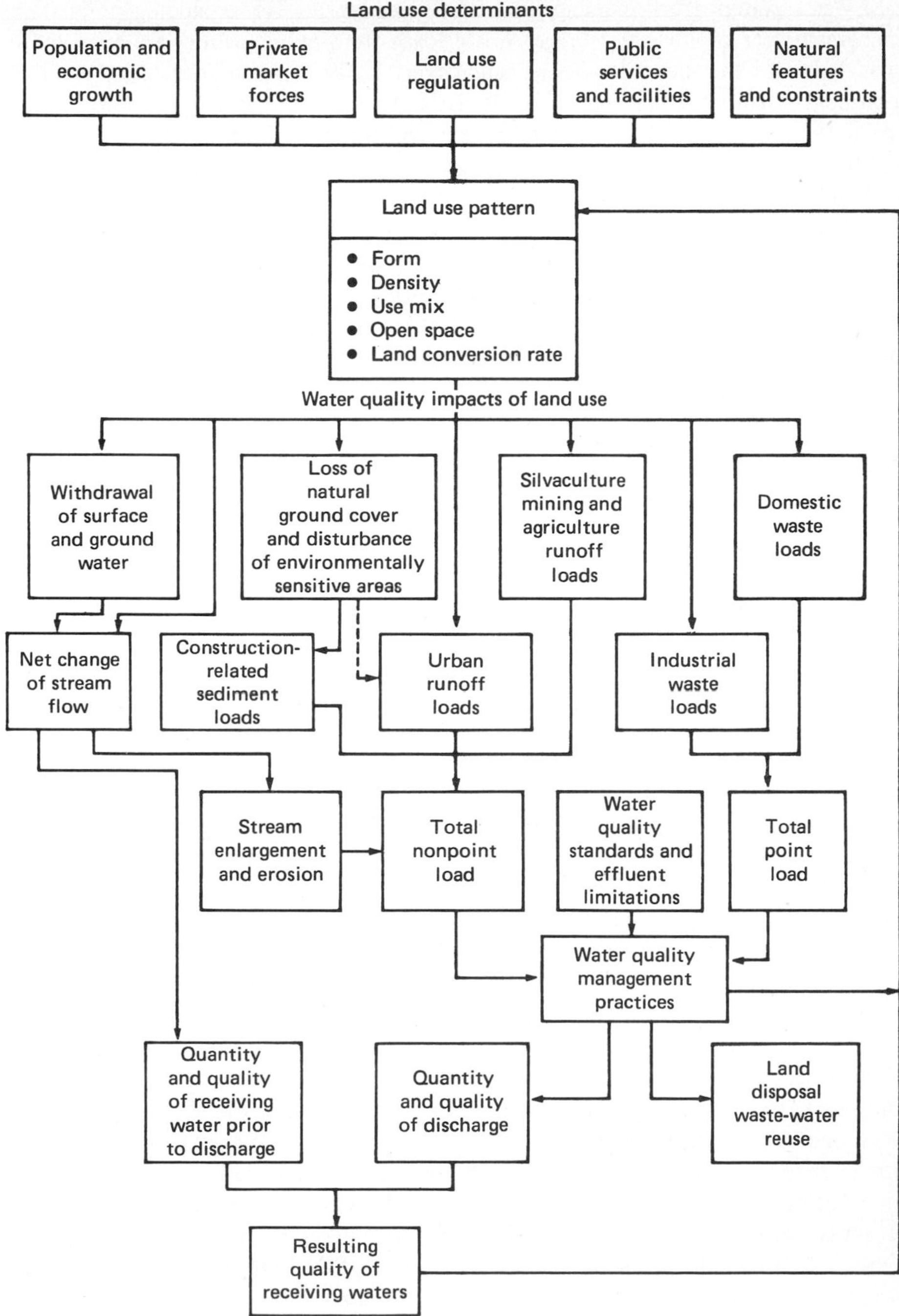

Figure 5-3 Schematic diagram of the land use/water quality relationship.

Waste waters from municipalities are increasing due to expanding populations and greater per capita water usage. Waste loads from municipalities are expected to quadruple in the next 50 years, and over 1,000 communities outgrow their treatment facilities each year (3). It is estimated that only one-third of the population is served with adequate sewers and waste-water treatment facilities. In addition, storm and combined sewers contain a collage of pollutant materials washed from streets and urban areas.

There are more than 300,000 water-using factories and industries in the United States (4). The growth rate of industrial waste-water discharges is greater than that for municipal discharges. Quality characteristics of industrial wastes vary considerably depending upon the type of industry. A useful parameter in describing industrial wastes is population equivalent:

$$\mathrm{PE} = \frac{A \times B \times 8.34}{0.17}$$

where PE = population equivalent based on organic constituents in the industrial waste
A = industrial waste flow, mgd
B = industrial waste BOD, mg/liter
8.34 = number of lb/gal
0.17 = number of lb BOD/person-day

Agricultural wastes include irrigation return flows as well as runoff from feedlots. Irrigation return waters comprise between 30 and 40 percent of the total U.S. water usage of approximately 400 billion gal/day. These waters exhibit salinities that are severalfold greater than unused irrigation water; also, hardness, total dissolved solids, and turbidity are at increased levels. Irrigation return flows may also exhibit increases in nitrogen, phosphorus, and pesticide contents. Runoff waters from feedlots have high organic, nutrient, and solids concentrations and contain microorganisms that are potentially pathogenic to animals and humans. As an illustration of the organic strength, a feedlot with 10,000 cattle produces runoff with a BOD content equivalent to the untreated waste water from a city of 45,000 persons.

Soil erosion is one of the major water pollutants in the United States in terms of quantity. Sediment from soil erosion is greatest in streams in the far west and southwest, for example, in the states of Nevada, Arizona, New Mexico, Texas, and Oklahoma. The total quantity of solids from soil erosion is approximately 700 times greater than the total from municipal waste-water discharges (5).

Accidental spillage of oil and other hazardous substances into watercourses is of fairly recent interest. Events such as the grounding of the Torrey Canyon in 1967 and the Santa Barbara offshore oil leak in 1969 have dramatized that these occurrences can cause devastating and extensive damage to the aqueous environment.

Acid mine drainage is of particular importance in the Appalachian region of the United States. It is estimated that acid mine drainage has deteriorated water quality along more than 10,000 mi of streams in the Appalachian area. Problems are generated when water and air react with sulfur-bearing minerals in mines or refuse piles to form sulfuric acid. Acid mine drainage also contains iron, copper, lead, zinc, and other metals that may be toxic to aquatic life. Sediment yields from strip-mining operations are also important since over 230 mi^2/yr are disturbed by surface mining operations in the total United States, and sediment yields may range up to 30,000 tons/mi^2-yr from strip-mine areas (6).

There are over 8 million watercraft that navigate in U.S. waters (7). These watercraft discharge sanitary wastes, oils, litter, ballast, and bilge waters. Although the total quantities of waste discharged from watercraft are small relative to other pollutant sources in the United States, they are important since most of the discharges are in high-use shoreline and harbor areas.

The effects of water pollutants on receiving water quality are manifold and dependent upon the type and concentrations of pollutants. Soluble organics, as represented by high BOD wastes, cause depletion of oxygen. Trace quantities of certain organics cause undesirable tastes and odors, and some may be biomagnified in the food web. Suspended solids decrease water clarity and hinder photosynthetic processes; if solids settle and form sludge deposits, changes in benthic ecosystems result. Color, turbidity, oils, and floating materials are of concern due to their aesthetic undesirability and possible influence on water clarity and photosynthetic processes. Excessive nitrogen and phosphorus can lead to algal overgrowth with concomitant water treatment problems resulting from algae decay and interference with treatment processes. Chlorides cause a salty taste to be imparted to water, and in sufficient concentration limitations on water usage can occur. Acids, alkalis, and toxic substances have the potential for causing fish kills and creating other imbalances in stream ecosystems. Thermal discharges can also cause imbalances as well as reductions in stream waste assimilative capacity. Stratified flows from thermal discharges minimize normal mixing patterns in receiving streams and reservoirs.

STEP 1: IDENTIFICATION OF WATER POLLUTANTS

The first step in prediction and assessment of water quality impacts involves identification of types and quantities of water pollutants emitted from construction and operation of each alternative under consideration. One approach for identifying water pollutants is to review EISs prepared for similar projects. Another approach is to utilize the unit waste generation factor, which is defined as the rate at which a pollutant is released to a drainage area or watercourse as a result of some activity, such as land clearing or production by industry, divided by that activity.

Table 5-1 shows erosion rates reported for various non-point sources of sediment (8). It is noted that highway construction and agricultural operations yield the highest erosion rates, whereas natural land drainage yields among the lowest erosion rates. Table 5-2 also provides information on erosion rates from various land uses (9).

Table 5-3 shows annual non-point source pollutional impacts by land use and

Table 5-1 Erosion Rates Reported for Various Sediment Sources

Sediment	Erosion rate, tons/mi^2-yr	Geographic location	Comment
Natural	15–20	Potomac River basin	Native cover
	32–192		Native cover
	200	Pennsylvania and Virginia	Natural drainage basin
	320	Mississippi River basin	Throughout geological history
	13–83	Northern Mississippi	Forested watershed
	25–100	Northwest New Jersey	Forest and underdeveloped land
	115		Soils eroding at the rate they form
Agricultural	12,800	Missouri Valley	Loess region
	13,900	Northern Mississippi	Cultivated land
	1,030	Northern Mississippi	Pasture land
	10,000–70,000		Continuous row crop without conservation practices
	200–500	Eastern U.S. Piedmont	Farmland
	320–3,840		Established as tolerable erosion
Urban	50–50,000	Kensington, Maryland	Undergoing extensive construction
	1,000–100,000		Small urban construction area
	1,000	Washington, D.C. area	750 mi^2 area average
	500	Philadelphia area	
	146	Washington, D.C. area watersheds	As urbanization increases
	280		
	690		
	2,300		
Highway construction	36,000	Fairfax Co., Virginia	Construction on 179 acres
	50,000–150,000	Georgia	Cut slopes

identifies unit waste generation factors for BOD, nitrogen, and phosphorus (10). Table 5-4 summarizes reported nutritional loadings from various land uses as well as rainfall, domestic sewage, and septic tanks (11). Table 5-5 is a summary of reported agricultural waste generation factors (12). The typical composition of domestic sewage in the United States is shown in Table 5-6 (13). Finally, Table 5-7 provides

Table 5-2 Representative Rates of Erosion from Various Land Uses

Land use	Erosion rate MT/km^2-yr	Erosion rate Tons/mi^2-yr	Erosion rate Relative to forest = 1
Forest	8.5	24	1
Grassland	85	240	10
Abandoned surface mines	850	2,400	100
Cropland	1,700	4,800	200
Harvested forest	4,250	12,000	500
Active surface mines	17,000	48,000	2,000
Construction	17,000	48,000	2,000

Table 5-3 Annual Non-Point Source Pollutional Impacts by Land Use

Land use	Quality parameters						
	Imperviousness, degree	BOD, lb/acre-yr	Nitrogen, lb/acre-yr	Phosphorus, lb/acre-yr	PO_4, lb/acre-yr	Erosion	Temperature change
Natural							
Forests (grasslands)	Low	Small	0.89	0.089	0.3	Low	
Agriculture			2.4	0.92			
(farms)	Low		24	3.88		High potential	Small
Feedlots	High	Varying with animal type, density, and management practices				High potential	Small
Single-family residential	Low-medium	5.9	0.6	0.2	0.6		Varies with degree of cover removed and surfaces heated (5°–15°)
Multifamily residential	Medium	14	2.5	0.67	2.1		
Commercial	High	43	2.4	1.3	3.9		
Industrial	Very high	23.4	3.3	0.56	1.7		
Resource extraction	Varies with methodology and management practices						
Recreation	Varies with intensity of use—extremely sensitive to overuse						
Urban and road construction						30,000–150,000 tons/acre	

Table 5-4 Summary of Reported Nutritional Loadings

Source	Nitrogen, lb/acre-yr	Phosphorus, lb/acre-yr
Farmland runoff	–	0.35 (total)
Good management	1.0 (NO_3)	0.10 (PO_4)
Poor management	5.3 (NO_3)	0.25 (PO_4)
4 lb/acre-yr applied	0.7–3.0 (total)	0.06–0.2 (total)
Irrigation return flow	2.45–24.0 (total)	0.95–3.88 (total)
Urban runoff	8.2 (total)	0.87 (total)
	2.95–15.97 (organic)	3.32-20-20 (PO_4)
Rainfall at 30 in./yr	4.8–32 (NO_3)	0.18–0.54 (PO_4)
	0.14–9.5 (inorganic)	–
	10 (N)	–
Forest runoff	1.30–296 (total)	0.32–0.77 (total)
	0.5 (total)	0.03–0.06 (total)
Leaf litter	6.5×10^{-7} (NO_3)	–
Domestic waste	6.9–10.7[a] (total)	1.3–3.8[a] (total)
Septic tanks	8[a] (NO_3, ground water seepage)	

[a]lb/capita-yr.

a summary of bacterial loads that can be anticipated from untreated and treated sewage and urban runoff (14).

The primary point of Tables 5-1–5-7 is to provide information to enable calculation of the total quantity of water pollution anticipated from a given activity. This information is basic to prediction of mesoscale and microscale water quality impacts.

STEP 2: DESCRIPTION OF EXISTING WATER QUANTITY AND QUALITY LEVELS

The second step involves assembling information on existing water quantity and quality levels in the area of the project, particularly focusing on quality parameters related to anticipated water pollutants to be emitted from construction and operational phases of the project. Sources of water quality information include relevant city, county, and state water resources agencies and private industries that

Table 5-5 Agricultural Pollutional Loads

	Total nitrogen, lb/acre-yr	Total phosphorus, lb/acre-yr
Surface	2.45–28.3	0.68–3.99
Subsurface	38–166	2.5–8.9

Table 5-6 Typical Characteristics of Domestic Sewage in the United States

Constituents	Weak	Medium	Strong
Physical characteristics			
Color, nonseptic	Gray	Gray	Gray
Color, septic	Gray-black	Blackish	Blackish
Odor, nonseptic	Musty	Musty	Musty
Odor, septic	Musty—H_2S	H_2S	H_2S
Temperature (average), °F	55-90	55-90	55-90
Total solids,[a] mg/liter	450	800	1,200
Total volatile solids, mg/liter	250	425	800
Suspended solids, mg/liter	100	200	375
Volatile suspended solids, mg/liter	75	130	200
Settleable solids, ml/liter	2	5	7
Chemical characteristics			
pH, units	6.5	7.5	8.0
Cl, SO_4, Ca, Mg, etc.[a]			
Total nitrogen, mg/liter	15	40	60
Organic nitrogen, mg/liter	5	14.5	19
Ammonia nitrogen, mg/liter	10	25	40
Nitrate nitrogen, mg/liter	—	0.5	1.0
Total phosphate, mg/liter	5	15	30
Biological characteristics			
Total bacteria, counts/100 ml	1×10^8	30×10^8	100×10^8
Total coliform, mpn[b]/100 ml	1×10^6	30×10^6	100×10^6
BOD	100	200	450

[a]Quite variable depending on natural water quality of region.
[b]mpn, most probable number.

Table 5-7 Summary of Bacterial Loads

Source	Coliform/100 ml	Coliform/capita-day
Raw sewage	1 million	3.7×10^9
Treated sewage		
Primary treatment	500,000	1.8×10^9
Activated sludge	60,000	2.3×10^8
Biological plus chlorination	15,000	5.7×10^7
Urban runoff, 34 in. rainfall/yr	1.275 million	5.5×10^{10}/acre-day
Irrigation returns	—	Insignificant

have monitoring programs. Another source is the STORET system of the EPA, with STORET meaning storage and retrieval of water quality data.

Table 5-8 shows water quality information obtained from STORET (15). This information has been extracted from the computer printout and is organized by physical, chemical, and bacteriological parameters. The historical record of information at a given sampling station varies in terms of the number of years of record, number of samples analyzed, and number of parameters reported per sample analyzed. It is noted that Table 5-8 identifies several points where erroneous data were contained in the STORET system. The procurement of water quality and quantity information from data storage systems must be coupled with an examination for reliability. A professional interface is necessary in order to effectively utilize the information obtained.

Since water quality standards vary with the beneficial uses assigned for particular streams or stream segments, it may be necessary to evaluate existing water quality relative to various standards. This step is important for projects that may have an impact over large distances in a single stream and for other projects, such as pipelines, that may cross numerous streams and several states. Interpretation of water quality data involves comparison with water quality standards. A rating system that has been used is based on assignment of negative values when standards for various parameters are not met and assignment of zero values for parameters in compliance with applicable standards. Table 5-9 includes a list of numerical assignments used for this stream quality rating system (15).

The rationale for Table 5-9 was that not meeting a bacteriological standard was three times as important as not meeting a physical standard, and not meeting a chemical standard was twice as important as not meeting a physical standard; therefore, the negative numerical assignments for the bacteriological classification were three times greater than those for the physical classification, and the chemical classification assignments were two times greater than those for the physical classifications. In addition, when the mean value of a parameter did not meet the applicable standard, it was considered three times as important as when the maximum or minimum did not meet the standard; therefore, the negative numerical assignments for the mean values were three times greater than those for the maximum or minimum values. Finally, stations from which 10 or more samples have been analyzed are twice as important as those stations from which fewer than 10 samples have been analyzed; therefore, the negative numerical assignments for stations with 10 or more samples were twice as great as the comparable assignments for stations with less than 10 samples. On these bases application of the stream water quality rating system to the STORET data in Table 5-8 is as follows: total dissolved solids = −2; dissolved oxygen = −4; ammonia = −4; NO_2 and NO_3 = −4; phosphates = −16; fecal coliform = −6; total = −36.

Determination of a numerical score for a station in the study was based on a composite summation of negative numerical assignments applied when standards were not met. The zero or negative scores for each station were assigned a rating on the following basis:

Table 5-8 Water Quality Information from STORET[a]

Parameter	Units	No. of samples	Begin date	End date	Mean	Max.	Min.	Standard
Physical								
Turbidity	JTU[b]	71	62/06/04	72/05/22	65	230	0	
Total dissolved solids	mg/liter	61	62/06/04	72/06/27	347	(570)	*c*	500
Temperature	°F	72	62/06/04	72/06/27	56	82	32	90
Suspended solids	mg/liter	—						
Chemical								
pH	units	69	62/06/04	72/06/27	*c*	*c*	*c*	6.5–9.0
Hardness	mg/liter	33	62/06/04	71/12/01	289	376	144	
Dissolved oxygen	mg/liter	67	62/06/04	72/06/26	*c*	*c*	(0)	5.0
BOD5[d]	mg/liter	26	62/06/04	67/07/06	6	10	2	30
Hydrocarbons	mg/liter	—						
Ammonia	mg/liter	43	62/06/04	62/06/27	1.1	(18)	0.0	1.5
NO_2 and NO_3	mg/liter	47	62/06/04	72/06/27	2.4	(70)	0.0	10
Phosphates	mg/liter	33	64/05/21	71/12/01	(11)[e]	(35)	0.0	0.15
Bacteriological								
Fecal coliform	mpn/100 ml	14	62/06/04	72/06/27	226	(1,900)	0	400

[a]STORET identification no. 47096. Stream: Rock River. Station location: Route 64 Town Bridge near Oregon, Illinois.
[b]JTU, Jackson turbidity units.
[c]Erroneous STORET data.
[d]BOD5, 5-day biochemical oxygen demand.
[e]Parentheses indicate standard not met.

Table 5-9 Numerical Assignments for Stream Quality Rating System

		Numerical assignments for parameter classifications		
No. of samples	Value	Physical	Chemical	Bacteriological
<10	Max.	−1	−2	−3
	Min.	−1	−2	−3
	Mean	−3	−6	−9
⩾10	Max.	−2	−4	−6
	Min.	−2	−4	−6
	Mean	−6	−12	−18

A–meets all existing standards; excellent water quality (score = 0).

B–meets most existing standards; good water quality (score = −1 to −10).

C–does not meet few existing standards; fair water quality (score = −11 to −30).

D–does not meet several existing standards; poor water quality (score = −31 and greater).

Based on these ratings, water quality at the station shown in Table 5-10 would be in the D category. Table 5-10 is a summary of the application of this rating system to several streams in Illinois. The use of a rating system is of value in graphically portraying existing water quality.

Stream flow information can be obtained from the U.S. Geological Survey as well as from local and state agencies dealing with water resources. Example data on stream flows for the Mountain Fork River near Eagletown, Oklahoma, are shown in Table 5-11 (16). Statistical analysis of the data allows the development of a frequency discharge relationship. Figure 5-4 shows that 10 percent of the time the flow in the Mountain Fork River is ⩾3,500 cfs, 50 percent of the time it is ⩾350 cfs, and 90 percent of the time it is ⩾10 cfs (17).

One of the key concerns with regard to stream flow is the flow frequency, which is utilized for calculation of compliance with water quality standards. In some instances the 7-day, 2-year low flow is utilized; in other cases the 7-day, 10-year low flow is required. The phrase 7-day, 2-year low flow indicates that this is the minimum flow that occurs over a 7-day period on a frequency of once every 2 years. Flow frequency information is available from local and state water resources agencies.

STEP 3: UNIQUE POLLUTION PROBLEMS

The primary purpose of this step is to identify any unique pollution problems that have occurred in the project area. This is necessary in order to adequately describe the environmental setting, to indicate a familiarity with the area, and to focus on environmentally sensitive parameters. Examples of unique pollution problems that should be identified include fish kills, excessive algal growth, and thermal discharges causing stratified flow. Many sources can be used to obtain information on unique pollution problems. Local and state water resources agencies constitute one source, and conservation groups, another. Local newspapers provide historical documentation of pollution concerns.

STEP 4: DESCRIPTION OF GROUNDWATER QUANTITY AND QUALITY

This particular step may not be required for all project types, but for alternatives that have potential for groundwater impact, this is a necessary step. The basic

Table 5-10 Illinois Stream Quality Rating

	No. of reported parameters			Values			
Stream	1–4	5–8	9–13	<10	≥10	Score	Rating
Apple River			X	X		–16	C
Plum River			X	X		–20	C
Vermilion River			X		X	–40	D
Rock River			X		X	–30	C
Rock River			X		X	–36	D
Rock River			X	X		–13	C
Rock River			X	X		–4	B
Rock River	X			X		0	A
Rock River			X		X	–20	C
Rock River			X		X	–64	D
Rock River			X		X	–20	C
Rock River			X		X	–32	D
Rock River			X		X	–32	D
Rock River			X		X	–30	C
Rock River			X		X	–44	D
Elkhorn River			X	X		–11	C
Kankakee River			X		X	–54	D
Kankakee River			X	X		–4	B
Kankakee River			X		X	–62	D
Kankakee River			X		X	–70	D
Kankakee River			X	X		0	A
Kankakee River			X		X	–24	C
Iroquois River			X		X	–36	D
Iroquois River			X	X		0	A
Iroquois River			X	X		–12	C
Mazon River		X		X		–10	B
Mazon River		X		X		–35	D
Illinois River	X			X		0	A
Illinois River	X			X		0	A
Illinois River			X	X		–29	C
Illinois River			X	X		–25	C
Illinois River			X		X	–42	D
Illinois River			X		X	–30	C

Table 5-11 Surface Water Records for Mountain Fork River near Eagletown, Oklahoma[a,b]

Water year	Oct.	Nov.	Dec.	Jan.	Feb.	Mar.	Apr.	May	June	July	Aug.	Sept.	Total
1951	17,220	9,610	5,260	36,050	235,500	78,600	67,900	42,890	131,400	151,400	5,930	10,460	792,200
1952	21,850	128,700	94,120	102,700	71,340	122,200	435,600	47,960	6,850	808	137	1.0	1,032,000
1953	0	69,420	79,300	83,460	99,250	209,000	319,500	256,300	4,600	151,000	8,310	1,090	1,281,000
1954	101	1,070	20,960	149,000	92,350	19,100	72,800	146,600	4,650	313	1.6	6,980	513,900
1955	171,600	27,720	57,510	61,690	117,000	153,000	109,400	66,830	11,920	5,170	5,690	16,590	804,100
1956	28,560	2,370	5,210	17,070	212,300	50,760	29,490	68,000	3,240	540	192	257	418,000
1957	4.8	16,150	43,880	124,000	114,700	177,100	487,700	302,400	138,500	5,350	2,660	34,520	1,447,000
1958	10,440	117,200	58,060	97,800	33,710	214,900	152,300	223,600	55,320	14,160	14,350	16,100	1,008,000
1959	21,650	157,100	26,880	31,150	64,150	131,400	80,490	42,700	17,610	54,410	9,910	10,220	647,700
1960	87,530	29,820	210,900	146,800	93,320	83,960	42,210	311,600	42,670	93,000	38,040	5,050	1,185,000
1961	5,970	11,050	240,500	63,140	94,930	184,600	91,490	161,900	17,810	54,420	24,290	18,960	969,100
1962	27,170	133,200	152,400	172,500	103,000	106,400	106,200	25,110	14,390	1,370	2,940	11,360	856,000
1963	111,600	43,530	30,060	40,780	10,640	135,900	56,870	25,540	2,240	4,190	2,750	1,160	465,300
1964	22	0	921	976	19,660	157,100	177,800	29,430	2,840	127	32,210	49,630	470,700
1965	20,400	76,390	38,010	77,320	208,300	77,860	48,240	165,500	110,700	8,980	841	24,510	857,100
1966	12,140	6,300	12,560	32,410	157,200	32,190	120,400	130,300	2,570	456	36,630	8,960	552,100
1967	1,050	1,090	6,920	11,270	13,110	58,540	157,200	194,500	78,460	48,060	1,640	14,410	586,200

[a]Drainage area: 787 mi^2. Average discharge: 39 yr, 915,800 acre-ft/yr. Extremes: max. discharge, 101,000 cfs, May 20, 1960; min., no flow at times.
[b]Compiled from U.S. Geological Survey surface water records. Values are monthly and yearly discharge, in acre-ft.

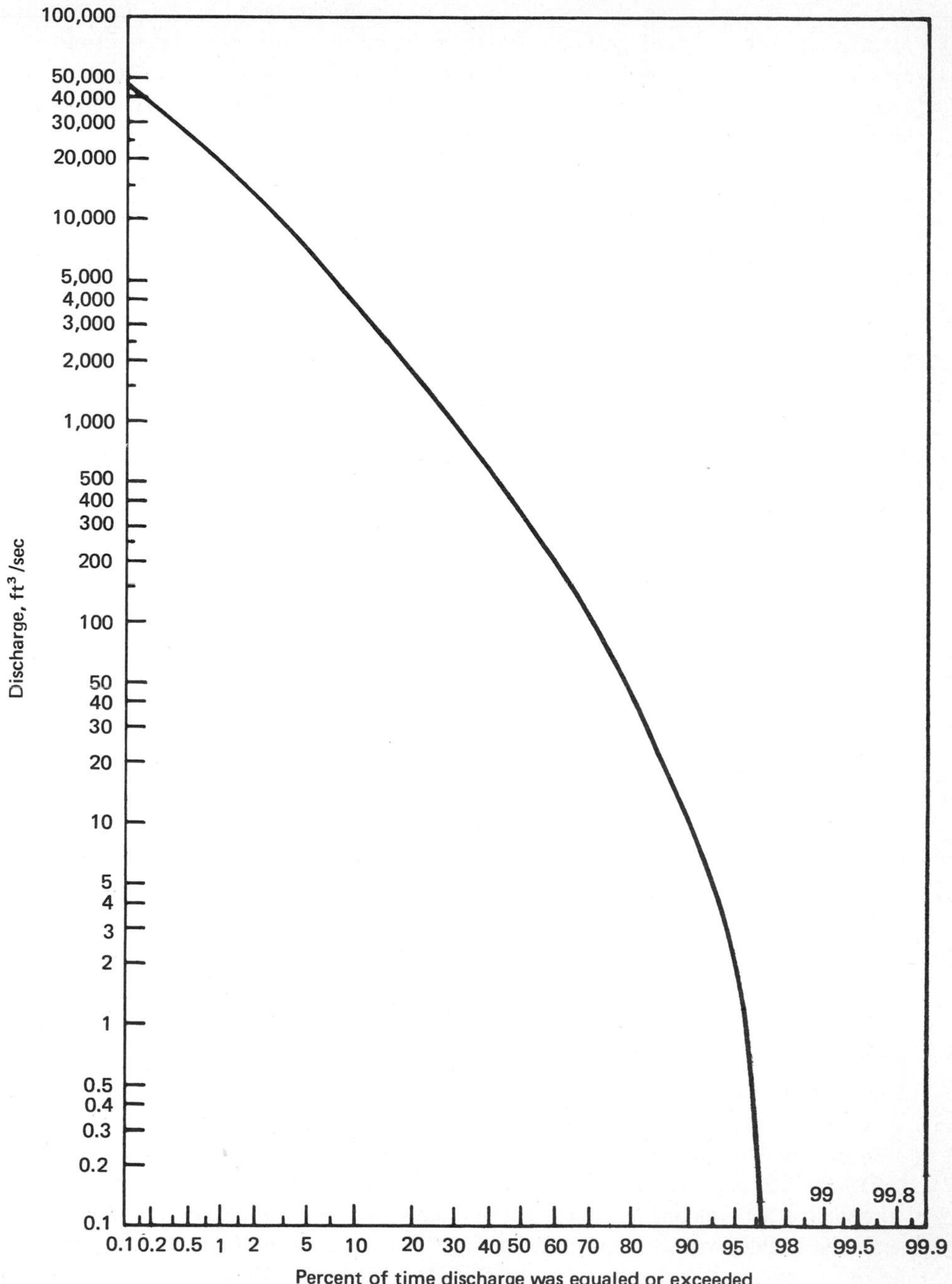

Figure 5-4 Duration curve of daily discharge, Mountain Fork River, Eagletown, Oklahoma. *Location:* latitude 34°02′30″, longitude 94°37′15″, on downstream side of pier of bridge on U.S. Highway 70, 2 miles west of Eagletown and at mile 8.9. *Drainage area:* 787 mi². ***Average discharge:*** 37 yr (water years 1925, 1930–1965) 1,291 cfs (934,600 acre-ft/yr). ***Daily discharge:*** max., 62,100 cfs; median, 350 cfs; min., no flow.

purpose is to determine the depth to the groundwater table in the area and to identify the direction of groundwater flow. Major users of groundwater from the area should be enumerated, as well as historical trends in groundwater depletion or quality deterioration. Sources of information include local and state agencies dealing with water resources and the U.S. Geological Survey.

STEP 5: SUMMARY OF METEOROLOGICAL INFORMATION

Meteorological data are required in order to predict and assess air quality impacts associated with proposed actions. In addition, certain climatological factors such as precipitation, evaporation, and air temperature are important in terms of predicting and assessing water quality impact. In addition, precipitation and temperature information may need to be considered for proper construction scheduling. The primary sources of information include local and state water resources agencies as well as the National Oceanographic and Atmospheric Administration.

STEP 6: WATER QUALITY STANDARDS

The Federal Water Pollution Control Act Amendments of 1972 (PL 92-500) established basic water quality goals and policies for the United States (18). Some of these are

1 The discharge of pollutants into navigable waters should be eliminated by 1985.

2 Wherever attainable, an interim goal of water quality, which provides for the protection and propagation of fish, shellfish, and wildlife and provides for recreation in and out of water, should be achieved by July 1, 1983.

3 The discharge of toxic pollutants in toxic amounts should be prohibited.

Strategies for point-source control have been developed, with every point source being subject to an effluent standard and a water quality standard, with the most stringent treatment requirements being applied. Control strategies for non-point-source pollution are being developed. Effluent standards for point sources represent requirements in terms of the quality characteristics of the effluent discharged from municipal and industrial waste-water treatment plants. Water quality standards are applicable to surface-water courses and represent the quality characteristics required to allow certain water uses.

A schedule has been established for point-source compliance with standards (19). By 1977 publically owned treatment works (municipalities) are to be in compliance with secondary treatment standards, and by 1983 they are to have the best practicable waste treatment technology. Industrial point sources are required to be in compliance with best practicable treatment by 1977 and best available treatment by 1983. New industrial point sources are to be planned in accordance with new source performance standards. Pretreatment standards are to be met by industrial sources prior to discharge into publically owned treatment works.

Secondary treatment standards describe treatment efficiency levels as well as required effluent qualities from publically owned treatment works (20). Table 5-12 summarizes these standards. All publically owned treatment works constructed with federal funds after June 30, 1974 must achieve best practicable waste treatment technology. Although no specific definition has been developed for this term, It must be considered relative to whether the receiving stream is classified as water quality limited or effluent limited. Water quality-limited segments cannot be expected to meet established water quality standards even if all point sources achieve effluent limitations such as secondary treatment for publically owned treatment works and best practicable treatment for industrial discharges. Effluent-limited segments are those where water quality standards can be achieved after all point sources meet effluent limitations. A 1973 inventory of stream segments in the United States indicated that 1,588 segments were water quality limited and 1,515 were effluent limited (21).

One of the key considerations in defining best practicable waste treatment technology is related to the most cost-effective treatment system. A cost-effective solution is one that will minimize total resources costs to the nation over time in order to meet federal and state water quality standards and treatment requirements. Total resources costs include capital cost (construction and land); operation, maintenance, and replacements; and social and environmental costs (22).

An example of effluent limitations for shrimp-processing plants is shown in Table 5-13 (23). Best practicable technology and best available technology that industries are required to use take into account such factors as age of equipment, facilities involved, processes and process changes, engineering aspects of control techniques, and environmental impact apart from water quality, including energy requirements. In assessing best practicable technology for a particular category of industry, a balance is struck between total costs and effluent reduction benefits. Best available technology is the highest degree of technology proven to be designable for plant scale operation. Pretreatment standards basically deal with industrial plant waste cleanup prior to discharge into publically owned treatment works (24). The focus of pretreatment standards is on pH control, minimization of toxic metals, and equalization of flow variations.

Water quality standards vary from state to state, river basin to river basin, and

Table 5-12 Secondary Treatment Standards for Publically Owned Treatment Works

Parameter	Effluent quality	Treatment efficiency, %
BOD	30 mg/liter[a]	85
Suspended solids	30 mg/liter[a]	85
pH	6-9	—
Fecal coliforms	200/100 ml	—

[a]30-day average, value not to exceed 45 mg/liter for any 7-day period.

Table 5-13 Effluent Limitations for Shrimp-Processing Plants

	BPT[a,b]		BAT[c]		New sources[d]	
Parameter	Daily	30 day	Daily	30 day	Daily	30 day
Nonremote Alaskan shrimp-processing subcategory						
BOD5, lb/1,000 lb			70	28		
Total suspended solids, lb/1,000 lb	320	210	45	18	270	180
Oil and grease, lb/1,000 lb	51	17	3.8	1.5	45	15
pH	6.0–9.0		6.0–9.0		6.0–9.0	
Remote Alaskan shrimp-processing subcategory						
Total suspended solids, lb/1,000 lb	No pollutants may be discharged that exceed 1.27 cm (0.5 in.) in any dimension		270	180	270	180
Oil and grease, lb/1,000 lb			45	15	45	15
pH			6.0–9.0		6.0–9.0	
Northern shrimp processing in the contiguous states subcategory						
BOD5, lb/1,000 lb			68	27.0	155	62
Total suspended solids, lb/1,000 lb	160	54	12	4.9	38	15
Oil and grease, lb/1,000 lb	126	42	9.5	3.8	14	5.7
pH	6.0–9.0		6.0–9.0		6.0–9.0	
Southern nonbreaded shrimp processing in the contiguous states subcategory						
BOD5, lb/1,000 lb			25	10	63	25
Total suspended solids, lb/1,000 lb	110	38	8.5	3.4	25	10
Oil and grease, lb/1,000 lb	36	12	2.8	1.1	4.0	1.6
pH	6.0–9.0		6.0–9.0		6.0–9.0	
Breaded shrimp processing in the contiguous states subcategory						
BOD5, lb/1,000 lb			43	17.0	100	40
Total suspended solids, lb/1,000 lb	280	93	19	7.4	55	22
Oil and grease, lb/1,000 lb	36	12	2.5	1.0	3.8	1.5
pH	6.0–9.0		6.0–9.0		6.0–9.0	

[a] In establishing the limitations set forth for BPT, EPA took into account all information it was able to collect, develop, and solicit with respect to factors (such as age and size of plant, raw materials, manufacturing processes, products produced, treatment technology available, energy requirements, and costs) that can affect the industry subcategorization and effluent levels established. It is, however, possible data that would affect these limitations have not been available and, as a result, these limitations should be adjusted for certain plants in this industry. An individual discharger or other interested person may submit evidence to the regional administrator (or to the state, if the state has the authority to issue National Pollutant Discharge Elimination System permits) that factors relating to the equipment or facilities involved, the process applied, or other such factors related to such discharger are fundamentally different from the factors considered in the establishment of the

various segments within river basins. As an example, Oklahoma water quality standards will be used (25). Most state standards include a statement regarding antidegradation. In Oklahoma the statement is as follows:

It is recognized that certain of the waters under consideration possess an existing quality, which is better than the minimum standards established. The quality of those waters will be maintained, unless and until it has been affirmatively demonstrated to the State through public hearings that other uses or different standards are justifiable as a result of necessary economic or social development. This will require that any industrial, public, or private project or development which would constitute a new source of pollution or an increased source of pollution to high quality waters will be required, as a part of the initial project design, to provide the highest and best degree of waste treatment (26).

State standards also include consideration of present and potential beneficial uses of water. Beneficial use designations in Oklahoma are as follows (27):

A–public and private water supplies.

B–emergency public and private water supplies.

C_1–fish and wildlife propagation.

C_2–fish and wildlife propagation to the extent allowed by specifically stated water quality parameters.

D–agriculture (includes livestock watering and irrigation).

E–hydroelectric power.

F_1–municipal and industrial cooling water.

F_2–receiving, transporting, and/or assimilating adequately treated waste.

G_1–recreation, primary body contact (includes recreational uses where the human body may come in direct contact with the water to the point of complete body submergence).

G_2–recreation, secondary body contact (includes recreational uses, such as fishing, wading, and boating, where ingestion of water is not probable).

---

guidelines. On the basis of such evidence or other available information, the regional administrator (or the state) will make a written finding that such factors are or are not fundamentally different for that facility compared to those specified in the development document. If such fundamentally different factors are found to exist, the regional administrator (or the state) shall establish for the discharger effluent limitations in the NPDES permit either more or less stringent than the limitations established herein, to the extent dictated by such fundamentally different factors. Such limitations must be approved by the administrator of the EPA. The administrator may approve or disapprove such limitations, specify other limitations, or initiate proceedings to revise these regulations.

[b]Effluent limitations guidelines for BPT are to be achieved no later than July 1, 1977 representing the degree of effluent reduction attainable by the application of the best practicable control technology currently available.

[c]Effluent limitations guidelines for BAT are to be achieved no later than July 1, 1983 representing the degree of effluent reduction attainable by application of the best available technology economically available.

[d]Effluent reduction is to be achieved by new sources through application of the best available demonstrated control technology, processes, operating methods, or other alternatives.

H–navigation.
I–aesthetics.
J–small-mouth bass fishery excluding lake waters.
K–trout fishery.

An example of water quality standards for Little River in central Oklahoma is shown in Table 5-14 (28).

STEP 7: WASTE LOAD ALLOCATION STUDY

The purpose of this step is to summarize the waste load allocation study for the particular surface watercourses in the vicinity of proposed alternatives. Most states are developing or have recently completed studies of waste loads in surface waters. One result of this step is the identification of known point sources of pollution in the vicinity of the study area. Attention should also be directed toward unique discharges or effluent constituents in the area.

It is also important to identify numbers and types of water users in the area, particularly those downstream from the project site. Water quantity concerns are of

Table 5-14 Water Quality Standards[a] for Little River in Central Oklahoma

Parameter	Standard
Chloride	1,382 mg/liter (sample[b]) 990 mg/liter (yearly mean)
Sulfate	60 mg/liter (sample[b]) 45 mg/liter (yearly mean)
Total dissolved solids	2,802 mg/liter (sample[b]) 2,046 mg/liter (yearly mean)
Fecal coliform	200/100 ml (monthly geometric mean) 400/100 ml (in not more than 10% of samples in 30-day period, min. of 5 samples)
Color	Limited to not interfere with beneficial uses[c]
Oil and grease	Essentially free
Taste- and odor-producing substances	Limited to not interfere with beneficial uses[c]
Solids	Face of floating materials
Turbidity (other than natural)	50 JTU
Temperature	Heat increase $\leqslant$ 5° F, max. temp. $\leqslant$ 90° F
Nutrients (phosphorus and nitrogen/ phosphorus)	Limited to prevent eutrophication
Dissolved oxygen	5 mg/liter
Toxic substances	Not allowed
Species diversity index	Minimum of 3 for benthic macroinvertebrates
pH	6.5–8.5

[a]Apply at all times with the exception of when the flow is equal to or less than the 7-day, 2-yr low flow.

[b]Not more than 1 in 20 samples should exceed sample standard.

[c]Beneficial uses = A, C_1, D, F_1, F_2, G_1, G_2, I.

major importance in water-deficient areas. The types of water uses are important since quality requirements vary for different uses. Table 5-15 summarizes specific quality characteristics of surface waters that have been used as sources for industrial water supplies (29).

STEP 8: MESOSCALE IMPACT CALCULATION

The purpose of this step is to examine the impact of alternatives in terms of their relative contributions to existing waste loads in streams. The approach consists of multiplying unit waste generation factors by their appropriate production quantities and then comparing these calculated waste loads with existing waste loads in the study area. One means of assessing the impact is to calculate percentage changes in pollutant loads resulting from the alternatives relative to existing pollutant loads in the study area. Waste loads should be considered for organic (nonconservative), inorganic (conservative), solid, nutrient, bacterial, and thermal pollutants.

STEP 9: CONSTRUCTION PHASE IMPACTS

The primary water quality impact during construction results from sediment that is eroded from the construction site, transported to local surface watercourses, and then dispersed or deposited (30). Many predictive methods have been developed to describe erosion, transport, and deposition. A summary of these methods is found in Table 5-16 (31).

STEP 10: MICROSCALE IMPACT CALCULATION

The next step involves calculation of downstream concentrations of various water pollutants from the project area for each alternative during both the construction and operational phases. It is beyond the scope of this presentation to derive various mathematical formulas used for microscale impact prediction; however, this section will be oriented to presentation of several mathematical models that can be used in microscale impact calculations.

Conservative Pollutants

Conservative pollutants are not biologically degraded in a stream, nor will they be lost from the water phase due to precipitation, sedimentation, or volatilization. The basic approach for prediction of downstream concentrations of conservative pollutant is to consider the dilution capacity of the stream. As an illustration, Fig. 5-5 represents a schematic drawing of a power plant located adjacent to a stream, with the power plant using a 205-acre cooling pond (32). Data about the stream at withdrawal point X are shown in Table 5-17. These data are expressed as a frequency distribution relative to flow and concentration of total dissolved solids. The total dissolved solids are comprised primarily of chlorides. Water withdrawal from the stream into the cooling pond is 3.5 mgd, and cooling water usage in the

Table 5-15 Summary of Specific Quality Characteristics of Surface Waters That Have Been Used as Sources for Industrial Water Supplies[a]

Characteristic	Boiler makeup water		Cooling water				Process water							
			Fresh		Brackish[b]									
	Industrial, 0–1,500 psig	Utility, 700–5,000 psig	Once through	Makeup recycle	Once through	Makeup recycle	Textile industry, SIC-22	Lumber industry, SIC-24	Pulp and paper industry, SIC-26	Chemical industry, SIC-28	Petroleum industry, SIC-29	Primary metals industry, SIC-33	Food and kindred products, SIC-20	Leather industry, SIC-31
Silica (SiO_3)	150	150	50	150	25	25	–	–	50	–	50	–	For the above two categories the quality of raw surface supply should be that prescribed by the National Technical Advisory Subcommittee on Water Quality Requirements for Public Water Supplies	
Aluminum (Al)	3	3	3	3	–	–	–	–	–	–	–	–		
Iron (Fe)	80	80	14	80	1.0	1.0	0.3	–	2.6	5	15	–		
Manganese (Mn)	10	10	2.5	10	0.02	0.02	1.0	–	–	2	–	–		
Copper (Cu)	–	–	–	–	–	–	0.5	–	–	–	–	–		
Calcium (Ca)	–	–	500	500	1,200	1,200	–	–	–	200	220	–		
Magnesium (Mg)	–	–	–	–	–	–	–	–	–	100	85	–		
Sodium and potassium (Na + K)	–	–	–	–	–	–	–	–	–	–	230	–		
Ammonia (NH_3)	–	–	–	–	–	–	–	–	–	–	–	–		
Bicarbonate (HCO_3)	600	600	600	600	180	180	–	–	–	600	480	–		
Sulfate (SO_4)	1,400	1,400	680	680	2,700	2,700	–	–	–	850	570	–		
Chloride (Cl)	19,000	19,000	600	500	22,000	22,000	–	–	200[c]	500	1,600	500		
Fluoride (F)	–	–	–	–	–	–	–	–	–	–	1.2	–		
Nitrate (NO_2)	–	–	30	30	–	–	–	–	–	–	8	–		
Phosphate (PO_4)	–	50	4	5	5	5	–	–	–	–	–	–		
Dissolved solids	35,000	35,000	1,000	1,000	35,000	35,000	150	–	1,080	2,500	3,500	1,500		
Suspended solids	15,000	15,000	5,000	15,000	250	250	1,000	*d*	–	10,000	5,000	3,000		
Hardness ($CaCO_3$)	5,000	5,000	850	850	7,000	7,000	120	–	475	1,000	900	1,000		
Alkalinity ($CaCO_3$)	500	500	500	500	150	150	–	–	–	500	–	200		
Acidity ($CaCO_3$)	1,000	1,000	0	200	0	0	–	–	–	–	–	75		
pH, units	–	–	5.0–8.9	3.5–9.1	5.0–8.4	5.0–8.4	6.0–8.0	5–9	4.6–9.4	5.5–9.0	6.0–9.0	3–9		
Color, units	1,200	1,200	–	1,200	–	–	–	–	360	500	25	–		
Organics														
Methylene blue active substances	2[e]	10	1.3	1.3	–	1.3	–	–	–	–	–	–		
Carbon tetrachloride extract	100	100	*f*	100	*f*	100	–	–	–	–	–	30		
Chemical oxygen demand (O_2)	100	500	–	100	–	200	–	–	–	–	–	–		
Hydrogen sulfide (H_2S)	–	–	–	–	4	4	–	–	–	–	–	–		
Temperature, °F	120	120	100	120	100	120	–	–	95[g]	–	–	100		

[a] Unless otherwise indicated, units are mg/liter and values are maximums. No one water will have all the maximum values shown.
[b] Water containing in excess of 1,000 mg/liter dissolved solids.
[c] May be ⩽ 1,000 for mechanical pulping operations.
[d] No large particles ⩽ 3 mm diameter.
[e] 1 mg/liter for pressures up to 700 psig.
[f] No floating oil.
[g] Applies to bleached chemical pulp and paper only.

Table 5-16 Summary of Sediment Prediction Methods

Prediction method	Process		
	Erosion	Transport	Deposition
Empirical			
Ellison	X	–	–
Musgrave	X	–	–
Universal soil loss equation	X	–	–
Einstein bedload function	–	X	–
Colby-modified Einstein	–	X	–
Toffaleti total load method	–	X	X
Lacey's silt theory	–	X	X
Pemberton-modified Einstein	–	X	–
Reservoir surveys	–	–	X
ARS			
SCS			
Corps of Engineers			
Bureau of Reclamation			
U.S. Geological Survey			
Statistical			
Flaxman	–	–	X
Sediment rating-flow duration	–	–	X
U.S. Geological Survey			
Bureau of Reclamation			
Corps of Engineers			
Woolhiser's deterministic watershed model	X	X	X
Simulation			
ARS upland erosion model	X	–	–
ARS USDAHL-73 watershed model	X	X	X
ARS "ACTMO" chemical transport model	–	X	–
Negev's watershed model	X	X	X
Stanford IV model	X	X	X
Hydrocomp simulation	X	X	X
Huff hydrologic transport model	–	X	–
Royal Institute (Sweden) hydrologic model	X	X	X
Snyder's parametric hydrologic model	–	X	X

power plant is 3.0 mgd. Water losses in the cooling pond occur from consumptive losses in the plant cooling cycle and from pond evaporation. Consumptive losses average 1.6 mgd, and evaporation losses vary from 0 to 1.4 mgd during May through October. On this basis, total water losses in the cooling pond vary from 1.6 to 3.0 mgd. If the concentration factor in the pond is defined as "water in" divided by "water out," the highest concentration factor is 7 (3.5 mgd/0.5 mgd) and the lowest is 1.8 (3.5 mgd/1.9 mgd). Using these concentration factors, calculations can be made for the total dissolved solids in the receiving stream, with the results summarized in Table 5-18. Comparison of the calculated values in the receiving stream can then be made relative to the applicable water quality standard.

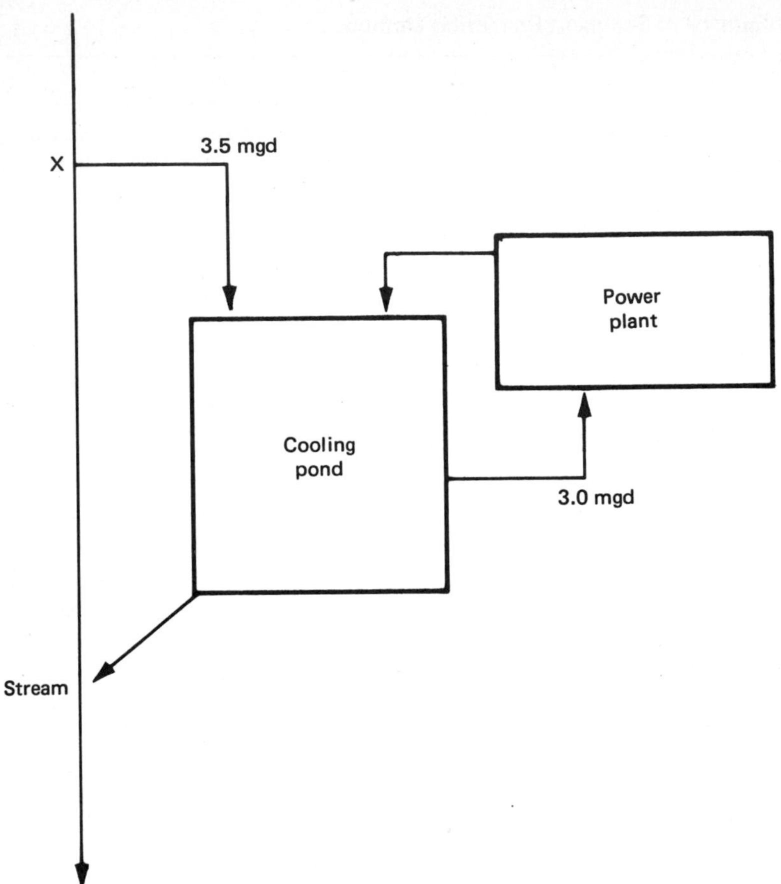

Figure 5-5 Schematic drawing of power plant and water usage.

Table 5-17 Summary of Total Dissolved Solids in Stream at Withdrawal Point

Stream flow	Frequency	Total dissolved solids, mg/liter
10.0	37	720
17.3	219	520
80.0	109	305

Table 5-18 Summary of Calculated Total Dissolved Solids Downstream from Cooling Pond Discharge

Condition[a]	Flow, mgd	Total dissolved solids, mg/liter
Worst	7.0	1030 (720)[b]
Average	14.7	610 (520)
Best	78.4	310 (305)

[a]Worst: stream flow of 6.4 mgd (3.5 mgd withdrawn from 10 mgd), pond effluent flow of 0.5 mgd; occurs 37 days/yr. Average: stream flow of 13.8 mgd (3.5 mgd withdrawn from 17.3 mgd), pond effluent flow of 0.9 mgd; occurs 219 days/yr. Best: stream flow of 76.5 mgd (3.5 mgd withdrawn from 80 mgd), pond effluent flow of 1.9 mgd; occurs 109 days/yr.

[b]Numbers in parentheses represent stream water quality just upstream from confluence with cooling pond discharge.

Nonconservative Pollutants

Nonconservative pollutants are organic materials that can be biologically decomposed by bacteria in aqueous systems. One of the most common examples of stream pollution results from oxygen deficiency caused by organic waste discharges. Dissolved oxygen in a stream is deficient when the actual concentration is less than the saturation concentration for existing conditions of temperature, pressure, and salt content. There are certain demand and supply forces relative to dissolved oxygen in a stream. Demand for oxygen is exerted by bacteria in the decomposition of organic materials, both in the liquid phase and in bottom deposits. Oxygen supply is from natural reaeration as well as from the net effect of photosynthesis.

Various mathematical models have been developed to describe dissolved oxygen relationships in streams. In 1925 Streeter and Phelps proposed a model for the "oxygen sag curve" (33). The assumptions for their model were that the BOD decrease is due only to bacterial oxidation, hence there is no sedimentation or volatilization; there is no benthal oxygen demand; there is no net photosynthetic effect in terms of oxygen supply, that is, algal photosynthetic oxygenation is counterbalanced by algal respiration; and reoxygenation is by natural reaeration only. The basic mathematical model developed by Streeter and Phelps is as follows:

$$D_t = \frac{K_1 L_a}{K_2 - K_1}(10^{-K_1 t} - 10^{-K_2 t}) + D_a 10^{-K_2 t}$$

where D_t = dissolved oxygen deficit at any flow time t downstream, the flow time t expressed in days

= saturation dissolved oxygen concentration − actual dissolved oxygen concentration

K_1 = coefficient of deoxygenation, day^{-1}

K_2 = coefficient of reaeration, day^{-1}

L_a = ultimate BOD in the stream following mixing, mg/liter
D_a = dissolved oxygen deficit upstream of waste discharge, mg/liter

It should be noted that K_1, K_2, and L_a are influenced by temperature. Specific mathematical relationships describing the temperature influence are as follows:

$$K_{1(T)} = K_{1(20)}\,(1.047)^{T-20}$$
$$K_{2(T)} = K_{2(20)}\,(1.016)^{T-20}$$
$$L_{a(T)} = L_{a(20)}\,(0.02T + 0.6)$$

where T = water temperature, °C
$K_{1(20)}, K_{2(20)}, L_{a(20)}$ = values at 20°C

The saturation dissolved oxygen concentration is a function of temperature, pressure, and salt content. Figure 5-6 illustrates the influence of temperature on the oxygen sag curve (34).

Other factors that may be important in predicting water quality impact are related to critical conditions in terms of the location and value of the minimum point on the oxygen sag curve, as well as the maximum permissible BOD loading that can be introduced and still maintain the dissolved oxygen standard. Equations for critical time and deficit are as follows:

$$t_c = \frac{1}{K_2 - K_1} \log_{10}\left(\frac{K_1 L_a - K_2 D_a + K_1 D_a}{K_1 L_a}\frac{K_2}{K_1}\right)$$

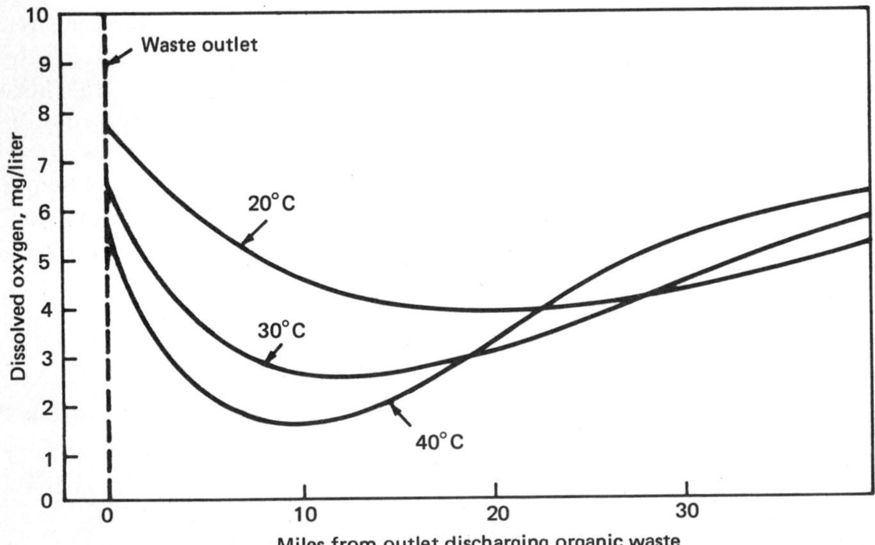

Figure 5-6 Relation between temperature and oxygen profile.

$$D_c = \frac{K_1}{K_2} L_a 10^{-K_1 t_c}$$

where t_c = critical time = time of flow (days) to point of occurrence of minimum dissolved oxygen concentration

D_c = critical deficit = maximum deficit (mg/liter); occurs at time of flow t_c

The equation that can be used for determining the maximum permissible BOD loading is as follows:

$$\log L_a = \log D_{\text{all}} + \left[1 + \frac{K_1}{K_2 - K_1}\left(1 - \frac{D_a}{D_{\text{all}}}\right)^{0.418}\right] \log \frac{K_2}{K_1}$$

where D_{all} = allowable deficit (mg/liter) = saturation dissolved oxygen concentration − dissolved oxygen standard

Modifications of the Streeter-Phelps equation developed since 1925 include those by Thomas, Camp, O'Connor, and Churchill. Many of these modifications have specific limitations on usage. There are excellent resource materials describing mathematical models that have been developed as modifications or alternatives to the Streeter-Phelps equation (35, 36). Mathematical models have been made that describe dissolved oxygen variations in reservoirs during both stratified and non-stratified periods (37). It is beyond the scope of this presentation to offer detailed aspects of these models.

Bacterial Pollution

In order to assess the potential environmental impact of a proposed action, it may be necessary to predict bacterial self-purification in streams. Bacterial self-purification can be defined as the decrease of bacteria of all types, and especially those of fecal origin, as a function of flow distance or flow time in a river (38). Bacteria may starve, be devoured by predators, or be otherwise inactivated.

The per capita contribution and seasonal variations of coliform bacteria in municipal waste water are shown in Table 5-19 (39). Seasonal variations are due to greater multiplications of intestional bacteria within sewage systems during summer months and also greater per capita discharge of these organisms in the summer months.

The basic mathematical relationship that describes bacterial self-purification is as follows:

$$B_t = B_0 10^{-Kt}$$

where B_t = bacterial residual after any time t, days

B_0 = initial number of bacteria in stream

K = bacterial death rate, day^{-1}

Table 5-19 Expected Seasonal Pattern of Monthly Average Coliform Contribution Based on an Annual Average of 200 Billion per Capita per Day

Month	Monthly avg., % of annual avg.	Monthly avg. coliform contribution/capita-day ($\times 10^9$)
January	45	90
February	46	92
March	53	106
April	65	130
May	92	184
June	135	270
July	172	344
August	182	364
September	165	330
October	110	220
November	80	160
December	55	110
Annual average	100	200

Table 5-20 Coliform Death Rates *K* Observed in Rivers

River	Reaction rate K, day^{-1}		Authority for survey data	Remarks
	Warm weather	Cool weather		
Ohio	0.50	0.45	Frost, Streeter, et al.	Generalized results of analysis of extensive data
Upper Illinois	0.90	0.32	Hoskins et al.	1-day decline
	0.67	0.29		2-day decline
Scioto	0.96	0.46	Kehr et al.	
Hudson	0.80		Hall, Riddick, Phelps	Freshwater reach below Albany
Upper Miami	0.80		Velz, Gannon, Kinney	Mean through reach above Dayton
Tennessee	0.46		Kittrell	1- and 2-day declines, below Knoxville
Tennessee	0.60		Kittrell	1-day decline
	0.57			2-day decline (below Knoxville)
Sacramento	0.77		Kittrell	1-day decline
	0.65			2-day decline (below Sacramento)
Missouri		0.30	Kittrell	1-day decline
		0.26		2-day decline (below Kansas City)

Increases in stream temperatures increase the K value, thus decreasing the B_t values at various distances downstream. Some typical K values are shown in Table 5-20, with values for large rivers being 0.5 ± 0.15 per day and for moderate-size rivers 0.8 ± 0.2 per day (40).

Thermal Pollution

Mathematical models of varying degrees of complexity have been developed to predict the persistence of heat in quiescent waters, flowing streams, estuaries, and the ocean (41, 42). It is beyond the scope of this presentation to analyze the detailed aspects of each of these models. An example has been published on the use of a simplified temperature-predictive equation for calculating downstream temperatures following a waste heat discharge into a flowing stream (41). Excellent reference materials are available on the subject of thermal pollution (43-45).

STEP 11: POLLUTION CONTROL MEASURES

If it is determined that water quality standards are exceeded by the proposed action, then abatement strategies or control measures should be presented. Excellent references are available to describe various control technologies that can be used for minimizing pollutant emissions (46-49).

STEP 12: DISCUSSION OF OPERATIONAL IMPACTS

This final step is included as a reminder to consider unique aspects of operational impacts relative to a variety of concerns. These concerns include the frequency distribution of decreased quality or quantity, fate of nutrients by incorporation into biomass, reconcentration of conservative pollutants into the food web, and chemical changes of certain inorganic chemicals within aqueous systems. Chromium serves as an example of the latter since it is a transition metal that exhibits various oxidation states and behavior patterns (50). Trivalent chromium is generally present as a cation, $Cr(OH)^{++}$, and is very chemically reactive, tending to sorb on suspended materials and subsequently settle from the liquid phase. Hexavalent chromium is anionic (CrO_4^{--}) and chemically unreactive, thus tending to remain in solution. Changes can occur in the chromium oxidation state due to stream quality. For example, hexavalent chromium can be chemically reduced to trivalent chromium under anaerobic conditions, whereas trivalent chromium can be oxidized to hexavalent chromium under aerobic conditions.

SUMMARY

In summary, this chapter is oriented to the types of information and practical steps for predicting the impacts of a given proposed action in terms of the mesoscale and microscale viewpoints relative to water quality. Suggestions are provided for assessment or interpretation of the predicted impacts, with these primarily related to an analysis of the resultant microscale concentrations relative to water quality standards.

SELECTED REFERENCES

1 Shubinski, R. P., and G. F. Tierney: Effects of Urbanization on Water Quality, *Proc. ASCE Urban Transportation Div. Specialty Conf. Environ. Impact,*, p. 180, 1973.

2 Federal Water Quality Administration: "Clean Water for the 1970's," pp. 4–13, Washington, D.C., June 1970.

3 *Ibid.*, pp. 4–5.

4 *Ibid.*, p. 5.

5 *Ibid.*, p. 10.

6 *Ibid.*, p. 9.

7 *Ibid.*, p. 13.

8 Shubinski and Tierney, *op. cit.*, p. 206.

9 Environmental Protection Agency: "Methods for Identifying and Evaluating the Nature and Extent of Non-Point Sources of Pollutants," p. 6, Publ. EPA-430/9-73-014, Washington, D.C., Oct. 1973.

10 Shubinski and Tierney, *op. cit.*, p. 193.

11 Pontier J.: Effect of Land Use and Water Use on Water Quality, p. 27, term paper, University of Oklahoma, Norman, 1973.

12 Shubinski and Tierney, *op. cit.*, p. 203.

13 Council on Environmental Quality: "Municipal Sewage Treatment–A Comparison of Alternatives," p. 41, Washington, D.C., 1974.

14 Pontier, *op. cit.*, p. 29.

15 Canter, L. W.: Internal report prepared for Williams Brothers Engineering Co., Tulsa, Okla., July 1973.

16 Oklahoma Water Resources Board: "Appraisal of the Water and Related Land Resources of Oklahoma–Regions Five and Six," p. 76, Publ. 27, Oklahoma City, 1969.

17 *Ibid.*, p. 77.

18 *Federal Water Pollution Control Act Amendments of 1972*, PL 92-500, 92d Cong., S. 2770, Oct. 18, 1972.

19 Environmental Protection Agency: "Water Quality Strategy Paper," p. 25, a statement of policy for implementing the requirements of the 1972 Federal Water Pollution Control Act Amendments and certain requirements of the 1972 Marine Protection, Research, and Sanctuaries Act, Washington, D.C., Mar. 15, 1974.

20 Environmental Protection Agency: "Secondary Treatment Information," *Fed. Reg.*, vol. 38, no. 159, pp. 22298–22299, Aug. 17, 1973.

21 Environmental Protection Agency, Mar. 1974, *op. cit.*, p. 17.

22 Smith, R.: Cost-Effectiveness Analysis for Water Pollution Control, *Proc. Symp. Upgrading Wastewater Stabilization Ponds to Meet New Discharge Standards*, p. 199, Utah State University, Logan, Nov. 1974.

23 Environmental Protection Agency: "Summary of Final Effluent Limitations Guidelines for Existing Sources and Standards of Performance for New Sources," pp. iv, 16, 17, Publ. EPA-330/9-74-001, National Field Investigations Center, Denver, Colo., Aug. 1974.

24 Environmental Protection Agency: "Pretreatment Standards," *Fed. Reg.*, vol. 38, no. 215, pp. 30982–30984, Nov. 8, 1973.

25 Oklahoma Water Resources Board: "Oklahoma's Water Quality Standards–1973," Publ. 52, Oklahoma City, Sept. 1973.

26 *Ibid.*, p. 2.
27 *Ibid.*, p. 3.
28 *Ibid.*, pp. 3–9, 37, 55.
29 Federal Water Pollution Control Administration: "Water Quality Criteria," p. 189, Washington, D.C., Apr. 1, 1968.
30 Clark, J. R.: "Environmental Assessment of Construction Activities," pp. 53–73, Master of Science thesis, University of Oklahoma, Norman, 1975.
31 Environmental Protection Agency, 1973, *op. cit.*, p. 47.
32 Canter, L. W.: Environmental Impact of Commanche Power Station in Lawton, Oklahoma, report prepared for Public Service Company of Oklahoma, Tulsa, Feb. 1973.
33 Velz, C. J.: "Applied Stream Sanitation," p. 142, Wiley-Interscience, New York, 1970.
34 Federal Water Pollution Control Administration: "Industrial Waste Guide on Thermal Pollution," p. 20, Washington, D.C., Sept. 1968.
35 Velz, *op. cit.*
36 Nemerow, N. L., "Scientific Stream Pollution Analysis," McGraw-Hill Book Company, New York, 1974.
37 Markofsky, M., and D. R. Harleman: "A Predictive Model for Thermal Stratification and Water Quality in Reservoirs," Publ. 16130 DJH 01/71, Environmental Protection Agency, Washington, D.C., Jan. 1971.
38 Phelps, Earle B.: "Stream Sanitation," pp. 201–221, John Wiley and Sons, New York, 1944.
39 Velz, *op. cit.*, p. 241.
40 *Ibid.*, p. 247.
41 Federal Water Pollution Control Administration, Sept. 1968, *op. cit.*
42 Edinger, J. E., and J. C. Geyer: "Heat Exchange in the Environment," Publ. 65-902, Edison Electric Institute, New York, June 1965.
43 Krenkel, P. A., and F. L. Parker: Biological Aspects of Thermal Pollution, *Proc. Natl. Symp. Thermal Pollution, Portland, Ore., June 3–5, 1968*, Vanderbilt University Press, Nashville, Tenn., 1969.
44 Krenkel, P. A., and F. L. Parker: Engineering Aspects of Thermal Pollution, *Proc. Natl. Symp. Thermal Pollution, Nashville, Tenn. Aug. 14–16, 1968*, Vanderbilt University Press, Nashville, Tenn., 1969.
45 Krenkel, P. A., and F. L. Parker: "Thermal Pollution: Status of the Art," Rept. 3, National Center for Research and Training in the Hydrologic and Hydraulic Aspects of Water Pollution Control, Vanderbilt University, Nashville, Tenn., Dec. 1969.
46 Liptak, Bela (ed.): "Environmental Engineer's Handbook," vol. 1, "Water Pollution," Chilton Book Company, Radnor, Pa., 1974.
47 Culp, Russell L., and Gordon L. Culp: "Advanced Waste Treatment," Van Nostrand Reinhold Company, New York, 1971.
48 Culp, Gordon L., and Russell L. Culp: "New Concepts in Water Purification," Van Nostrand Reinhold Company, New York, 1974.
49 Lund, H. F. (ed.): "Industrial Pollution Control Handbook," McGraw-Hill Book Company, New York, 1971.
50 Canter, L. W., and E. F. Gloyna: Transport of Chromium-51 in an Organically Polluted Environment, *Proc. 23d Purdue Industrial Waste Conf.*, pp. 374–387, Purdue University, Lafayette, Ind., 1968.

Chapter 6

Prediction and Assessment of Impacts on the Noise Environment

Another of the major impacts of many actions is on the noise environment in and adjacent to the project area. Construction of power plants, highways, airports, and pipelines generates noise intrusions. Utilization of airports and highways and the operation of compressor stations lead to persistent noise impacts in the environmental setting. This chapter is addressed to data needs and associated technology for predicting and assessing the impact of proposed actions on the noise environment.

BASIC STEPS FOR PREDICTION AND ASSESSMENT

Basic steps associated with prediction of changes in the noise environment and assessment of the impact of these changes are as follows:

1 Identify noise levels for the alternatives under consideration during both the construction and operational phases.
2 Determine existing noise levels for the project area. This may involve field measurements or the determination of land-use patterns. Identify unique noise sources in the area as well as unique places where noise levels must be minimized.

3 Obtain applicable noise standards and criteria for the area.
4 Determine the microscale impact by predicting anticipated noise levels for each alternative during both construction and operational phases. Compare predicted noise levels with applicable standards or criteria in order to assess impact.
5 If standards or criteria are exceeded, consider noise abatement methods to minimize impact on the noise environment.

These five steps are directed toward determining the noise impacts of alternatives and the proposed action on the microscale noise environment. Steps 2 and 3 should be summarized in the environmental setting and the remainder in the environmental impact section of the EIS.

The organization of this chapter is primarily oriented to the five steps identified above. However, prior to step-by-step discussion and analysis, brief information on noise pollution considerations is presented.

BASIC INFORMATION ON NOISE POLLUTION

Noise can be defined as unwanted sound or sound in the wrong place at the wrong time. Noise can also be defined as any sound that is undesirable because it interferes with speech and hearing, is intense enough to damage hearing, or is otherwise annoying (1). The definition of noise as unwanted sound implies that it has an adverse effect on human beings and their environment, including land, structures, and domestic animals. Noise can also disturb natural wildlife and ecological systems.

Sound is mechanical energy from a vibrating surface, transmitted by cycling series of compressions and rarefactions of molecules of the materials through which it passes (2). Sound can be transmitted through gases, liquids, and solids. A vibrating source producing sound has a total power output, and the sound results in sound pressure that alternately rises to a maximum pressure of compression and drops to a minimum pressure of rarefaction. The number of compressions and rarefactions of the air molecules in a unit of time is described as its frequency. Frequency is expressed in Hertz (Hz), which is the same as the number of cycles per second. Humans can identify sounds with frequencies from about 16 to 20,000 Hz (3).

Sound power or sound pressure do not provide practical units for sound or noise measurement for two basic reasons (3). First, a tremendous range of sound power and sound pressure can be produced. Expressed in microbars (one-millionth of 1 atm pressure), the range is from 0.0002 to 10,000 μbars for peak noises within 100 ft from large jet and rocket propulsion devices. Second, the human ear does not respond linearly to increases in sound pressure. The nonlinear response is essentially logarithmic. Therefore noise measurements are expressed by the term "sound pressure level" (SPL), which is the logarithmic ratio of the sound pressure to a reference pressure and is expressed as a dimensionless unit of power, the decibel (dB). The reference level is 0.0002 μbar, the threshold of human hearing. The equation for sound pressure level is as follows (2):

$$SPL = 20 \log_{10} \left(\frac{P}{P_0}\right)$$

where SPL = sound pressure level, dB
P = sound pressure, μbar
P_0 = reference pressure, 0.0002 μbar

Table 6-1 contains a summary of various sound pressures and the corresponding decibel levels, with examples of recognized noise sources cited (2). Conventional speech has a sound pressure of 0.2 μbar and a SPL of 60 dB, while sound pressure from light trucks at 20 ft is 2 μbars (10 times greater than conventional speech), and the SPL is 80 dB. Since the SPL scale is logarithmic, decibel values are not additive. For example, a SPL of 70 dB from one source superimposed on a SPL of 72 dB does not result in 142 dB. A SPL of 74.1 dB results. To determine the resultant dB level, it is necessary to convert decibel readings to sound pressures, add these pressures, and then reconvert the resultant ratio to the decibel value. Table 6-2 is provided as an aid for determining the cumulative decibel level of two or more individual noise sources.

In most noise considerations, the A-weighted sound level is used. This scale is appropriate because the human ear does not respond uniformly to sounds of all frequencies, being less efficient in low and high frequencies than it is at medium or speech frequencies (2). To obtain a single number representing a sound level containing a wide range of frequencies and yet representative of the human response, it is necessary to weight the low and high frequencies with respect to medium frequencies. The resultant SPL is "A weighted," and the units are dBA. The A-weighted sound level is also called the noise level. Sound-level meters have an A-weighting network, thus yielding A-weighted dB readings.

Table 6-1 SPL, Sound Pressure, and Recognized Sources of Noise in Our Daily Experiences

Sound pressure, μbar	SPL, dBA	Example
0.0002	0	Threshold of hearing
0.00063	10	
0.002	20	Studio for sound pictures
0.0063	30	Studio for speech broadcasting
0.02	40	Very quiet room
0.063	50	Residence
0.2	60	Conventional speech
0.63	70	Street traffic at 100 ft
1.0	74	Passing automobile at 20 ft
2.0	80	Light trucks at 20 ft
6.3	90	Subway at 20 ft
20	100	Looms in textile mill
63	110	Loud motorcycle at 20 ft
200	120	Peak level from rock and roll band
2,000	140	Jet plane on the ground at 20 ft

Table 6-2 Aid for Determining the Cumulative Decibel SPL When the Differences between Two or More Levels Are Known

Difference between levels, dBA	No. of dBA to be added to higher level
0	3.0
1	2.6
2	2.1
3	1.8
4	1.5
5	1.2
6	1.0
7	0.8
8	0.6
10	0.4
12	0.3
14	0.2
16	0.1

The first comprehensive federal legislation for abating noise emissions from a broad range of sources was contained in the Noise Control Act of 1972 (4). One factor cited for passage of the act was that inadequately controlled noise presents a growing danger to the health and welfare of the human population, particularly in urban areas. In the past several years, noise has been increasing in some urban areas by as much as 1 dB per year, or 10 dB per decade (1). Some reasons for these increases include growth in urban freeways, increases in commercial air traffic, shifts in aircraft from propeller-type to jets, increases in construction activity, and increases in noise-generation devices such as power lawn mowers and motorcycles. It should be noted that while primary responsibility for control of environmental noise rests with state and local governments, federal action is essential to deal with major noise sources in commerce, the control of which requires national uniformity. The general objective of noise control is to promote an environment free from noise that jeopardizes human health or welfare.

Table 6-3 contains a summary of hearing losses in the overall U.S. population, with these hearing losses ranging from moderate to major. Approximately 10 million persons have experienced hearing losses, with some 3 million having losses due to noise-associated activities (1). In addition to the hearing losses shown in Table 6-3, other effects of noise range from disruption of normal human patterns to temporary hearing damage. It is estimated that 16 million persons are presently exposed to aircraft noise levels with effects ranging from moderate to very severe (5).

STEP 1: IDENTIFICATION OF NOISE LEVELS

The first step in prediction and assessment of noise impacts involves identification of the noise levels to be anticipated from construction and operation of each alternative under consideration for a proposed action. One approach for identifying

Table 6-3 Hearing Loss (Moderate to Profound) in United States

Age range, yr	Population totals, thousands	Loss of hearing totals, thousands	Noise-associated hearing loss, thousands
0–5	17,000	850	?
5–10	20,000	1,000–1,400	200[a]
10–18	32,500	650–975	150[b]
18–65	113,000	2,260	2,000 (approx.)
Over 65	20,000	4,000	400–600
Totals	202,500	8,700–11,135	2,750–2,950

[a]The most common cause is explosions from toy caps (20% sensory-neural hearing loss).

[b]Firearms and toy caps (based on approximately 20% sensory-neural hearing loss).

anticipated noise levels is to review other EISs prepared on projects of similar type. Another approach involves conduction of noise surveys in the vicinity of existing projects of similar type. The best approach is to utilize noise generation factors related to various activities.

Construction activities generally generate noise levels in excess of those in the project environs. Construction sites can be categorized into four major types: domestic housing, including residences for one to several families; nonresidential buildings, including offices, public buildings, hotels, hospitals, and schools; industrial, including industrial buildings, religious and recreational centers, stores, and repair facilities; and public works, including roads, streets, water mains, and sewers (1). Noise from construction of major civil works such as dams generally affects

Table 6-4 Typical Ranges of Energy-Equivalent Noise Levels in dBA at Construction Sites

	Domestic housing		Office building, hotel, hospital, school, public works		Industrial parking garage, religious amusement and recreations, store, service station		Public works roads and highways, sewers, and trenches	
Phase	I[a]	II[b]	I	II	I	II	I	II
Ground clearing	83	83	84	84	84	83	84	84
Excavation	88	75	89	79	89	71	88	78
Foundations	81	81	78	78	77	77	88	88
Erection	81	65	87	75	84	72	79	78
Finishing	88	72	89	75	89	74	84	84

[a]I, all pertinent equipment present at site.

[b]II, minimum required equipment present at site.

relatively few people other than those employed at or near construction sites; thus they are not included in these construction-site categories.

Noise at a construction site varies relative to the particular operation in progress. Operations can be divided into five consecutive phases: ground clearing, including demolition and removal of prior structures, trees, and rocks; excavation; placing foundations, including reconditioning of old roadbeds and compacting trench floors; erection, including framing, placing of walls, floors, windows, and pipe installation; and finishing, including filling, paving, and cleanup (1). Table 6-4 shows typical energy-equivalent noise levels at construction sites. Energy-equivalent noise level (L_{eq}) refers to the equivalent steady noise level that, in a stated period of time, would contain the same noise energy as the time-varying noise during the same time period (6). Maximum levels range from 77 to 89 dBA for all categories and have an average value of approximately 85 dBA. Minimum values have a wider range, extending from 65 to 88 dBA, with an average of 78 dBA.

Table 6-4 also shows that the ground-clearing and excavation phases are the noisiest, foundation placement and erection phases are somewhat quieter, and the final finishing phase tends to produce considerable noise annoyance. Table 6-5 contains information on noise levels observed 50 ft from various types of construction equipment. These levels range from 72 to 96 dBA for earth-moving equipment, from 75 to 88 dBA for materials-handling equipment, and from 70 to 87 dBA for stationary equipment. Impact equipment may generate noise levels up to 115 dBA (1).

Noise from project operations includes highway vehicles, aircraft, rail systems, recreation vehicles, internal combustion engines, industrial machinery, and building equipment. Noise produced by highway vehicles can be attributed to three major generating systems: rolling stock such as tires and gearing; propulsion systems related to engine and other accessories; and aerodynamic and body systems (7). Noise levels produced by highway vehicles are dependent upon vehicle speed, as illustrated in Fig. 6-1. General characteristics of highway vehicles are shown in Fig. 6-2, including information on automobiles, trucks, utility and maintenance equipment, and buses.

Commercial and private aircraft are among the most recognized noise sources in society. Information regarding general noise characteristics of commercial aircraft is presented in Fig. 6-3 for 2-3-engine turbofan, 4-engine turbofan, 4-engine widebody turbofan, 3-engine widebody turbofan, and propeller aircraft (7). Noise levels from vertical and short takeoff and landing aircraft are shown in Fig. 6-4. Figure 6-5 identifies noise characteristics of general aviation aircraft, including single-engine propeller, multiengine propeller, and executive jets.

Figure 6-6 identifies noise characteristics of rail systems, including railroads and rail transit systems (7). The noise characteristics of recreation vehicles including motorcycles, snowmobiles, and pleasure boats are summarized in Fig. 6-7. Figure 6-8 includes the noise characteristics of devices powered by internal combustion engines such as generators, lawn-care equipment, and others.

Table 6-6 shows the noise levels to be anticipated at the operator positions for a range of industrial machinery and equipment (1). Information in Table 6-6 is of

Table 6-5 Construction Equipment Noise Ranges

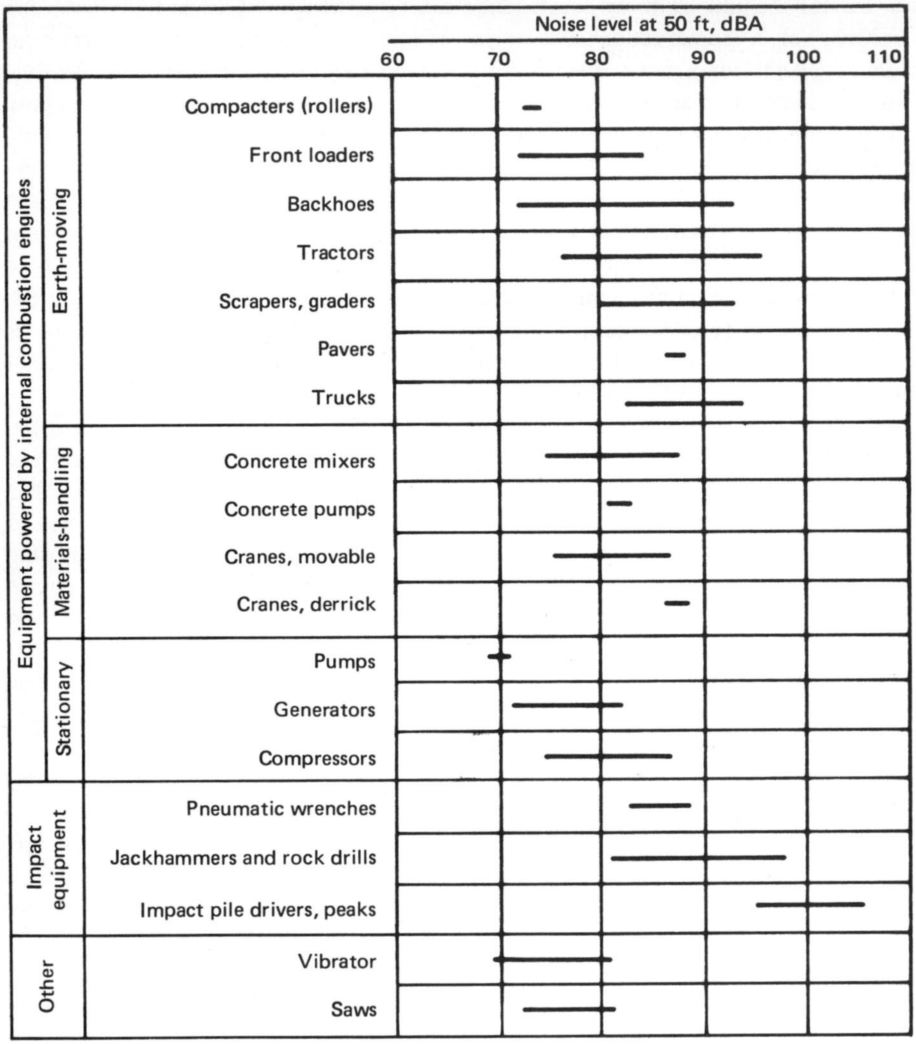

Category	Type	Equipment	Noise level at 50 ft, dBA (scale 60–110)
Equipment powered by internal combustion engines	Earth-moving	Compacters (rollers)	
		Front loaders	
		Backhoes	
		Tractors	
		Scrapers, graders	
		Pavers	
		Trucks	
	Materials-handling	Concrete mixers	
		Concrete pumps	
		Cranes, movable	
		Cranes, derrick	
	Stationary	Pumps	
		Generators	
		Compressors	
Impact equipment		Pneumatic wrenches	
		Jackhammers and rock drills	
		Impact pile drivers, peaks	
Other		Vibrator	
		Saws	

Note: Based on limited available data samples.

primary concern from the work environment standpoint; however, it is also relevant for ambient noise considerations. Table 6-7 shows various types of building equipment and noise levels associated with usage.

STEP 2: DESCRIPTION OF EXISTING NOISE LEVELS

The second step involves aggregation of information on existing noise levels in the project area. Sources of information include relevant county and state environmental agencies and private industries that have noise monitoring programs.

There is a dearth of information on noise levels occurring in most environmental settings. One means of establishing existing noise levels in various areas is to consider the land-use patterns. Noise levels for urban areas are shown in Table 6-8 (6). The day-night noise level (L_{dn}) is defined as the weighted average of the noise of the nighttime hours as compared to that occurring during daytime hours of greater activity. Specifically, L_d refers to the L_{eq} from 0700 to 2200, and L_n is for the nighttime L_{eq} from 2200 to 0700 (8). Quiet suburban residential areas have an average L_{dn} of 50 dBA, while very noisy urban residential areas exhibit L_{dn} values of 70 dBA. Typical noise levels in rural settings are 30-35 dBA, and in wilderness locations they are on the order of 20 dBA. The outdoor daytime residual noise level at the Grand Canyon rim is 16 dBA (9).

Another approach for determining existing noise levels in an area is to conduct

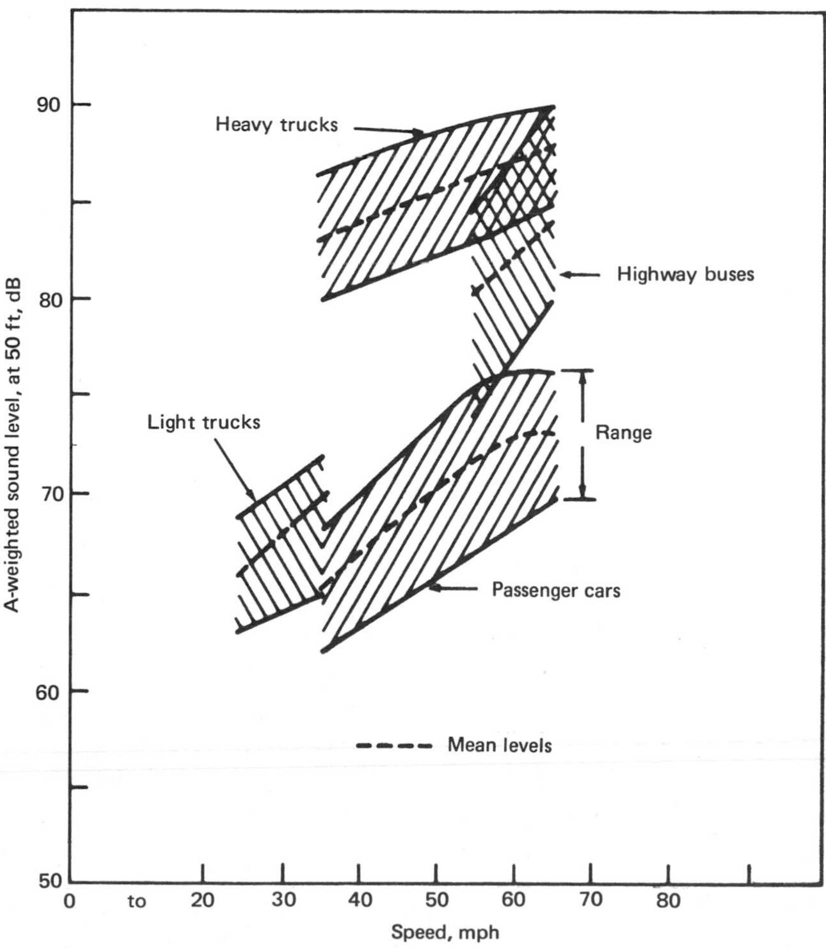

Figure 6-1 Single vehicle noise output as a function of vehicle speed.

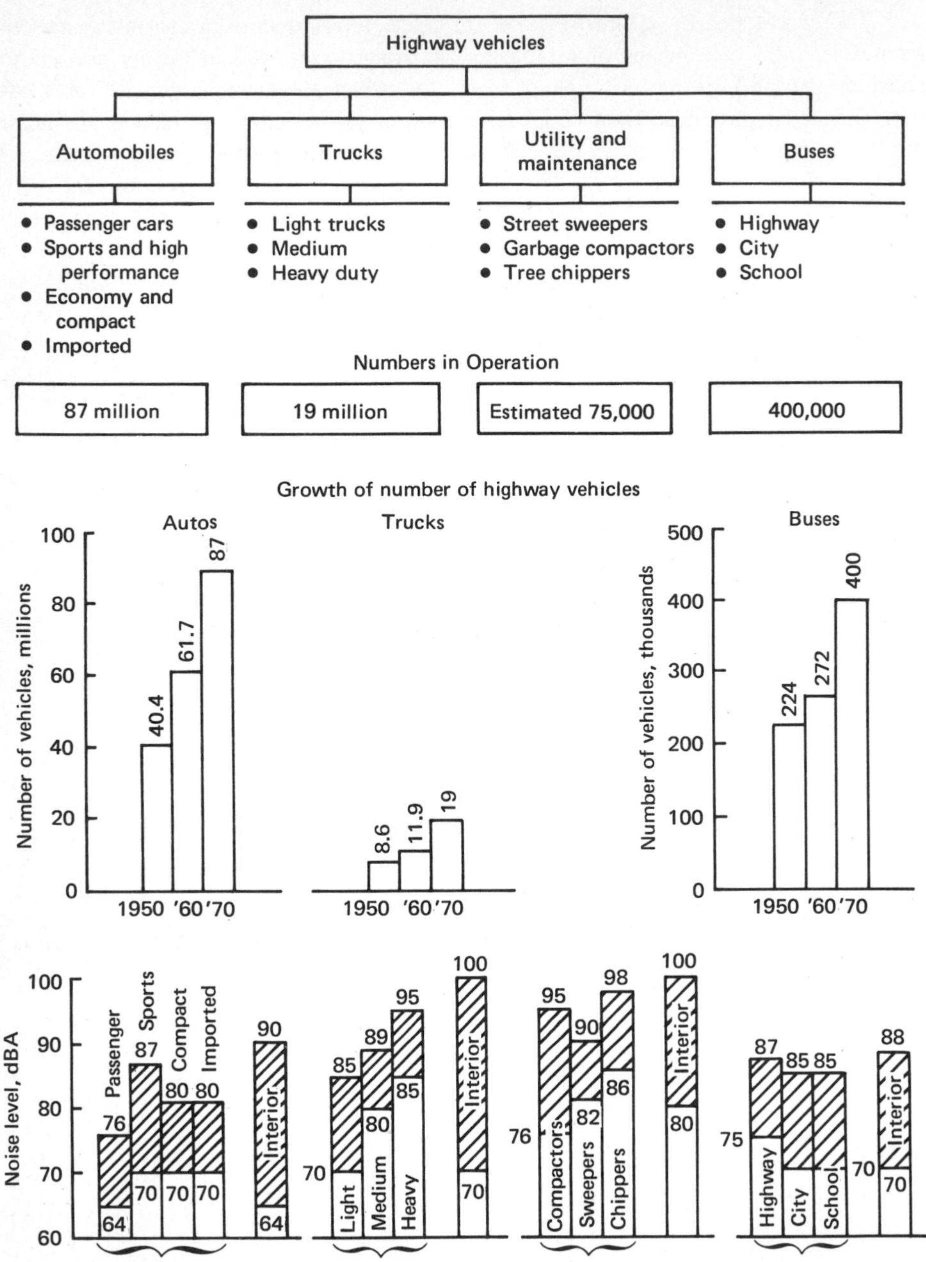

Figure 6-2 Characteristics of highway vehicles.

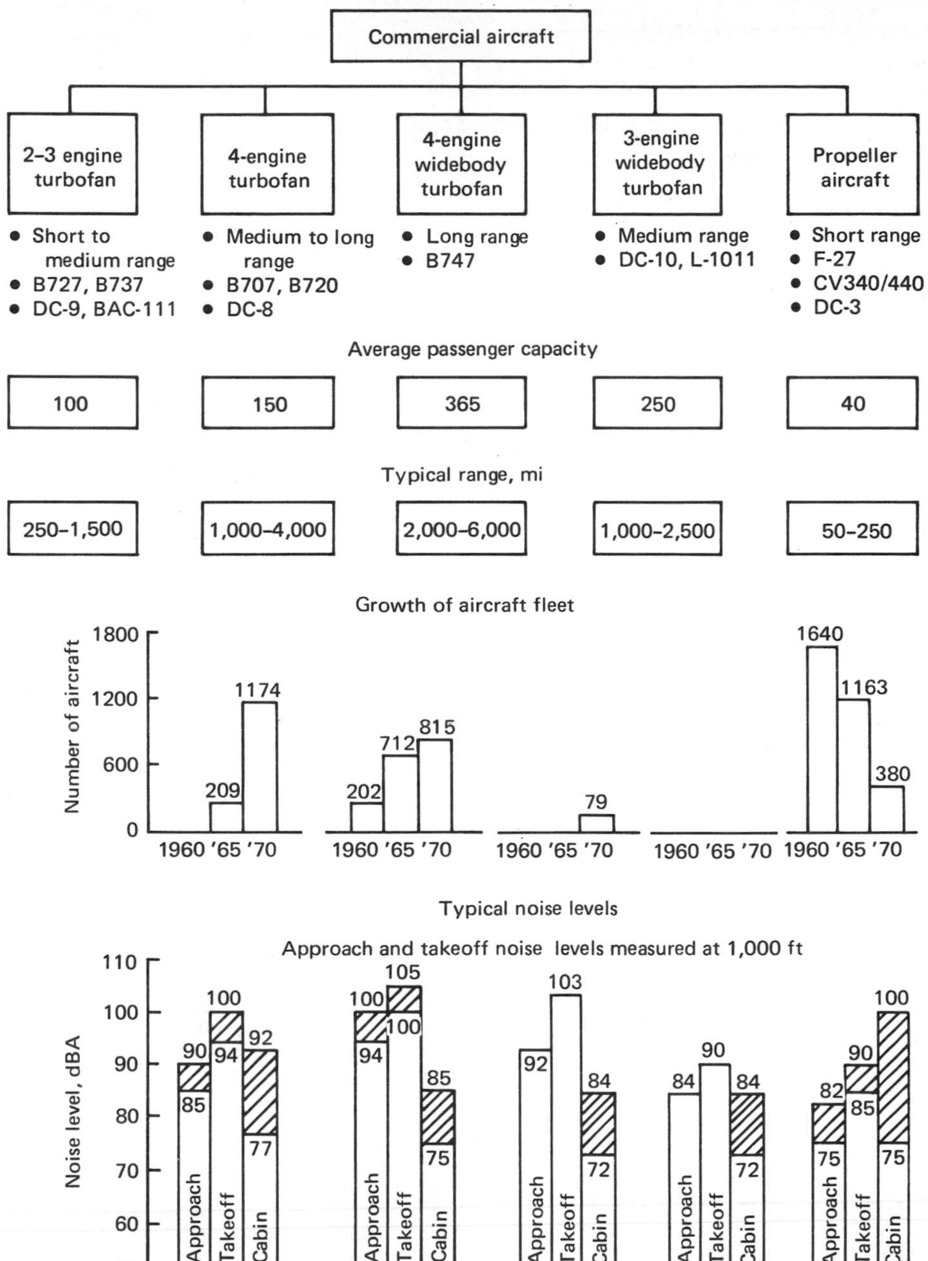

Figure 6-3 Characteristics of commercial aircraft.

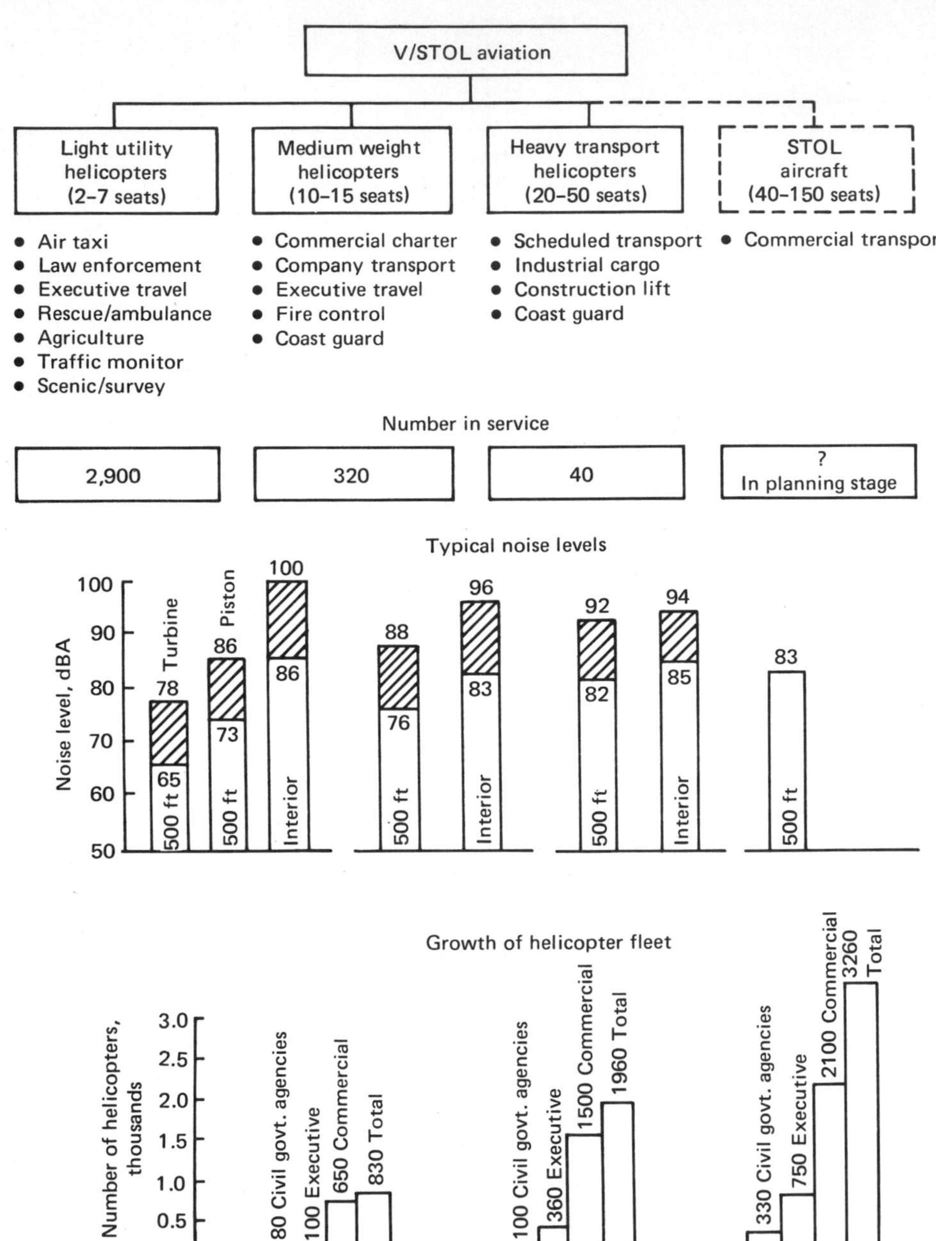

Figure 6-4 Characteristics of vertical (V) and short takeoff and landing (STOL) aircraft.

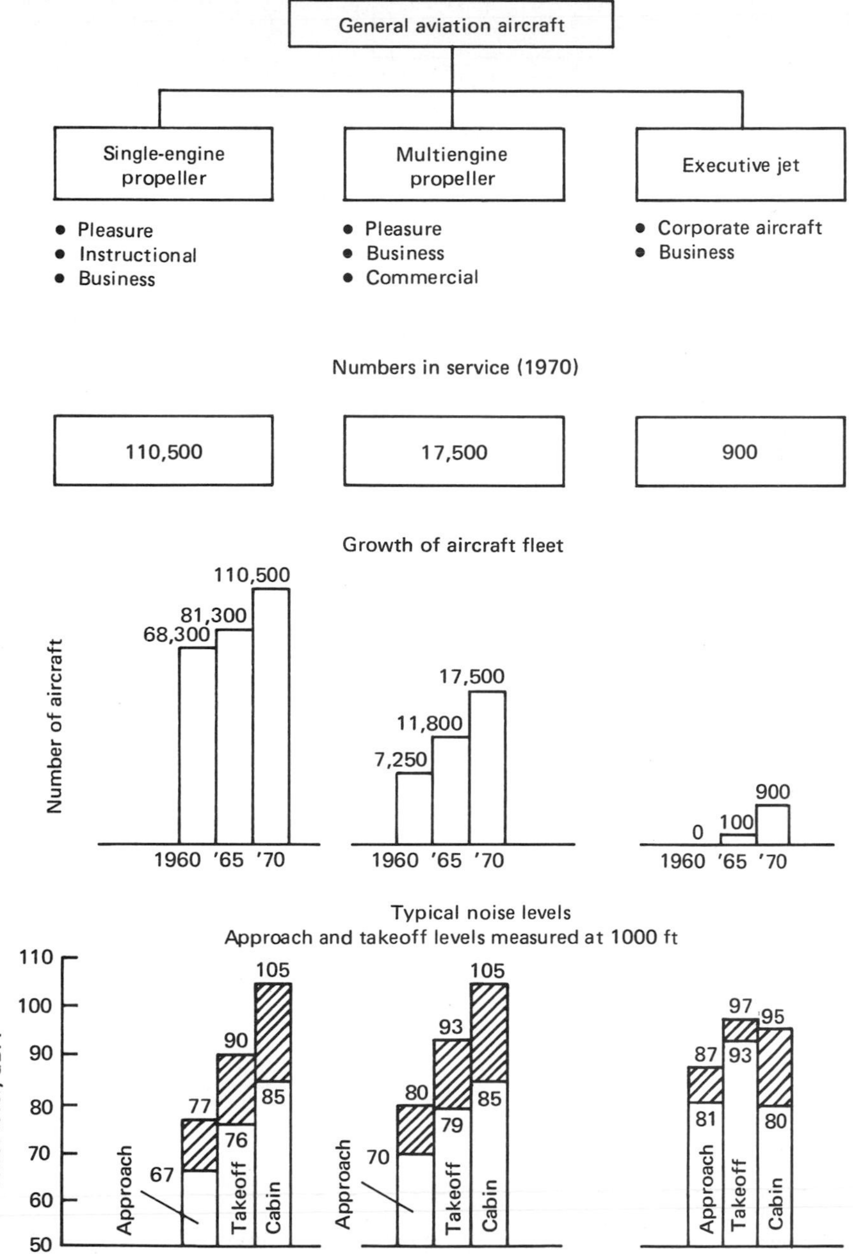

Figure 6-5 Characteristics of general aviation aircraft.

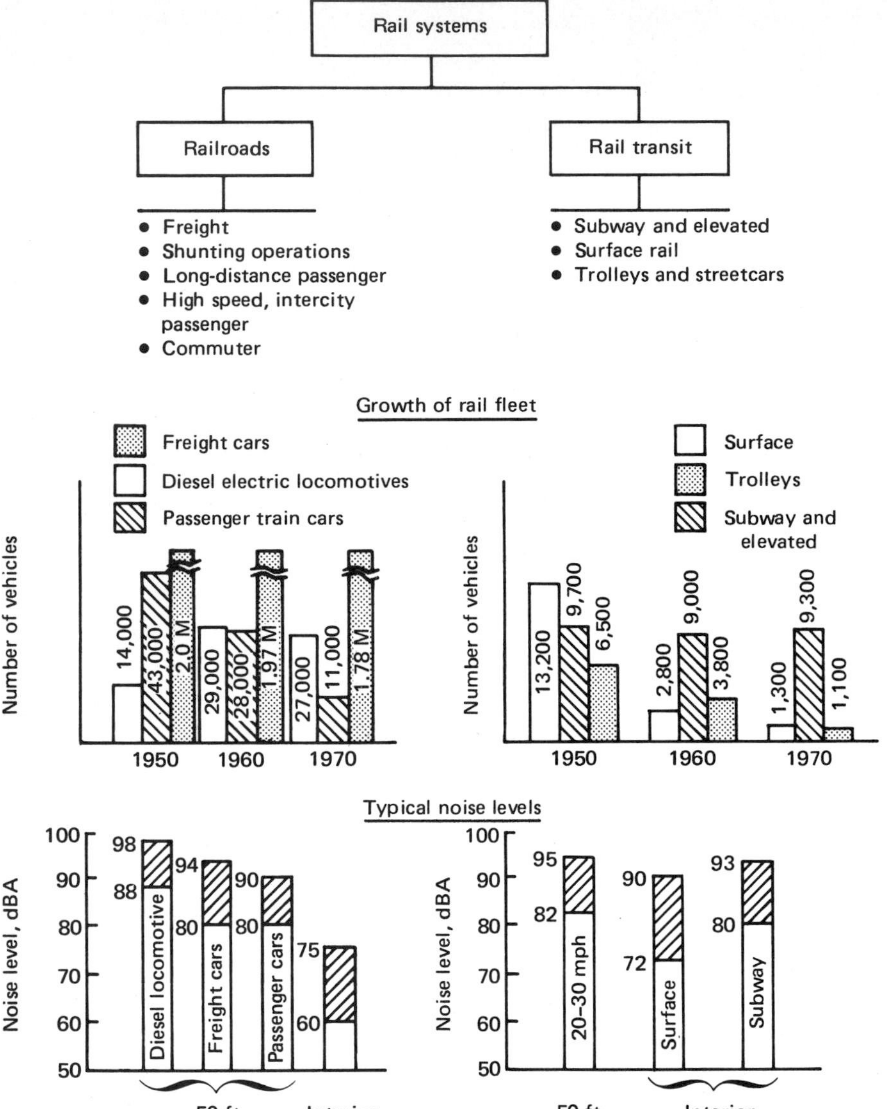

Figure 6-6 Characteristics of rail systems.

a noise survey. Details associated with the planning and conduction of surveys are beyond the scope of this discussion. Examples of noise surveys are available (9, 10).

EFFECTS OF NOISE (BASIS FOR STEP 3)

Information on the physiological effects of noise has been developed primarily from industrial exposures (2) and on the psychological effects of noise, as a result of legal

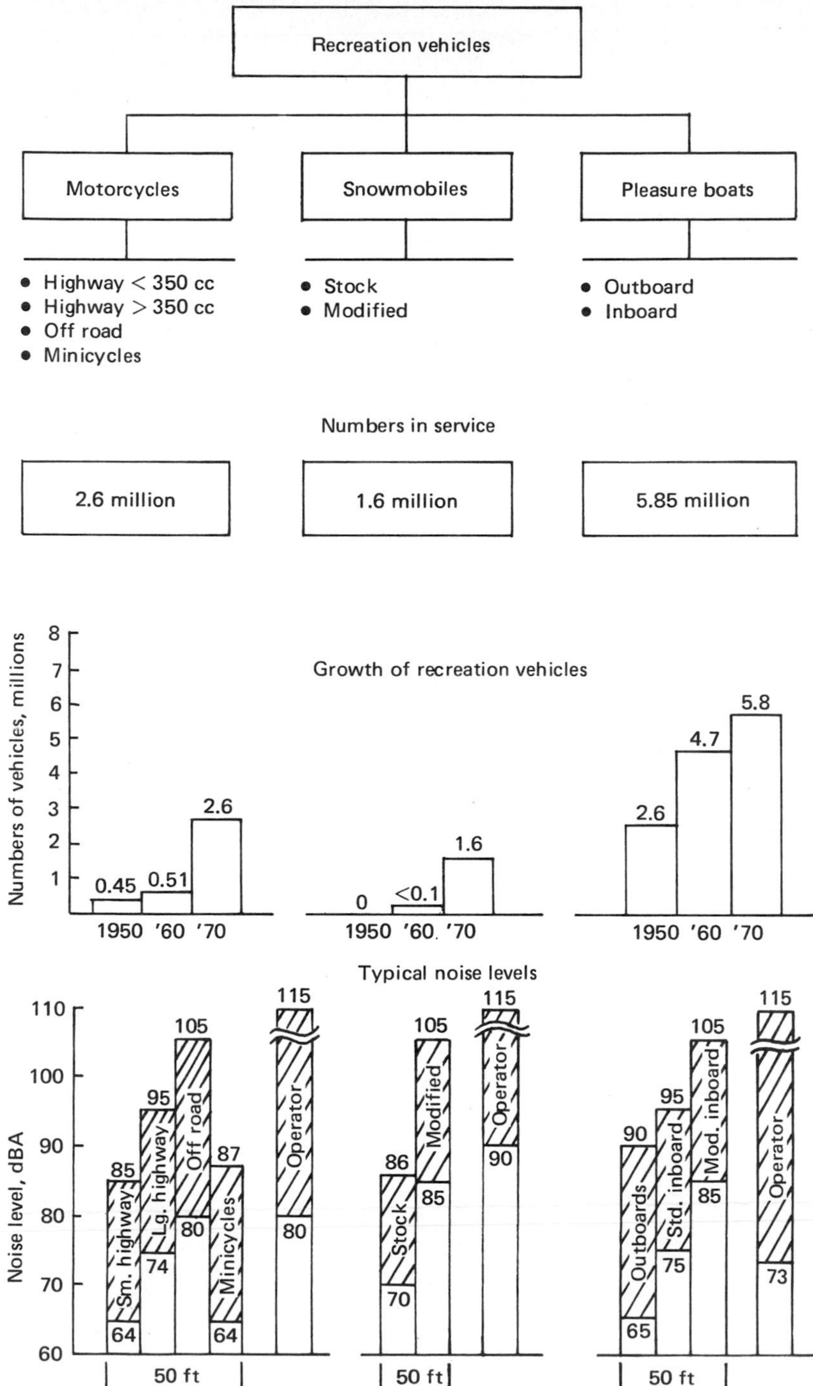

Figure 6-7 Characteristics of recreation vehicles.

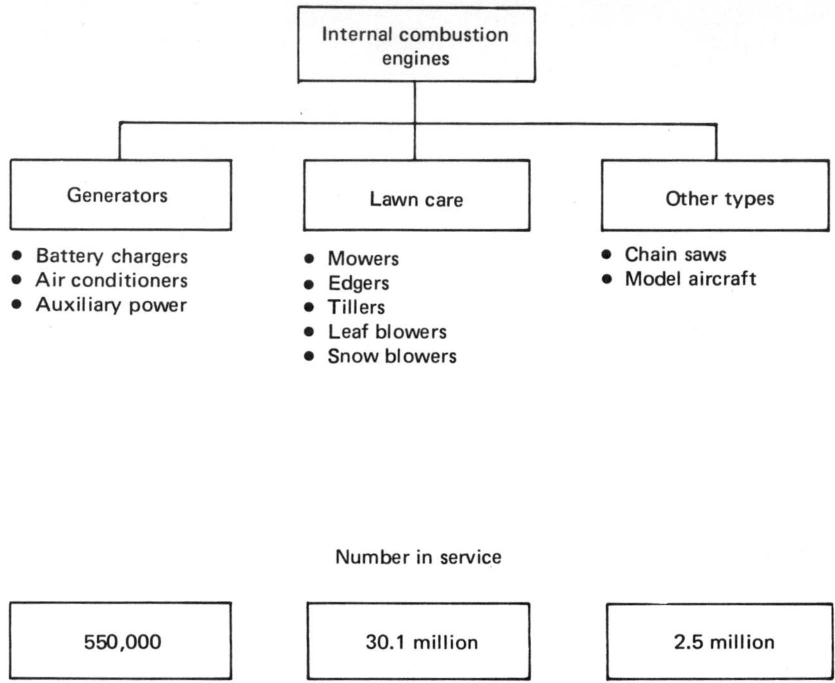

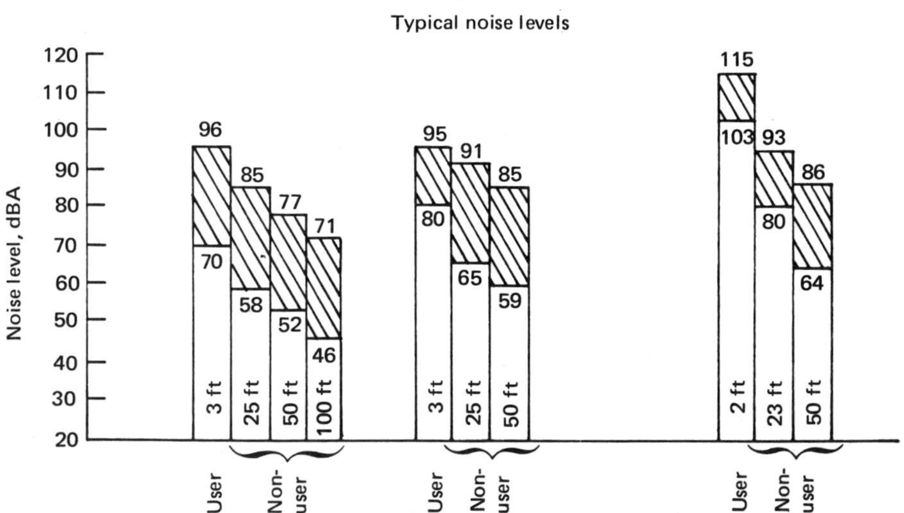

Figure 6-8 Characteristics of devices powered by internal combustion engines.

Table 6-6 Range of Industrial Machinery, Equipment, and Process Noise Levels[a]

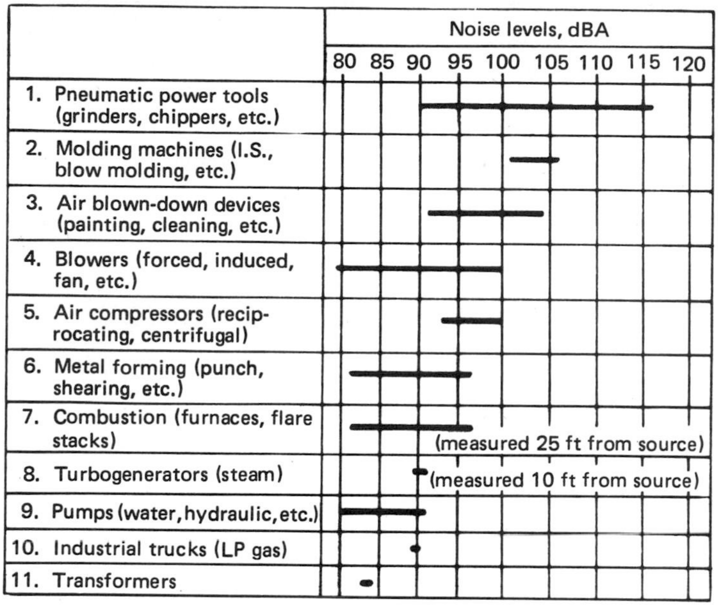

	Noise levels, dBA								
	80	85	90	95	100	105	110	115	120
1. Pneumatic power tools (grinders, chippers, etc.)									
2. Molding machines (I.S., blow molding, etc.)									
3. Air blown-down devices (painting, cleaning, etc.)									
4. Blowers (forced, induced, fan, etc.)									
5. Air compressors (reciprocating, centrifugal)									
6. Metal forming (punch, shearing, etc.)									
7. Combustion (furnaces, flare stacks)				(measured 25 ft from source)					
8. Turbogenerators (steam)				(measured 10 ft from source)					
9. Pumps (water, hydraulic, etc.)									
10. Industrial trucks (LP gas)									
11. Transformers									

[a]Measured at operator positions, except for 7 and 8.

actions taken in conjunction with airports (3, 11). Noise effects can be categorized into hearing changes and losses, interference with speech communication, annoyance, and other effects.

Two types of hearing changes are caused by noise exposure (2). Temporary threshold shift (TTS) is a lessened ability to hear weak auditory signals; however, recovery occurs in a few hours up to 4 wk. Noise-induced permanent threshold shift (NIPTS) represents hearing loss from which there is no recovery. TTS increases linearly with the average noise level from about 80 to 130 dBA. It is proportional to the length of exposure; thus steady noise is the major offender. Some facts about NIPTS, which is a form of deafness, are as follows:

1 Unprotected exposures of 8 hr/day for several years to noise above 105 dBA produces NIPTS.

2 In the occupational setting, NIPTS will appear in almost all men exposed 8 hr/day to noise above 105 dBA. It will appear in about 50 percent of those exposed similarly to a level of 95 dBA. It will not appear in anyone at a level below 80 dBA.

3 Regular exposures to moderate noise levels do not make the ear more resistant to occasional exposures to high noise levels.

Noise can also interfere with speech communication, which impedes activities and human relationships. Figure 6-9 illustrates noise levels causing speech interference (1). If two people are standing 10 ft apart, normal speech communication can occur as long as the interference noise is less than 55 dBA; the voice will have to be raised if the noise is between 55 and 75 dBA; verbal communication can occur only with shouting when the noise is between 75 and 92 dBA; and above 92 dBA speech communication is impossible.

Table 6-7 Range of Building Equipment Noise Levels to Which People Are Exposed

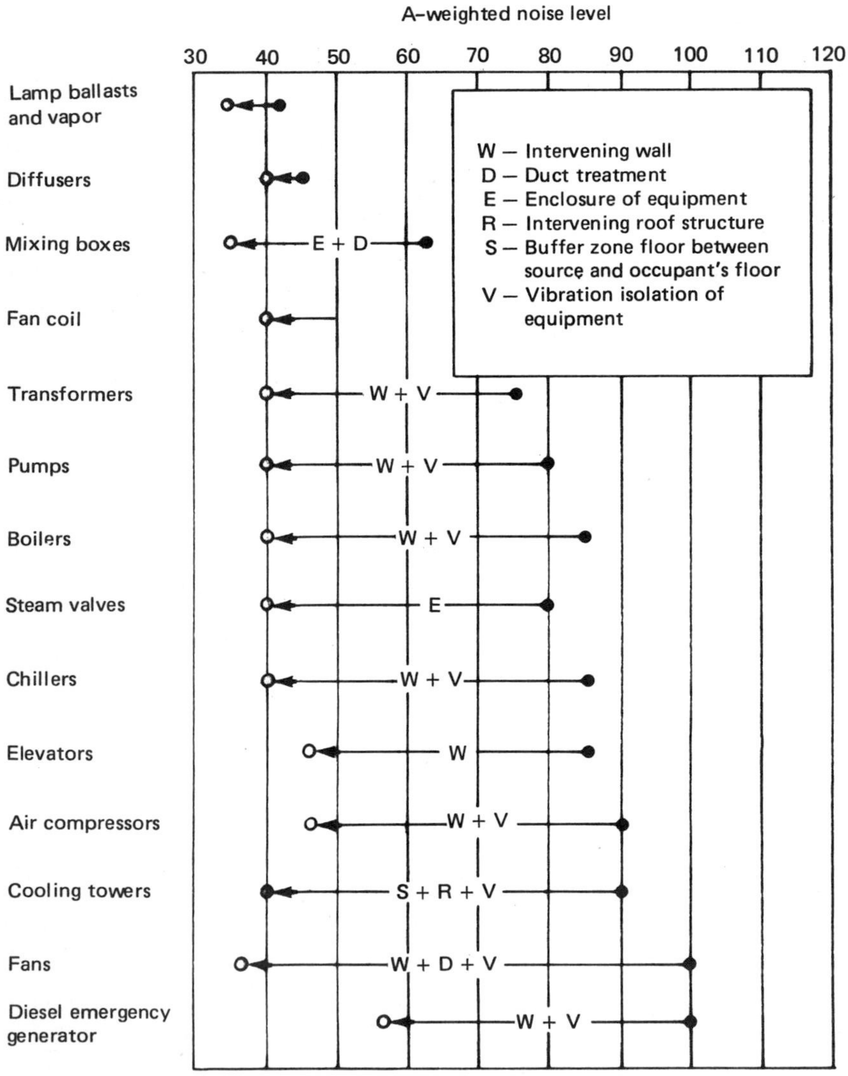

● Noise level at 3 ft from source
○ Noise level at occupant's position

Table 6-8 Estimated Percentage of Urban Population (134 Million) Residing in Areas with Various Day-Night Noise Levels Together with Customary Qualitative Description of the Area

Description	Typical range of L_{dn}, dB	Average L_{dn}, dB	Estimated percentage of urban population	Average census tract population density, no. of people/mi^2
Quiet suburban residential	48–52	50	12	630
Normal suburban residential	53–57	55	21	2,000
Urban residential	58–62	60	28	6,300
Noisy urban residential	63–67	65	19	20,000
Very noisy urban residential	68–72	70	7	63,000

Annoyance resulting from noise involves the subjective response of people. Figure 6-10 shows typical community responses to noise levels (1). In general, as noise levels increase, community reaction increases in magnitude and intensity. Many studies have been conducted on noise annoyance associated with airports (3).

Additional effects of noise include disruption of sleep and rest, reduction in work performance, property devaluation resulting from sonic booms, and

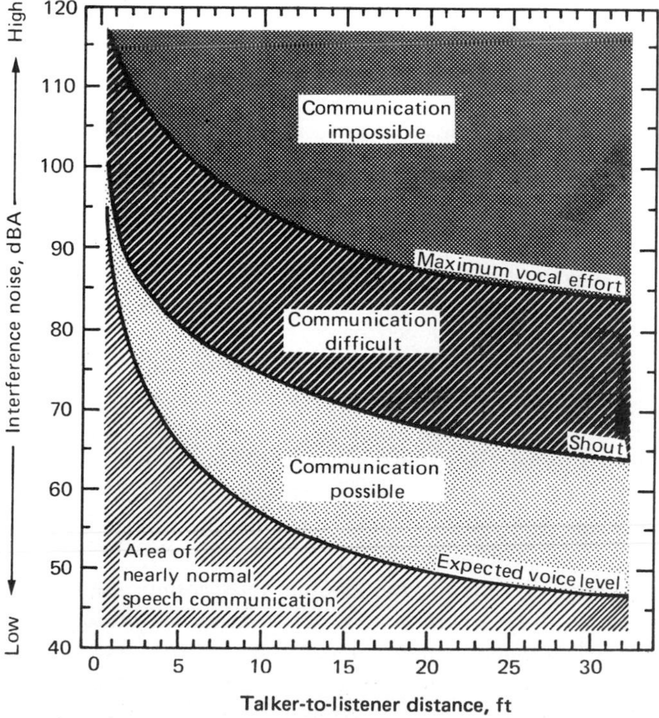

Figure 6-9 Speech interference levels.

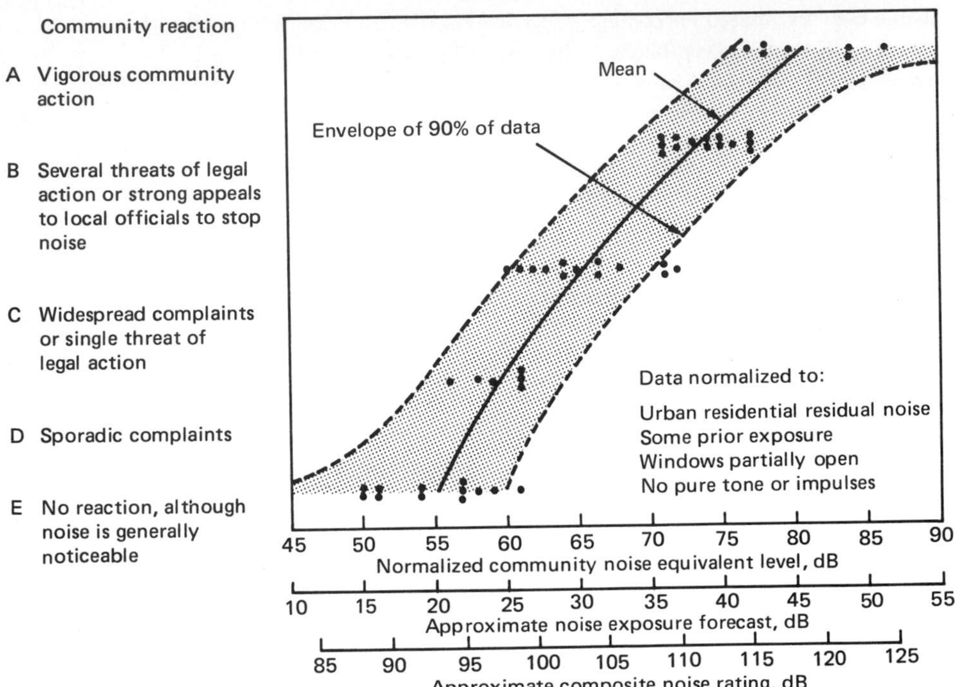

Figure 6-10 Community reaction to intrusive noises of many types as a function of the normalized community noise equivalent level.

interference with normal patterns of behavior of domestic and wild animals. Reference materials with detailed presentations on these manifold effects of noise are available (1, 3, 6).

STEP 3: NOISE STANDARDS AND CRITERIA

Noise standards and criteria have been developed based on consideration of various effects of noise. There is a scientific terminology on noise levels in conjunction with standards and criteria. One category is the percentage of time a given noise level is exceeded. In order to present the information given in Table 6-9, statistical analyses of noise levels measured over time are necessary. For the statistical distribution of Table 6-9, the noise level exceeds 60 dBA 1 percent of the time, 55 dBA for 10 percent, 50 dBA for 50 percent, and 45 dBA for 90 percent. These noise levels are abbreviated as L_1, L_{10}, L_{50}, and L_{90}, respectively.

Table 6-10 shows the design noise levels and land-use relationships used by the Federal Highway Administration (12). Exterior noise levels apply to outdoor areas that have regular human use and where a lowered noise level is of benefit. These design L_{10} values are to be applied at those points within the sphere of human activity (at approximate ear-level height) where outdoor activities actually occur. The values do not apply to an entire tract upon which the activity is based, but

Table 6-9 Example of Statistical Distribution of Outdoor Noise Analyzed in Intervals of 5-dB Widths

Interval, dBA	Percent of total time	Cumulative percent of total time	Noise level
61-65	1	1	L_1
56-60	9	10	L_{10}
51-55	40	50	L_{50}
46-50	40	90	L_{90}
41-45	10	100	

only to that portion where the activity occurs. The design values need not be applied to areas having limited human use or where lowered noise levels would produce little benefit. Such areas include but are not restricted to junk yards, industrial areas, railroad yards, parking lots, and storage yards. The interior design noise level and land-use category E applies to indoor activities for those situations where no exterior noise sensitive land use or activity is identified.

Table 6-11 summarizes criteria for protection of public health and welfare with an adequate margin of safety (6). The phrase "public health and welfare" is defined as complete physical, mental, and social well-being, and not merely the absence of disease and infirmity. Table 6-11 is useful for noise impact assessment in the absencc of specific noise standards for a given area.

Detailed noise standards have been developed for airports and residential areas. The scope of these standards, their criteria, and their associated terminology will

Table 6-10 Design Noise Level/Land-Use Relationships

Land-use category	Design noise level (L_{10}), dBA	Description of land-use category
A	60 (exterior)	Tracts of lands in which serenity and quiet are of extraordinary significance and serve an important public need and where the preservation of those qualities is essential if the area is to continue to serve its intended purpose; such areas could include amphitheaters, particular parks or portions of parks, or open spaces that are dedicated or recognized by appropriate local officials for activities requiring special qualities of serenity and quiet
B	70 (exterior)	Residences, motels, hotels, public meeting rooms, schools, churches, libraries, hospitals, picnic areas, recreation areas, playgrounds, active sports areas, and parks
C	75 (exterior)	Developed lands, properties, or activities not included in categories A and B
D	–	For requirements on undeveloped lands see paragraphs (5) and (6) in Ref. (12)
E	55 (interior)	Residences, motels, hotels, public meeting rooms, schools, churches, libraries, hospitals, and auditoriums

Table 6-11 Yearly Average[a] Equivalent Sound Levels Identified as Requisite to Protect the Public Health and Welfare with an Adequate Margin of Safety

		Indoor		To protect against both effects[c]	Outdoor		To protect against both effects[c]
	Measure	Activity interference	Hearing loss consideration[b]		Activity interference	Hearing loss consideration[b]	
Residential with outside space and farm residences	L_{dn}	45		45	55		55
	$L_{eq(24)}$		70			70	
Residential with no outside space	L_{dn}	45		45			
	$L_{eq(24)}$		70				
Commercial	$L_{eq(24)}$	*d*	70	70[e]	*d*	70	70[e]
Inside transportation	$L_{eq(24)}$	*d*	70	*d*			
Industrial	$L_{eq(24)}$[f]	*d*	70	70[e]	*d*	70	70[e]
Hospitals	L_{dn}	45		45	55		55
	$L_{eq(24)}$		70			70	
Educational	$L_{eq(24)}$	45		45	55		55
	$L_{eq(24)}$[f]		70			70	
Recreational areas	$L_{eq(24)}$	*d*	70	70[e]	*d*	70	70[e]
Farm land and general unpopulated land	$L_{eq(24)}$				*d*	70	70[e]

[a] Refers to energy rather than arithmetic averages.

[b] The exposure period that results in hearing loss at the identified level is 40 yr.

[c] Based on lowest level.

[d] Since different types of activities appear to be associated with different levels, identification of a maximum level for activity interference may be difficult except in those circumstances where speech communication is a critical activity.

[e] Based only on hearing loss.

[f] An $L_{eq(8)}$ of 75 dB may be identified in these situations so long as the exposure over the remaining 16 hr/day is low enough to result in a negligible contribution to the 24-hr average, i.e., no greater than an L_{eq} of 60 dB.

not be covered here. Reference materials are available that enable a more detailed study (3).

STEP 4: PREDICTION OF NOISE LEVELS

Prediction of anticipated noise levels resulting from construction and operation is the most critical step in assessment of the impact of alternatives on the noise environment. Two simple models for noise level prediction (point source and line source) will be considered (13). Sound travels through air in waves with the characteristics of frequency and wavelength. As shown in Fig. 6-11, if a sound is created at a point, a system of spherical waves propagates from that point outward through the air at a speed of 1,100 ft/sec, with the first wave making an ever-increasing sphere with time. As the wave spreads, the height of the wave or the intensity of sound at any given point diminishes as the fixed amount of energy is spread over an increasing surface area of the sphere. This phenomenon is known as geometric attenuation of sound. Point source propagation can be defined as follows:

$$\text{Sound level}_1 - \text{sound level}_2 = 20 \log_{10} \frac{r_2}{r_1}$$

The sound level at station 1 minus the sound level at station 2 is equal to 20 times the log of the ratio of the radii (r). This means that for every doubling of distance, the sound level decreases by 6 dBA. This point-source relationship is called the inverse square law and is applicable for single vehicles and aircraft when sound is propagating in free air, either from the airplane to the ground in a complete

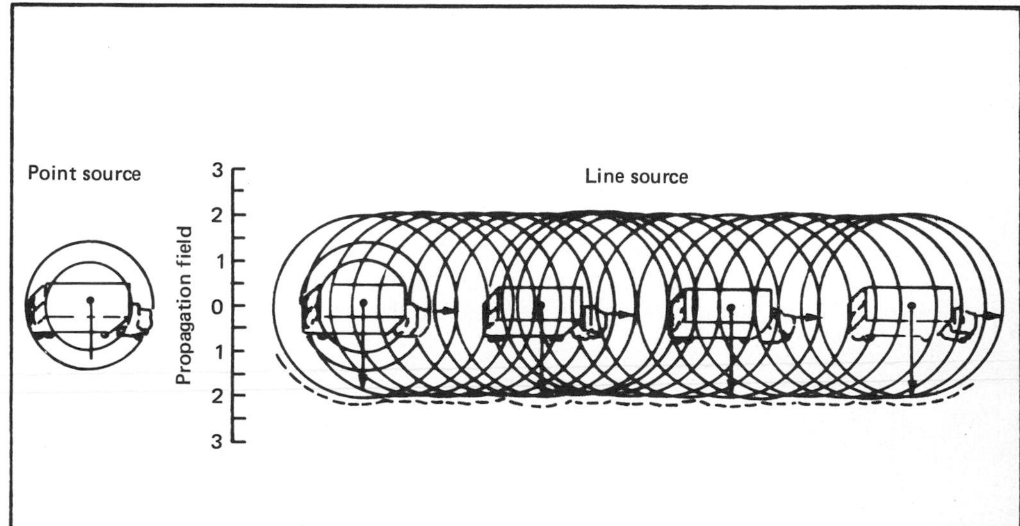

Figure 6-11 Sound propagation comparison.

spherical sense or in the case of an automobile on the ground when the propagation field is only half a sphere. It would also apply for construction equipment.

Line-source propagation occurs when there is a continuous stream of noise sources. As shown in Fig. 6-11, the propagation is no longer characterized by a spherical or hemispherical spreading of sound, but rather the reinforcement by the line of point sources makes the propagation field like a cylinder or half cylinder. Line source propagation prediction is as follows:

$$\text{Sound level}_1 - \text{sound level}_2 = 10 \log_{10} \frac{r_2}{r_1}$$

The decrease in sound level for each doubling of distance from a line source is 3 dBA. When noise levels from a busy highway are considered, it is appropriate to utilize the highway as an infinite line source and consider a 3-dBA doubling of distance propagation rate.

The Federal Highway Administration uses two highway noise prediction approaches (12)—the method developed by the National Cooperative Highway Research Program (14) and the one developed by the Transportations Systems Center (15). These computerized methods involve consideration of the characteristics of highway segments as one input variable. Pertinent characteristics include the traffic using the highway (quantities and speeds of both automobiles and trucks), physical dimensions of the facility (the elevation, depressions, grades, and surface types), and aspects of the environment bordering the facilities that have an effect on noise levels (landscaping, structures, and barriers). Both models calculate the noise level at a particular perpendicular distance away from a point along the highway. Once this noise level is calculated, the model moves outward an incremental distance and calculates another noise level. This process is repeated until the model reaches a maximum prescribed distance away from the highway. At this point the model moves further down the highway and calculates another group of noise levels. This is repeated until the model has covered the entire length of the highway section of interest. Model printouts include contour maps of noise levels over the entire length of the facility (16).

Predictions of noise levels from aircraft may involve three noise-prediction systems: aircraft sound description system, noise exposure forecast, and composite noise rating. It is beyond the scope of this presentation to review details of these methods of prediction. References are available that allow more detailed study (16).

STEP 5: NOISE CONTROL PRACTICES

If the predicted noise levels from step 4 indicate that certain noise standards or criteria are exceeded, consideration must be given to noise minimization in order to accomplish compliance. There are three basic principles of noise control, namely, reduction of vibrating sources, enclosure of the source, and attenuation by absorption (2). Noise abatement programs have been developed for controlling industrial noise and include plant planning; substitution of quieter equipment, processes, and materials; reduction at the source; and reduction of transmission by air. Description of specific noise abatement procedures and practices is beyond the scope of this presentation. Two general examples will be cited to indicate options for highway noise abatement and subsonic aircraft noise abatement.

Noise control measures for highways include constructed barriers to obstruct or dissipate sound emissions, elevated or depressed highways, and the absorption effects of landscaping (trees, bushes, and shrubs). Constructed barriers can be an effective approach for reducing noise from highways. Important factors include the relative height of the barrier, the noise source and the affected area, and the horizontal distances between the source and the barrier and the barrier and the noise-affected area (13). Elevating or depressing highways in urban areas provides differences in grade, which shield traffic noise and reduce noise levels on adjacent properties. Plantings of trees, bushes, and shrubs adjacent to a highway generally produce little physical reduction in noise unless the plantings are very dense and have significant depth. Additional noise control measures for highways include limitations on allowable grades, maintenance of proper road surface repairs, route locations planned to ensure maximum separation between roadways and existing noise-sensitive areas, and provisions for compatible land uses adjacent to highway rights-of-way.

Subsonic aircraft noise abatement involves consideration of aircraft design or modification, changes in aircraft operations and route locations, more frequent aircraft maintenance, and landscape architecture and acoustic insulation at facilities located on or near airports (13). In addition, land use zoning for compatible uses in the vicinity of airports is desirable.

SUMMARY

This chapter is oriented to the types of information and practical steps that can be used in predicting the impacts of alternatives on the noise environment. Suggestions are provided for assessment or interpretation of the predicted impacts, with these primarily based on comparisons of predicted noise levels with applicable standards and criteria for the given area.

SELECTED REFERENCES

1 Environmental Protection Agency: *Report to the President and Congress on Noise*, 92d Cong., 2d Sess., Document 92-63, Washington, D.C., Feb. 1972.
2 Chanlett, E. T.: "Environmental Protection," pp. 523–544, McGraw-Hill Book Company, New York, 1973.
3 Environmental Protection Agency: "Public Health and Welfare Criteria for Noise," Publ. 550/9-73-002, Office of Noise Abatement and Control, Washington, D.C., July 27, 1973.
4 *Noise Control Act of 1972*, PL 92-574, 92d Cong., H.R. 11021, Oct. 27, 1972.
5 Council on Environmental Quality: "Environmental Quality–the Fifth Annual

Report of the Council on Environmental Quality," pp. 167–171, Washington, D.C., Dec. 1974.

6 Environmental Protection Agency: "Information on Levels of Environmental Noise Requisite to Protect Public Health and Welfare with an Adequate Margin of Safety," Publ. 550/9-74-004, Office of Noise Abatement and Control, Washington, D.C., Mar. 1974.

7 Wyle Laboratories: "Transportation Noise and Noise From Equipment Powered by Internal Combustion Engines," Publ. NTID 300.13, prepared for Office of Noise Abatement and Control, U.S. Environmental Protection Agency, Washington, D.C., Dec. 31, 1971.

8 Clark, J. R.: "Environmental Assessment of Construction Activities," M.S. thesis, University of Oklahoma, Norman, 1975.

9 Wyle Laboratories: "Community Noise," Publ. NTID 300.3, prepared for Office of Noise Abatement and Control, U.S. Environmental Protection Agency, Washington, D.C., Dec. 31, 1971.

10 Putnicki, G. J., C. S. Riddel, and H. Watson: "Environmental Noise Assessment, Waco, Texas Metropolitan Area," region VI, U.S. Environmental Protection Agency, Dallas, Tex., Apr. 5, 1974.

11 Hildebrand, J. L.: Noise Pollution: An Introduction to the Problem and an Outline for Future Legal Research, *Columbia Law Rev.*, vol. 70, pp. 652–692, Apr. 1970.

12 Federal Highway Administration: Noise Standards and Procedures, Policy and Procedure Memorandum 90-2, U.S. Department of Transportation, Washington, D.C., Feb. 8, 1973.

13 Department of Transportation: "Transportation Noise and Its Control," Publ. DOT P 5630.1, Washington, D.C., June 1972.

14 Gordon, C. G., et al.: Highway Noise–A Design Guide for Highway Engineers, *Natl. Coop. Highway Res. Progr. Rept.* 117, 1971.

15 Wesler, J. E.: "Manual for Highway Noise Prediction," Rept. DOT-TSC-FHWA-72-1, prepared by Transportation Systems Center for U.S. Department of Transportation, Washington, D.C., Mar. 1972.

16 Nelson, K. E., and T. D. Wolsko: "Transportation Noise: Impacts and Analysis Techniques," Rept. ANL/ES-27, prepared by Argonne National Laboratory for Illinois Institute for Environmental Quality, Chicago, Oct. 1973.

Chapter 7

Prediction and Assessment of Impacts on the Biological Environment

Major impacts from many actions occur on floral and faunal species that are components of the biological environment within and adjacent to project areas. General impacts on the biological environment are related to changes in community types and their geographical distribution. Specific impacts may occur in the life cycles of rare and endangered species inhabiting the area of concern. This chapter is addressed to data needs and associated technology for predicting and assessing impacts of proposed actions on the biological environment.

BASIC STEPS FOR PREDICTION AND ASSESSMENT

1 Prepare a description of the flora and fauna comprising the biological environmental setting. Describe community types and their geographical distribution, and develop species descriptions for each community type.
2 Identify rare and endangered species inhabiting the area of interest, and discuss relevant characteristics of these species.
3 If appropriate, discuss past and current management practices as related to floral and faunal species, as well as special activities associated with protected species.
4 Discuss natural succession as it relates to the alteration of communities with time. In essence, this is an attempt to describe what the environment will become without implementation of the proposed action.

5 Predict the impacts of alternatives on the biological environmental setting. Quantify the impacts where possible, and qualitatively discuss the implications of the remainder.
6 Summarize the critical impacts associated with various alternatives. Do not only consider individual species, but rather describe general impacts on the overall ecosystem.

BASIC INFORMATION ON ECOLOGY

Ecology is the study of the relation of organisms or groups of organisms to their environment. In essence, this study involves consideration of the structure and function of various organisms and groups of organisms within the environment; thus ecology consists of the study of ecosystems. An ecosystem is defined as the abiotic, physicochemical, and biotic assemblage of the plants, animals, and microbes in which an ecological relationship is demonstrated. The characteristics of an ecosystem include abiotic (nonliving) and biotic (living) components, interaction of the community and the environment, and exchange of materials (energy and nutrients) between nonliving and living parts. Energy flow and nutrient cycling involve interactions between the physicochemical environment and the biotic assemblage, thus constituting the keys to ecosystem dynamics (1).

A community includes all organisms in a given area interacting with the abiotic environment. A group of organisms of a single species is a population. Within a community and the environment, energy flow is noncyclic, and nutrient movement is cyclic. Therefore, energy flow is unidirectional, and nutrient flow can be bidirectional. The primary energy source for the biological environment is the sun.

Ecological succession describes the orderly process of development or change that involves species structures and community processes. It is usually directional and thus predictable. Succession results from modification of the physical environment by the community, that is, succession is community controlled, although the physical environment determines the pattern and rate of change. Ecological succession terminates in a stable ecosystem with maximum biomass and species symbiosis. Transitory or temporary communities are called seral stages, and the stabilized system is known as a climax community. Changes in species composition and community structure may result from natural phenomena.

It is beyond the scope of this chapter to provide a detailed presentation on ecological principles and concepts. The selected terms that have been defined are representative of those appropriate for consideration in describing the biological environmental setting and predicting and assessing impacts of proposed actions.

STEP 1: DESCRIPTION OF BIOLOGICAL SETTING

The first step in the process of prediction and assessment of impacts on the biological environment involves the preparation of an adequate description of the environmental setting from the biological perspective. One of the first considerations is to define the general community types and to describe these types in accordance with their geographical distribution. In a recent study of eight alternative waterway navigation routes from Tulsa, Oklahoma to Wichita, Kansas, community types were defined in accordance with vegetation, including floodplain forests, postoak-blackjack forests, and bluestem prairies (2).

The second part of this step is to develop species descriptions for each community type. The most effective way to present species information is to organize the data by community type and, where possible, provide quantitative information indicating the population density of the particular species. If population density information is unavailable, the likelihood of occurrence of a given species can be indicated as ranging from common to occasional to rare (3). Common occurrence denotes that the species occurs in many localities within the community type in large numbers. Occasional denotes occurrence in several localities in small numbers, and rare denotes highly localized occurrence that is restricted by scarcity of habitat or low numbers.

The primary categories of flora and fauna utilized in a waterway navigation study (2) are as follows:

Floral components

General vegetation patterns of entire area.
Plant species for bluestem prairie.
Tree species in upland forests.
Tree species in lowland forests.
Shrubs and vines of postoak-blackjack forest.
Shrubs and vines of floodplain forest.
Herbaceous plants of postoak-blackjack forest.
Herbaceous plants of floodplain forest.
Rare plant species in entire area.

Faunal components of entire area

Amphibians–frogs, toads, salamanders.
Reptiles–turtles, lizards, snakes.
Naiads (freshwater mussels or freshwater clams).
Fishes.
Sport fisheries.
Birds.
Mammals.
Rare faunal species.

Table 7-1 presents information on amphibians in the nine-county region associated with the waterway navigation study. The occurrence of these amphibians by alternative routes and the necessary habitats and breeding areas associated with the amphibians, as well as comments highlighting items of unique interest with regard to the individual species, are all included in Table 7-1.

Sources of information for preparing a description of the biological environmental setting include published papers in technical journals and conference proceedings and reports from local and state agencies such as wildlife commissions,

Table 7-1 Partial List of Amphibians of the Mid-Arkansas River Basin Region[a]

Amphibian	Occurrence[b,c]		Habitat[d,e]	Comments
Scaphiopus hurteri (Hurter's spadefoot)	C	C	T (B)	Inhabits woodlands and grasslands
Bufo americanus (American toad)	C	C	T (B)	Inhabits moist woodlands
Bufo woodhousei (Rocky Mountain toad)	O	O	T (B–R)	Frequents moist woodlands
Acris crepitans (Blanchards cricket frog)	C	C	PR	
Rana catasbeiana (bullfrog)	C	C	PR	
Rana aerolata (northern crayfish frog)		R	T (B)	Lowland thickets/waterways; associated with crayfish burrows
Rana clamitans (green frog)	O	O	P–R–B	
Rana pipiens (leopard frog)	C	C	P–R–B	
Gastrophryne olivacea (Great Plains narrow-mouthed toad)	O		T (B)	Mainly subterranean

[a]After (2).
[b]The two listings under occurrence represent two alternative routes.
[c]C = common; occurring in many localities in large numbers. O = occasional; occurs in several localities in small numbers. R = rare; highly localized, restricted by scarcity of habitat or low numbers.
[d]T = terrestrial; land areas not associated with water. R = running water; streams and springs. P = permanent; stationary bodies of water. B = temporary bodies of water.
[e]Breeding habitats are listed in parentheses.

fish and game commissions, or departments of natural resources. Federal sources of information include the Bureau of Land Management, Fish and Wildlife Service, National Forest Service, and National Park Service. University documents can also provide previously unpublished information and data.

STEP 2: IDENTIFICATION OF RARE AND ENDANGERED SPECIES

One of the critical aspects of environmental impact assessment is the potential impact on any rare and endangered plant or animal species within the area of interest. Information on rare and endangered species can be obtained from the U.S. Department of the Interior. Once the potential existence of rare and endangered species is established for an area, a description of the characteristics of these species should be prepared. Characteristics include breeding and nesting requirements, life-cycle features, and other unique requirements that may be of importance in considering the impacts of alternatives.

STEP 3: DISCUSSION OF MANAGEMENT PRACTICES

Many areas under federal, state, or local control are subjected to practices directed toward managing the biological environment. Examples include spraying for pest control, introducing new species, controlled burning, stocking water bodies with fish species, and limiting hunting seasons in terms of time and number of animals killed. If the areas of interest for any alternatives under consideration include subareas having management practices, these subareas as well as the practices themselves need to be described. Conduction of management practices indicates that those components of the biological environment for which practices exist are of significance and concern.

STEP 4: DISCUSSION OF NATURAL SUCCESSION

Natural succession is descriptive of normal changes within the biological environment that lead to alteration of community types and of species within community types over extended periods of time. If an adequate analysis of succession can be accomplished, the information developed will provide the basis for describing the biological environment that will exist in the future without implementation of the proposed action.

STEP 5: PREDICTION OF IMPACTS OF ALTERNATIVES

The biological impacts of the alternatives under consideration should be described, using quantitative techniques where possible and qualitative discussions of the remainder. Biological impacts associated with many alternatives are related to resultant land-use changes. Land-use changes interfere with community types and, in turn, interfere with the individual species within community types.

One way of identifying the multiple impacts of a proposed action on the biological environmental setting is to utilize a checklist approach that considers the relevance of individual potential impacts on various flora and fauna in the area of interest. A list of possible impacts that can occur on the biological environment (3) includes but is not limited to the following:

1 Effects on the resiliency and fitness of ecosystem types, for example, lowland forest, upland forest, grassland, marsh, bog, and streams.
2 Effects on the total standing crop of organic matter.
3 Effect on annual plant productivity.
4 Effects on mulch or litter removal as related to top-soil stripping.
5 Effects on animal production.
6 Effects on sediment load carried by streams.
7 Effects on aquatic macroinvertebrate populations.
8 Effects on drift rate of aquatic macroinvertebrates.
9 Effects on population density of fish.
10 Effects of sediment load on fish growth.
11 Effects of sediment load on fish spawning.
12 Effects on species diversity of the aquatic biota.
13 Effects on undesirable proliferations of biota.
14 Effects on localized survival of rare plant and animal species.
15 Effects on the habitat carrying capacity of both aquatic and terrestrial systems.

16 Effects on abandonment of habitat.
17 Effects on endemic populations of plants and animals.
18 Effects on wildlife breeding and nesting sites.
19 Effects on endangered plant and animal species.
20 Effects on vegetation communities of denuded areas.
21 Effects on wildlife refuges and sanctuaries.
22 Effects on scientific and educational areas of biological interest.
23 Effects on vegetation recovery rates.
24 Effects on forage areas for both upland and lowland game species.
25 Effects on migratory game bird species.
26 Effects on terrestrial microbial communities.
27 Effects on the amount of forest removed.
28 Effects on population density of past species.
29 Effects on domestic animal species.
30 Effects on the amount of grassland removed.
31 Effects on natural drainage systems.
32 Effects on natural animal corridors.
33 Effects on eutrophication.
34 Effects on expansion of population range for both plant and animal species.
35 Effects on cropland removal.
36 Effects on the potential for wildlife management.
37 Effects on food-web index, including herbivores, omnivores, and carnivores.
38 Effects on species diversity of the terrestrial biota.
39 Effects on nutrient supply available to terrestrial biota.
40 Effects on sport fishing and hunting.
41 Effects of resultant air pollution on crop yield.
42 Effects on relict vegetation types.
43 Effects on the responses of sensitive native plants to air pollutants, both particulates and gases.
44 Effects on unnatural dispersion and subsequent overutilization of habitat.
45 Effects of noise levels on reproductive inhibition of small mammals.
46 Effects of air pollutants on tree canopy.
47 Effects of noise levels on broodiness of upland and lowland game birds.
48 Effects on water-temperature stability.
49 Effects on areas of high brush-fire potential.
50 Effects on water quality and dependent biota.
51 Effects of noise levels on insect maturation and reproduction.
52 Effects on natural biological character loss.

This list can be used to identify potential impacts and, by qualitative or quantitative means, their magnitude.

An example of the types of biological impacts associated with a project is shown in Table 7-2, which lists the biological impacts considered for the waterway navigation project from Tulsa, Oklahoma to Wichita, Kansas (2). A total of 31 biological parameters were identified, and the potential changes resulting from each of eight alternatives to each parameter were determined in this study. Table 7-2 also

Table 7-2 Parameters and Weights Used in the Biological Section of the Mid-Arkansas River Basin Analysis[a]

	Parameter	Weight
1	Biological effect of water quality change	11.9
2	Biological effect of salinization change	11.9
3	Biological effect of eutrophication change	11.9
4	Biological implications of siltation	11.9
5	Browser populations (deer)	7.8
6	Change from lotic to lentic habitat	11.9
7	Changes in lowland feeding and breeding habitats	11.9
8	Changes in lowland nesting habitat	11.9
9	Expansion of population ranges	3.9
10	Fishing pressure	11.9
11	Food-web index	11.9
12	Grazers (cattle)	7.8
13	Interruption of wildlife refuges	11.9
14	Forest land removed	11.9
15	Cropland removed	11.9
16	Grassland removed	11.9
17	Migration of waterfowl	7.8
18	Miles of new shoreline (channel)	11.9
19	Miles of new shoreline (lake)	11.9
20	Number of oxbow lakes	11.9
21	Pest species	3.9
22	Potential areas for fish and wildlife management	3.9
23	Potential for recreation areas	7.8
24	Potential for sport and commercial fishing	7.8
25	Rare and endangered species (plants and animals)	11.9
26	Recruitment from tributaries	11.9
27	Size of oxbow lakes	11.9
28	Terrestrial ecosystem stability or diversity	11.9
29	Unique habitats	11.9
30	Weedy aquatic vegetation	3.9
31	Weedy terrestrial vegetation	3.9

[a]After (2).

identifies the relative importance weight assigned to the 31 parameters. Of the 31 biological parameters used, 21 were considered of high importance, 5 of medium importance, and 5 of low importance. High-importance parameters were weighted as three times more important than the low-importance parameters, with the medium parameters being intermediate between these extremes.

Many environmental impact statements are characterized by having a very complete species list in the environmental setting section but a very cursory presentation of the impacts of the alternatives on these species. It may be impossible to completely describe every potential impact; however, it is important that some attempt be made to describe the general changes that would result from the implementation of the proposed action on as many species as possible. Of particular importance is a discussion of the potential changes on any rare and endangered plant or animal species identified in step 2.

A detailed discussion of the implications of changes on the biological environment, as well as interrelationships among various parameters, is considered beyond the scope of this chapter.

STEP 6: SUMMARY OF CRITICAL IMPACTS

The final step in prediction and assessment of impacts on the biological environment is a discussion of any critical changes that have been identified in conjunction with step 5. In essence, this final step represents the results of assessment of the magnitude and importance of changes resulting from project implementation. This step is a means of summarizing biological impact information for agency decision makers.

SUMMARY

One of the most important categories of impacts associated with project development is related to potential changes in the biological environment. These changes can result from changes in land-use patterns and are of particular importance relative to any rare and endangered plant or animal species present in the area of interest. Numerous sources of information are available for assembling species lists for areas as well as for describing general community types. Impacts of various alternatives on the biological environment are potentially manifold. One technique for addressing these impacts consists of using a checklist of potential impacts and determining their relevance for the particular study. The impacts should be quantified where possible, with qualitative descriptions prepared for the balance.

SELECTED REFERENCES

1 Odum, E. P.: "Fundamentals of Ecology," 3d ed., chaps. 1–7, W. B. Saunders, Philadelphia, Pa., 1971.
2 Canter, L. W., P. G. Risser, and L. G. Hill: "Effects Assessment of Alternative Navigation Routes from Tulsa, Oklahoma to Vicinity of Wichita, Kansas," pp. 53–122, 227–250, University of Oklahoma, Norman, June 1974.
3 Hill, L. G.: personal communication, Oklahoma Biological Station, University of Oklahoma, Norman, 1975.

Chapter 8

Prediction and Assessment of Impacts on the Cultural Environment

One of the major concerns associated with many actions is relative to their potential impact on cultural resources, which include architectural, historical, and archeological sites, as well as areas of unique importance due to their ecological, scientific, or geological information. The sphere of cultural resources includes not only the precise limits of the project area, but also all surrounding lands on which the project may have a reasonably direct impact by modifying land-use patterns or by opening areas for agriculture or for public use, thus increasing potential vandalism (1). Possible impacts on cultural resources include inundation, destruction, disruption, or disturbance. This chapter is addressed to data needs and approaches for predicting and assessing impacts of proposed actions on cultural resources.

BASIC STEPS FOR PREDICTION AND ASSESSMENT

Basic steps associated with prediction of changes in the cultural environment and assessment of the impact of these changes are as follows:

1 Identify known cultural resources in the area of interest. These sources should include historical and archeological sites; areas of ecological, scientific, or geological significance; and areas of ethnic importance. Prepare a discussion of

the cultural overview of the area, including prehistorical as well as historical patterns in the area.

2 Identify potential cultural resources in the area of interest.
3 Determine significance of known and potential cultural resources relative to local, regional, and national concerns.
4 Delineate possible impacts of alternatives on known and potential cultural resources in the area of interest. Impacts should be determined for preconstruction, construction, operation, and postoperation phases.
5 Depending upon the findings of steps 3 and 4, do one of the following: proceed with selection of proposed action from the alternatives, or eliminate one or more alternatives and then proceed with selection of the proposed action. Following selection of the proposed action, conduct a detailed reconnaissance of the project area for the selected action and develop mitigation measures for impact minimization and cultural resources preservation.
6 Develop procedures that will be used during the construction phase if previously unidentified cultural resources are uncovered.

BASIC INFORMATION ON CULTURAL RESOURCES

Cultural resources are becoming more important with the growing realization that our environment and civilization are the products of history. Cultural resources are nonrenewable, and this feature in itself is one reason that these resources are important. Finally, information on cultural resources, particularly archeological resources, can yield important environmental data since past ecological conditions often are reflected in archeological sites (1).

Federal, state, and local laws dealing with cultural resources have been evolving within the past several decades. Federal laws and regulations related to cultural impact include the following:

The Antiquity Act of 1906 (PL 59-209, 34 STAT. 225; 16 U.S.C. 431–433): This act provides for the protection of all historic and prehistoric ruins of monuments on federal lands. It prohibits any excavation or destruction of such antiquities without permission of the secretary of the department having jurisdiction. It authorizes the secretaries of the interior, agriculture, and war to give permission for excavation to reputable institutions for increasing knowledge and for permanent preservation in public museums. It also authorizes the president to declare areas of public lands as national monuments and to reserve lands for that purpose (2).

The Historic Sites Act of 1935 (PL 74-292, 49 STAT. 666; 16 U.S.C. 461–467): This act declares as national policy the preservation for public use of historic sites, buildings, and objects. It has led to the establishment of the Historic Sites Survey, the Historic American Building Survey, and the Historic American Engineering Record, by giving the Secretary of the Interior the power to make historic surveys, to secure and preserve data on historic sites, and to acquire and preserve data on historic sites. The National Historic Landmarks program and its Advisory Board were also established under this act to designate properties having exceptional value as commemorating or illustrating the history of the United States. The National Historic Landmarks program was the beginning of the National Register program (2).

The Reservoir Salvage Act of 1960 (PL 86-523, 74 STAT. 220; 16 U.S.C. 469–469c): This act requires that before any agency of the United States undertakes the construction of a dam or issues a license for construction of a dam (greater than 5,000 acre-ft or 40 surface acres of capacity), it should provide written notice to the Secretary of the Interior. (The provisions of the act apply regardless of the size of the reservoir if the constructing agency finds or is presented with evidence that archeological resources are affected.) The Secretary of the Interior should cause a survey to be made and, if deemed necessary, should cause the necessary research to be conducted as expeditiously as possible (1).

The National Historic Preservation Act of 1966 [PL 89-665; 16 U.S.C. 470–470m (1970) as amended 16 U.S.C.A. 470h, 470i, 470l–470n (Supp. 1973)]: This act provides for an expanded National Register of Historic Places to register districts, sites, buildings, structures, and objects significant in American history, architecture, archeology, and culture. It provides for a program of matching grants-in-aid to the states for historical surveys and planning and for preservation, acquisition, restoration, and development projects. The act also established the Advisory Council on Historic Preservation, appointed by the president, to advise the president and the Congress on matters relating to historic preservation (2).

Under this law, if a site is on the National Register, this fact must be taken into consideration when any project using federal funds or under federal permit might adversely affect it. Section 106 of this act requires that the President's Advisory Council on Historic Preservation be afforded an opportunity to comment on any undertaking that adversely affects properties listed on the National Register (1).

Procedures of the Advisory Council on Historic Preservation (36 CFR 800): The Advisory Council on Historic Preservation has established procedures for compliance with the National Historic Preservation Act of 1966. If a federal agency finds that any undertaking will have any adverse effect on a property on or determined by the Secretary of the Interior to be eligible for the National Register, that agency must allow the Advisory Council opportunity to comment on the undertaking. Although Advisory Council comments and recommendations are not binding on federal agencies, the consultation and negotiation procedure and full Advisory Council hearings may prevent preservation controversies from arising by bringing the interests to light early in project planning stages and providing a high level forum for ventilation of more complex controversies. The prestigious composition of the Advisory Council and its position as advisor to the president and Congress on historic preservation matters provides an incentive to federal agencies to comply with Advisory Council recommendations (2).

The National Environmental Policy Act of 1969 (PL 91-190, 31 STAT. 852; 42 U.S.C. 4321–4347): This act declares that it is the policy of the federal government to use all practical means, consistent with other essential considerations of national policy, to–among other things–improve and coordinate federal plans, functions, programs, and resources to the end that the nation may preserve important historic, cultural, and natural aspects of our heritage (1).

Executive Order 11593, Protection and Enhancement of the Cultural Environment, 16 U.S.C. 470 (Supp. 1, 1971): This order requires federal agencies to take a leadership role in preservation in two particular ways. First, for all property under federal jurisdiction or control, the responsible agencies must survey and nominate all historic properties to the National Register. These historic properties must also be maintained and preserved by the responsible agency. Second, for every action funded, licensed, or executed by the federal government, the agency involved must ask the Secretary of the Interior to determine if any property in the environmental impact area is eligible for the National Register. The determination of eligibility process is faster than the nomination process and gives the same protection as nomination to the National Register. If the federal action will substantially alter or destroy a historic property, the agency must have the property recorded by the Historic American Buildings Survey or the Historic American Engineering Record (2).

The Archeological and Historic Preservation Act of 1974 (PL 93-291): Enacted May 24, 1974 the act is directed to preservation of historic and archeological data that would otherwise be lost as a result of federal construction or other federally licensed or aided activities. It authorized the Secretary of the Interior, or the agency itself, to undertake recovery, protection, and preservation of such data. Where the federal government financially aids in activities that may cause irreparable damage, the Secretary of the Interior may survey the data and undertake recovery and preservation. Archeological salvage or recording by the Historic American Buildings Survey or the Historic American Engineering Record are among the alternatives available to the secretary. When the activity takes place on private land, the secretary must compensate the owner for any resultant delays or loss of use of the land. This act presents two innovations over previous laws: (1) only dams were covered in the Reservoir Salvage Act of 1960, now all federal projects are; and (2) up to 1 percent of project funds may be used for these purposes (2).

Most states have legislation directed toward archeological and cultural resources (1). Laws dealing with cultural resources regulate their disturbance on state lands and, in some instances and to different degrees, on private lands. State environmental policy acts usually include consideration of cultural resources. In some states aggressive programs for investigation, protection, and recovery of archeological resources have been established. State laws dealing with cultural resources vary considerably, and local inquiry is necessary in order to determine pertinent laws and regulations.

STEP 1: IDENTIFICATION OF KNOWN CULTURAL RESOURCES

The first step in prediction and assessment of impacts on cultural resources consists of identifying known resources in the area of interest. Information on known historical and archeological sites can be obtained from state historic preservation officers and the National Register of Historic Places (3). Archeological resources can be defined as objects and areas made or modified by humans, as well as the data associated with these artifacts and features (1). These objects include such artifacts as Indian arrowheads, stone axes, and broken or whole pottery vessels. Areas made or modified by humans include hunting stations, temporary camps, permanent settlements, and habitation sites. Careful excavation of these areas can reveal the location and dimensions of postholes and storage pits and can allow recovery of pollen, seeds, bones, cracked and flaked stones, and other debris indicative of food habits, manufacturing practices, and other details of the life patterns of inhabitants of the sites. Contextual information consists of the location (horizontal and vertical) in or on the ground of the artifacts or features and observations about the natural or constructed strata in or on which they are found. Persons trained in archeology can interpret the contextual information relative to the archeological objects or areas (1).

Cultural resources related to areas of ecological, scientific, or geological importance are of recent interest. Examples include wildlife refuges, caves, and unique areas such as the Painted Desert in Arizona. Local professional societies, as well as universities and colleges, can provide information on known nonarcheological resources in the area of interest.

Local resources of importance to ethnic groups, such as burial grounds and cemeteries or areas of unique religious importance, can be identified through contact with local government officials as well as organized ethnic groups in the area of interest or having responsibility therein.

STEP 2: IDENTIFICATION OF POTENTIAL CULTURAL RESOURCES

Even though the concept of a national register of historic places was initiated with the Historic Sites Act of 1935, there has never been a complete survey conducted. Most of the United States has not been thoroughly or professionally surveyed for cultural resources, and in many areas known resources include only those previously discovered by cursory studies or accident. Lack of completeness is the reason that step 1 above is not sufficient, in and of itself, to satisfy the attention that needs to be directed toward cultural resources in environmental impact assessments. Potential historic sites in the area of interest can be determined through contact with the state historic preservation officer, as well as state and local historical commissions, societies, or clubs, which can also provide information regarding the eligibility of certain sites for the National Register, as well as the status of sites that have been previously recommended or are in the process of being recommended for inclusion in the Register.

In order to identify previously unknown archeological resources in the area of interest, a preliminary archeological reconnaissance is necessary (1). This reconnaissance involves surface examination of selected portions of the area to be affected. It should be adequate to describe the general nature of archeological resources present, to assess the probable impacts of each alternative, and to estimate the costs of mitigating impacts. Field research associated with preliminary archeological reconnaissance will probably constitute no more than 25% of the total research time required to make an adequate assessment. Review of available information associated with step 1 and analysis of field data gathered in this step, along with report preparation, will require the major portions of the total cultural resources efforts.

Assistance for conduction of archeological surveys can be obtained through contact with the National Park Service or the relevant state archeological program. The National Park Service is the primary federal agency charged by Congress with developing and maintaining expertise, but the extent of staffing varies widely. Some counties, cities, and special districts have employed archeologists and thus have the capability for conducting archeological surveys.

Potential cultural resources of ecological, scientific, or geological interest or of ethnic importance can be identified through contact with university and college personnel in the vicinity, as well as liaison with local governmental organizations, ethnic groups, and professional societies.

STEP 3: SIGNIFICANCE OF CULTURAL RESOURCES

Every archeological site is of importance to our understanding of human history (1). Other cultural resources are important from past as well as future historical perspectives. However, not every cultural resource can be preserved or carefully and completely excavated. Decisions must often be made with regard to which sites should be preserved, which sites should be investigated, and the nature and intensity of investigation. Proper assessment is required to identify those sites with sufficient significance to warrant preservation, since over 5,000 sites are lost on an annual basis within the United States (2). Significance determinations for archeological and other cultural resources require the professional judgment of trained specialists.

In a recent study of eight alternative waterway navigation routes from Tulsa, Oklahoma to Wichita, Kansas, 19 parameters were used to assess the significance of cultural resources (4). Each parameter represented a matter of importance for each archeological site, and each waterway route was evaluated in terms of these parameters. A list of the parameters and their importance weights is given in Table 8-1. Importance weights reflect the fact that some of the parameters have greater priority than others. The considerations associated with each of the 19 factors are as follows:

1 Age or occupation period of the site—This is concerned with the approximate age and period of the individual site. Some period sites are more rare than others, and archeological work should make a serious effort to gather data from the entire range of time periods. Consequently, some sites from each period should be examined. As the number of sites available for each period increases, the priority in terms of losses becomes reduced.

Table 8-1 Parameters and Weights Used in the Archeological Section of the Mid-Arkansas River Basin Analysis[a]

	Parameter	Weight
1	Age or occupational period of the site	6.0
2	Concern by local population	3.0
3	Cost of conducting site survey	9.0
4	Depth of the occupational area	8.0
5	Ecological setting	6.0
6	Eligibility of the site for state or national register	3.0
7	Estimated number of sites to be obtained by the survey	7.0
8	Importance in terms of local, state, and national level	3.0
9	Minimum salvage costs for estimated sites	8.0
10	Nature of the site	7.0
11	Number of known sites to be damaged	7.0
12	Presence of single or multiple occupations	8.0
13	Preservation of archeological data	5.0
14	Previous knowledge of the area	8.0
15	Site frequency within the area concerned	3.0
16	Site importance in terms of geographic area and problems	8.0
17	Site preservation or damages	5.0
18	Size of occupational area	8.0
19	Value of site for nonarcheological fields	5.0

[a]After (4).

2 Concern by local population–The local inhabitants are often concerned about destruction or conservation of cultural resources.

3 Cost of conducting on-foot archeological survey–Cost of archeological surveys can be used as a determinant in the selection from alternatives, since these costs are proportional to the area involved for each alternative.

4 Depth of the occupational area–The depth of the deposit is important in estimating excavation costs. It is also important in terms of potential recoverable information for occupation, duration, and stratigraphy. In general, shallow deposits are less likely to be undisturbed and produce less cultural data than deep deposits. On the other hand, shallow occupations may provide activity areas or data that can be confused in thicker deposits.

5 Ecological setting–This factor is concerned with the ecological setting with reference to river valleys, that is, whether the site occupies bottom land, terrace ridge, hilltop, or rock shelter.

6 Eligibility of the site for state or national register–Known sites on state and national registers are of importance. Sites eligible for inclusion on either state or national registers also need to be considered in evaluation of alternatives.

7 Estimated number of sites to be obtained by survey–Reasonable estimates can be made for potential archeological sites by considering published data for similar types of projects or land areas. For example, a pipeline survey in Oklahoma across 330 mi produced 191 sites, thus averaging 1 site per 1.7 mi. Archeological surveys in areas to be occupied by reservoirs have revealed the existence of approximately 1 site per 0.5 mi.

8 Importance in terms of local, state, and national level–The importance of archeological sites varies in terms of local, regional, state, and national interest. Professional judgment can be used to estimate relative importance for these geographical concerns.

9 Minimum salvage costs for estimated sites–This factor includes an estimate of the archeological salvage costs for the total probable number of sites for each alternative, including both known and potential sites. An average salvage cost of $7,000 has been determined for sites in Oklahoma, with this figure including field excavations, laboratory preparations, and laboratory analyses.

10 Nature of the site–This is concerned with the character of the individual site and whether it represents a hunting camp, village site, special activity area, burial ground, quarry, workshop area, or other use. Sites likely to contain greater diversity in the cultural activities represented will have greater value in reconstructing the culture involved. A village site, for example, will provide information about village size, dwellings and architecture, house arrangements, special activity areas, associated features, and other items.

11 Number of known sites to be damaged–This information represents the sites previously identified in the area of interest. This information would be obtained from an archeological records check.

12 Presence of single or multiple occupations–The existence of single or multiple occupations at a site will affect excavation costs and requirements for sampling. Multiple-component sites may provide data for chronology and are normally of greater value as an information source.

13 Preservation of archeological data–This parameter is concerned with the preservation of materials at each archeological site. Some sites are more likely to

have materials preserved than others because of soil conditions (acid or alkaline), protection from the weather as in a rock shelter, and other factors. Consequently, sites more likely to have preserved various archeological materials have greater potential importance than those where this is not the case.

14 Previous knowledge of the area–This factor takes into account the overall knowledge of the archeological resources in the entire area. In general, a well-known region would require less investigation than unknown regions.

15 Site frequency within the area concerned–This deals with the general nature of each site and the relative number of similar sites in the area to be affected. Some sites are more rare, sometimes unique, while others may be well represented by numerous examples. Obviously sites that are rare are of greater importance than those that are plentiful.

16 Site importance in terms of geographic area and problems–Some archeological sites are more important than others in terms of providing answers to problems of concern not only locally, but also possibly on a statewide or even regional level.

17 Site preservation from damages–Some archeological sites have been damaged by vandalism, contour plowing, deep plowing, construction, erosion, and other factors. Others may be relatively intact and without obvious disturbance. Consequently, sites that are better preserved are likely to be of greater importance than those that have been disturbed.

18 Size of occupational area–The total area covered by the occupation will partly determine the cost of excavation; also, larger sites are more likely to provide greater amounts of information.

19 Value of site for nonarcheological fields–Information collected by archeological research may be of value to a number of other disciplines. Materials such as animal bones, plant materials, tree-ring specimens, pollen samples, soil samples and profiles, baked clay, shells, and charcoal may provide raw data for numerous fields concerned with past events.

STEP 4: IMPACTS ON CULTURAL RESOURCES

Impacts that can occur on cultural resources include inundation, destruction, damage, and disruption. Impacts can directly result from construction-phase disturbances or indirectly occur from treasure hunting by persons during any phase of the project. Preconstruction impacts occur primarily through acts of vandalism on known cultural resources sites. Indirect impacts include those that occur as a result of land-use changes and secondary growth and development. These secondary impacts can also be a result of direct construction activities or indirect acts of vandalism and disturbance. Attempts should be made to quantify the nature and extent of impacts during the various temporal phases of a project, with qualitative discussion provided where quantification is impossible. A basic approach that can be used in this step involves the concept of overlay mapping, with a base map showing known and potential sites and overlayed by maps identifying the nature and extent of the impacts from various alternatives under study to meet project needs.

STEP 5: SELECTION OF PROPOSED ACTION AND IMPACT MITIGATION

Selection of the proposed action from alternatives can be based upon elimination of one or more alternatives due to their potential impact on cultural resources, or selection of the alternative least likely to cause an impact on potential cultural resources. Once the proposed action is selected, it is then necessary to conduct an intensive archeological reconnaissance (1). This survey might be appropriate as a final step prior to selection of the proposed action, with the reconnaissance being conducted for two or more alternatives. An intensive archeological reconnaissance involves an on-the-ground surface survey and testing of an area sufficient to permit determination of the number and extent of the resources present, their scientific importance, and the time factors and cost of preserving them or otherwise mitigating any adverse affects on them (1). This level of investigation is appropriate once a specific region or area to be affected has been determined or the choice has been narrowed to one or a few prime alternatives.

Following an intensive archeological reconnaissance, the archeologist should be able to indicate which sites are particularly worth preservation and/or investigation, suggest areas where caution should be exercised because of possible buried sites, and provide an estimate of the cost of protecting or recovering an adequate amount of information from the area to be affected (1). Salvage may be determined to be appropriate for one or more sites in the area of interest. Archeological excavation associated with scientifically controlled recovery or salvage can be designed to yield maximum information about the life of the inhabitants, their ways of solving human problems, and their ability to adjust to and modify their natural environment. Archeological excavations should be programmed during final planning stages or at least during early stages of project construction.

STEP 6: PROCEDURES FOR CONSTRUCTION PHASE FINDINGS

One of the common problems of many proposed actions is that unidentified cultural sites are encountered during the construction phase. Prior to initiation of construction, the proposing agency should develop agreements with appropriate state and local archeological agencies regarding procedures to be utilized should sites be identified during the construction phase. These agreements should aid in precluding lengthy delays in project construction.

SUMMARY

Proper attention should be given to cultural resources in conjunction with project decision making. The geographical area to be considered for cultural resources should not be limited to the specific area where project construction will occur, but rather should encompass an area sufficient to include both primary and secondary consequences. Numerous federal, state, and local laws and regulations have been passed for protection of cultural resources. If an adequate assessment of cultural resources is accomplished at each step in the planning process and if proper funding is provided, it is generally possible to provide for protection or, where necessary, preservation for all significant sites, with minimal or no effect on project time schedules.

SELECTED REFERENCES

1 McGimsey, C. R., III: "Archeology and Archeological Resources," Society for American Archeology, Washington, D.C., 1973.
2 Neal, L.: personal communication, Oklahoma Archeological Survey, University of Oklahoma, Norman, 1975.
3 National Park Service: "The National Register of Historic Places–1972," U.S. Government Printing Office, Washington, D.C., 1972.
4 Canter, L. W., P. G. Risser, and L. G. Hill: "Effects Assessment of Alternate Navigation Routes from Tulsa, Oklahoma to Vicinity of Wichita, Kansas," pp. 295-341, University of Oklahoma, Norman, June 1974.

Chapter 9

Prediction and Assessment of Impacts on the Socioeconomic Environment

Many major impacts associated with certain proposed actions are evidenced by changes in socioeconomic factors in the project area and surrounding region. Socioeconomic changes may be beneficial or detrimental. Emphasis on this category of the environment is more recent than the focus on the physical-chemical, biological, and cultural environments, and results from the realization that the total environment includes factors associated with human concerns, which are described by socioeconomic parameters. This chapter is addressed to the data needs and associated technology for predicting and assessing impacts of proposed actions on the socioeconomic environment.

BASIC STEPS FOR PREDICTION AND ASSESSMENT

Basic steps associated with prediction of changes in the socioeconomic environment and assessment of the impact of these changes are as follows:

1 Describe the environmental setting in terms of socioeconomic factors. The area of interest for each factor will be dependent upon the relationship of the factor to the alternatives under consideration, as well as upon the available data base.
2 Identify critical environmental concerns relative to the socioeconomic factors

described above. Primary emphasis should be given to those factors that would be deemed marginal or inadequate in terms of societal standards.

3 Predict changes in the socioeconomic factors as a function of various alternatives under consideration, including the no-action alternative. Changes should be quantified where possible and qualitatively described as a minimum.

4 Discuss the implications of the changes relative to critical or marginal items defined in step 2. Identify factors that will be changed from satisfactory to marginal or critical.

BASIC INFORMATION ON THE SOCIOECONOMIC ENVIRONMENT

Factors that describe the socioeconomic environment represent a composite of numerous interrelated and nonrelated items. On the one hand, this category represents a catchall group since it includes factors not associated with the physical-chemical, biological, or cultural environments. On the other hand, this category is the one most descriptive of human relationships and interactions. Table 9-1 includes a summary of socioeconomic factors that have been used in EISs. Also included are typical changes following construction and operation of proposed actions. Many of the changes described in Table 9-1 represent impacts of unique significance or importance to the human population.

Factors descriptive of the socioeconomic environment were not included in most EISs written in the first several years following the passage of the NEPA. In 1973 the CEQ directed specific attention to this category of environmental concerns and thus provided the impetus for consideration of these items in project decision making.

As secondary, or indirect, consequences of projects are being given greater attention, increased emphasis is also directed to those environmental components influenced by these secondary effects. Many secondary effects are associated with land-use changes, population increases, and shifts in employment trends. These factors are components in the mosaic of the socioeconomic environment. Secondary effects often represent the major effects of proposed actions, thus adding increased importance to consideration of the socioeconomic environment in decision making.

STEP 1: DESCRIPTION OF SOCIOECONOMIC ENVIRONMENTAL SETTING

The first step is to assemble pertinent data and information that will enable description of the environmental setting in terms of various selected socioeconomic factors. Table 9-1 provides a general list of socioeconomic factors, and Table 9-2 identifies factors used in an environmental impact assessment methodology for transportation projects (1). Table 9-3 is a listing of socioeconomic impacts identified for gas pipeline projects (2).

Sources of information for socioeconomic factors include the Bureau of Census, statistical abstracts for various governmental levels, Chambers of Commerce, planning agencies and departments, educational programs, and recreational organizations. Aggregation of data and pertinent information for socioeconomic factors

Table 9-1 Examples of Socioeconomic Factors and Their Potential Changes Resulting from Project Implementation

Factor	Potential changes
General characteristics and trends in population for state, substate region, county, and city	Increase or decrease in population
Migrational trends in study area (study area is a function of alternatives and available data base)	Increase or decrease in migrational trends
Population characteristics in study area, including distributions by age, sex, ethnic groups, educational level, and family size	Increase or decrease in various population distributions; people relocations
Distinct settlements of ethnic groups or deprived economic/minority groups	Disruption of settlement patterns; people relocations
Economic history for state, substate region, county, and city	Increase or decrease in economic patterns
Employment and unemployment patterns in study area, including occupational distribution and location and availability of work force	Increase or decrease in overall employment or unemployment levels and change in occupapational distribution
Income levels and trends for study area	Increase or decrease in income levels
Land-use patterns and controls for study area	Change in land usage; may or may not be in compliance with existing land-use plans
Land values in study area	Increase or decrease in land values
Tax levels and patterns in study area, including land taxes, sales taxes, and income taxes	Changes in tax levels and patterns resulting from changes in land usage and income levels
Housing characteristics in study area, including types of housing and occupancy levels and age and condition of housing	Changes in types of housing and occupancy levels
Health and social services in study area, including health manpower, law enforcement, fire protection, water supply, waste-water treatment facilities, solid waste collection and disposal, and utilities	Changes in demand on health and social services
Public and private educational resources in study area, including K–12, junior colleges, and universities	Changes in demand on educational resources
Transportation systems in study area, including highway, rail, air, and waterway	Changes in demand on transportation systems; relocations of highways and railroads
Community attitudes and life-styles, including history of area voting patterns	Changes in attitudes and life-styles
Community cohesion, including organized community groups	Disruption of cohesion
Tourism and recreational opportunities in study area	Increase or decrease in tourism and recreational potential
Religious patterns and characteristics in study area	Disruption of religious patterns and characteristics
Areas of unique significance such as cemeteries or religious camps	Disruption of unique areas

Table 9-2 Socioeconomic Factors for Transportation Projects[a]

Factor	Comment
I Sociological	Social relationships
A Community (local area)	
1 Neighborhood severance	Violation of neighborhood boundaries
2 Cultural patterns	Ethnic cohesion, stability, life style
3 Crime	Assault, robbery, breaking, etc.
a Rate	Change in opportunity for
b Police protection	Availability and speed
4 Fire hazard	Type and density of land uses
a Hazards	Dwellings, trash, etc.
b Fire protection	Available equipment and time
5 Health	
a Health factors	Sanitation, dangerous spots, etc.
b Medical services	Time to reach health facilities or obtain services
6 Religious services	Opportunity to attend
a Loss of places	Removal of churches
b Access to	Isolation of members
7 Educational	Loss of or effect on access to
a Elementary	
b Junior high school	
c High school	
d Trade and college	
8 Recreational facilities	Other than parks and playgrounds
9 Social services	Gathering places other than previously considered
10 Public utilities	
11 Neighborhood livability	Pleasantness of surroundings
a Construction period	Disruption
b Long run	Cleanliness, repairs, etc.
B Metropolitan area	Loss of and effects on access to
1 Police protection	
2 Fire protection	
3 Medical services	
4 Educational services	
5 Parks	
6 Recreation	
7 Historical sites	
8 National defense	
a Evacuation	As a link in system
b Military movements	As a link in system
c Hazards to critical industry	
II Economic impact	
A Community (local area)	
1 Employment	
a Construction period	Change in place or access
b Long run	Change in place or access
2 Shopping facilities	
a Construction period	Change in place or access
b Long run	and loss of customers
3 Residential values	

Table 9-2 *(continued)* **Socioeconomic Factors for Transportation Projects**[a]

Factor	Comment
4 Other property values	
5 Property tax base	
a Construction period	Loss of taxable values
b Long run	Potential for change
6 Displaced residents	
a Owners	
1 DSS housing	DSS—decent, safe, and sanitary
2 Non-DSS housing	
b Renters	
1 DSS housing	
2 Non-DSS housing	
c Ease of replacement	
7 Displaced businesses	
a Small businesses	
1 Number	
2 Number of jobs	
3 Ease of relocation	
b Other businesses	
1 Number	
2 Number of jobs	
3 Ease of relocation	
8 Remaining businesses	Effects on jobs
a Small businesses	and solvency
1 Construction period	
2 Long run	
b Other businesses	
1 Construction period	
2 Long run	
9 New business	Potential for
10 Multiple use of right-of-way	
B Metropolitan area	Outside of local area
1 Access to employment	
2 Access to shopping	
3 Commercial activity	
4 Property values and tax base	

[a]After (1).

involves contacts with numerous groups. Socioeconomic information is more likely to be existent than is information for many specific items in the physical-chemical, biological, and cultural environments.

In a recent study of eight alternative waterway navigation routes from Tulsa, Oklahoma to Wichita, Kansas, numerous socioeconomic factors were included in the analysis (3). A brief description of the 11 pertinent factors is as follows:

1 Income—The potential navigation routes included four counties in Oklahoma and five in Kansas; information on total personal income in these counties

Table 9-3 Socioeconomic Impacts of Gas Pipeline Projects[a]

Project phase	Impacts
Construction	Land features and uses—Assess the impact on present or future land use, including commercial use, mineral resources, recreational areas, public health and safety, and the aesthetic value of the land and its features; describe any temporary restriction on land use due to construction activities; state the effect on construction-related activities upon local traffic patterns, including roads, highways, ship channels, and aviation patterns
	Socioeconomic considerations—Discuss the effect on local socioeconomic development in relation to labor, housing, local industry, and public services; discuss the need for relocations of families and businesses; describe the beneficial effects, both direct and indirect, of the action on the human environment, such as benefits resulting from the services and products and other results of the action (include tax benefits to local and state governments, growth in local tax base from new business and housing development, and payrolls); describe the impact on human elements, including the need for increased public services (schools, health facilities, police and fire protection, housing, waste disposal, markets, transportation, communication, energy supplies, and recreational facilities)
Operation and maintenance	Land features and uses—Outline restrictions on existing and potential land use in the vicinity of the proposed action, including mineral and water resources; state the effect of operation-related activities upon local traffic patterns including roads, highways, ship channels, and aviation patterns and the possible need of new facilities
	Socioeconomic considerations—Discuss the effect on the local socioeconomic development in relation to labor, housing and population growth trends, relocation, local industry and industrial growth, and public service; describe the beneficial effects, both direct and indirect, of the action on the human environment such as economic benefits resulting from the services and products, energy, and other results of the action (include tax benefits to local and state governments, growth in local tax base from new business and housing developments, and payrolls); describe impacts on human elements, including any need for increased public service (schools, police and fire protection, housing, waste disposal, markets, transportation, communication, and recreational facilities); indicate the extent to which maintenance of the area is dependent upon new sources of energy or the use of such vital resources as water
Termination and abandonment	Land uses and aesthetics—Discuss the impact on land use and aesthetics of the termination and/or abandonment of facilities resulting from the proposed action

[a]After (2).

was accumulated from 1950 to 1960. Data were also obtained on the per capita personal income over the same time period.

2 Physical resources—Physical resources were associated with assessed value of property on a county-by-county basis, as well as the ratio of county-assessed value per capita to the states of Kansas and Oklahoma. In addition, this category included consideration of county-by-county land proportions between urban, federal, agricultural, and water usage. Agricultural usage was further subdivided into crop land, pasture land, and range land.

3 Human resources—The human resources category included consideration of labor force skills as a proportion of the total labor force, as well as the actual numbers of persons in each category of skills. Categories of labor force skills included professional, managers, sales, clerical, craftsmen, operatives, non-farm labor, farmers, farm labor, service, and private households.

4 Employment—The nine-county region was also described in terms of the number of persons employed by sector of the economy, as well as the percentages employed in each sector. Sectors of the economy included agriculture; manufacturing; public utilities; wholesale and retail; government; insurance, financial services, and real estate; and all other.

5 Interregional competition—This factor addressed the extent of regional interaction between the nine-county study area and the surrounding areas and nation. Interregional competition was described by considering the value added by manufacturers over the period from 1958 to 1967 on a county-by-county basis. In order to determine the extent of interregional competition, a location quotient for value added by manufacturing was developed, with the location quotient being defined as county value added per county land value divided by national value added per national land value. A location quotient of one indicated a concentration equal to the national average. A value of more than one means a concentration greater than the national average, with values less than one describing the converse. The total value of agricultural products in the period 1949-1969 was also examined along with location quotients for the value of wheat production. In the category of manufacturing, location quotients for the nine counties varied from 0.2 to 1.2, while location quotients for wheat production ranged from 0.8 to 25.0.

6 Intraindustry structure—Detailed consideration was given to the different types of industrial activities within the nine-county region. Employment sizes by county for the following major industrial groups were developed: printing and publishing, petroleum and coal products, stone clay and glass products, primary metal industries, fabricated metal products, machinery except electrical, and transportation equipment.

7 Transportation—Accessibility to various transportation systems within the nine-county study area was addressed. Attention was also given to the quality and quantity of transportation facilities such as roads and highways, railroads, and airports.

8 Population—The nine-county study area was described in terms of population trends from 1950 to 1970, population density (persons per square mile), urban population as a percentage of total population, and components of population change, including natural increase and net migration. County population density information was compared to the average for the United States and the average for the nine-county study area.

9 Welfare—Several social welfare indicators were utilized, including median

school years completed by the 1970 population of persons 25 yr and older, the infant mortality rate per 1,000 persons, the number of tuberculosis cases per 1,000 persons, and the number of persons per physician. Unemployment trends for a 6-yr period from 1967 to 1972 were examined and compared to national averages for the same period. Unique areas of high unemployment were noted for certain study area counties heavily populated by American Indians.

10 Community cohesion–Community cohesion is descriptive of the degree of stability within a given area. On the basis of the nine-county study area, housing turnover was considered relative to the percentage of housing occupied by owners who moved in before 1960. Additional items descriptive of community stability included the annual percentage housing units for sale or rent, the number of suicide cases per 1,000 population, the number of homicide cases per 1,000 population, and the number of divorces per 1,000 population. The latter three items were compared to national averages.

11 Mobility–Mobility characteristics in the study area included consideration of the percentages of persons over the 5-yr period 1965–1970 that lived in the same house, different house but same county, different county but same state, different state but same country, and other country.

STEP 2: IDENTIFICATION OF CRITICAL SOCIOECONOMIC FACTORS

The purpose of this step is to identify any socioeconomic factors that represent critical items relative to the human environment. Examples of critical items include locations where water and waste-water treatment plants are operated in excess of design capacity or where existing school systems are overcrowded. Although no overall standards can be cited, many approaches can be used to define critical factors. Examples include consideration of the relationship of the factor to original design standards; the relationship of the factor to regional, state, or national averages; and actual recommended standards such as teacher-to-student ratios in grade schools.

STEP 3: PREDICTION OF CHANGES IN SOCIOECONOMIC FACTORS

Table 9-1 shows examples of changes that can occur in socioeconomic factors. The primary purpose of step 3 is to quantitatively predict, or at least qualitatively describe, changes in socioeconomic factors outlined in step 1. Particular attention should be addressed to critical factors identified in step 2. Anticipated changes in socioeconomic factors should be developed for the various alternatives under consideration, including the no-action alternative.

Table 9-2 suggests changes in socioeconomic factors resulting from transportation projects (1). The methodology used involves relative ranking of the changes resulting from each of the alternatives under consideration. Quantification is advantageous for ranking alternatives, although ranking can be accomplished by using professional judgment.

Table 9-3 lists changes that must be addressed in projects associated with gas pipelines (2). These changes can be expressed on a qualitative basis, although it is suggested that quantitative information be included where relevant.

In the study of eight alternative routes for extending waterway navigation from Tulsa, Oklahoma to Wichita, Kansas, changes in socioeconomic factors were predicted for the without-project situation (no-action alternative), as well as for each of the alternatives (3). Quantitative projections were made for a 50-yr period, with the projections including specific data for 1970, 1990, 2000, and 2020. Without-project projections were made for population; per capita income; total personal income; total agricultural earnings; total earnings in manufacturing; total earnings in finance, insurance, and real estate; total employment; total employment in agriculture; and total employment in manufacturing. Available forecasts were used to develop projections for these nine variables for the nine-county study area. Qualitative descriptions of anticipated changes were included for the other socioeconomic factors.

With-project projections were made by assuming that a navigable waterway would be built (3). The general socioeconomic impacts anticipated from waterway projects were identified. This information, coupled with general indicators of socioeconomic impact, was used to project, on a relative scale, anticipated changes that would result from the development of one of the eight alternatives being studied. Quantitative projections were made through the use of an impact susceptibility index, which is defined as a function of waterway frontage in each county, accessibility of the waterway in a county to other transportation networks, and other factors. Quantitative projections for with-project conditions were made for the years 2000 and 2020. Projections were made for population; per capita income; total personal income; total agricultural earnings; total earnings in manufacturing; total earnings in finance, insurance, and real estate; total employment; total employment in agriculture; and total employment in manufacturing. Additional items included in the socioeconomic category in this study were described from the standpoint of anticipated qualitative changes.

STEP 4: DISCUSSION OF IMPLICATIONS OF CHANGES

The final step in prediction and assessment of impacts on the socioeconomic environment is a discussion of the implications of the changes for each of the socioeconomic factors, considering both without- and with-project conditions. Changes could be presented relative to percentage differences based on current conditions or relative to current standards for certain socioeconomic factors. Particular attention should be paid to socioeconomic factors identified as critical items in step 2 above. In addition, socioeconomic factors that will become marginal or critical as a result of the implementation of the selected alternative should be discussed.

SUMMARY

Socioeconomic factors are of more recent interest in environmental impact assessment than factors associated with the physical-chemical, biological, or cultural environments. Socioeconomic factors are important since they represent specific aspects of the human environment, and changes in these factors often represent the most critical changes associated with project implementation. The majority of the information necessary to describe the socioeconomic environment is available from a variety of informational sources. Projections of anticipated changes can be made through applications of both qualitative and quantitative techniques.

SELECTED REFERENCES

1 Adkins, W. G., and D. Burke, Jr.: "Social, Economic, and Environmental Factors in Highway Decision Making," pp. 69–74, Res. Rept. 148-4, Texas Transportation Institute, Texas A&M University, College Station, Nov. 1974.

2 Federal Power Commission: Implementation of the National Environmental Policy Act of 1969, Order 485, Order Amending Part 2 of the General Rules to Provide Guidelines for the Preparation of Applicants' Environmental Reports Pursuant to Order 415-C, Washington, D.C., June 7, 1973.

3 Canter, L. W., P. G. Risser, and L. G. Hill: "Effects Assessment of Alternative Navigation Routes from Tulsa, Oklahoma to Vicinity of Wichita, Kansas," pp. 10–52, 168–213, University of Oklahoma, Norman, June 1974.

Chapter 10

Methods of Impact Analysis

Parts A and B of section 102 in the NEPA require agencies to use systematic and interdisciplinary approaches and to develop methods and procedures to ensure that presently unquantified environmental amenities and values may be given appropriate consideration in decision making along with economic and technical considerations (1). In response to these requirements, numerous environmental impact assessment methodologies have been developed since 1970. An impact can be defined as any change in the physical-chemical, biological, cultural, and/or socioeconomic environmental system that can be attributed to human activities relative to alternatives under study for meeting a project need (2). Impact methodologies provide an organized approach for predicting and assessing these impacts. In this chapter various impact methodologies developed in response to NEPA requirements are discussed.

PURPOSES OF ENVIRONMENTAL ASSESSMENT METHODS

Several purposes are served by impact analysis methods. One is to ensure that all environmental factors that need to be considered are included in the analysis. This purpose is relevant since the environment is a complex system of physical-chemical, biological, cultural, and socioeconomic resources, and various types of actions can create complex impacts and interrelationships on these resources. Methods whose approach to considering environmental factors is systematic are desirable.

Impact analysis methods should provide a means for evaluation of alternatives on a common basis. Many impact statements adequately describe the environmental impacts of proposed actions; however, they consider only the relative economic evaluation of alternatives to the proposed action. Methods of impact analysis provide the approach for evaluating absolute or relative impacts of alternatives. In conjunction with impact evaluation, it may be determined that there are data deficiencies in terms of either the description of the environmental setting, factors associated with the proposed action, or technology available for impact prediction and assessment. Methods for impact analysis can aid in identifying data needs and planning special studies or field studies.

Another important purpose of methods of impact analysis is the evaluation of mitigation measures. Attention should be directed toward measures that will minimize the environmental impact of alternatives and the proposed action. Methods for impact analysis aid in evaluation of the effectiveness of proposed mitigation measures.

Another purpose for assessment methodologies is to provide information in summary form for public participation. Utilization of a systematic, interdisciplinary, and organized approach gives credence to the validity of the impact analysis. Care must be exercised in any public distribution of information resulting from the application of an impact methodology that the information does not appear to mislead the public or misrepresent or confuse the results. Information presented to the public should be provided in summary form only.

Finally, methods of impact analysis are required to ensure compliance with the spirit and intent of the NEPA.

COMPARATIVE STUDIES OF METHODOLOGIES

As methods for impact analysis have developed, periodic comparisons of methodologies have been made in accordance with certain predetermined criteria. Dickert (3), Drobny and Smith (4), Warner (2), Warner and Bromley (5), Warner and Preston (6), and Smith (7) have conducted comparative analyses of environmental impact assessment methodologies. In each of these studies, selected criteria for methodology groupings and comparisons were presented, and selected methodologies were compared based on their degree of compliance with these criteria.

Dickert (3) considered three analytical functions associated with environmental impact assessment: identification, prediction, and evaluation (Table 10-1). Methods for identification of environmental impacts can assist in specifying the range of impacts that may occur, including their spatial dimensions and time period. Generally, identification methods answer questions concerning the components of the project and what elements of the environment may be affected by these components. Dickert identified two types of identification methods, namely, checklists and matrices/networks. Checklists contain environmental factors that need to be addressed relative to the impact of alternatives. A matrix is a two-dimensional

Table 10-1 Classification of Methodology for Environmental Impact Assessment

Function	Methodology
Identification	Description of the existing environmental system
	Determination of the components of the project
	Definition of the environment modified by the project (including all components of the project)
Prediction	Identification of environmental modifications that may be significant
	Forecasting of the quantity and/or spatial dimensions of change in environment identified
	Estimation of the probability that the impact (environmental change) will occur (time period)
Evaluation	Determination of the incidence of costs and benefits to user groups and populations affected by the project
	Specification and comparison of the trade offs (costs or effects being balanced) between various alternatives

checklist that identifies various types of project actions and their potential impact on environmental items. Networks emphasize interrelationships between affected environmental items.

Predictive methodologies involve the greatest application of technology. This area of impact analysis is the least developed in terms of specific methodologies that can be directly applied in environmental impact assessment. Predictive methodologies for air quality impacts, water quality impacts, and noise impacts are discussed in Chapters 4, 5, and 6, respectively. Quantitative predictive methods for the biological, cultural, and socioeconomic environments are not as available. Methods associated with evaluation include the Battelle environmental evaluation system (8) and the Georgia optimum pathway matrix (9). The major result of these methods is an aggregate index of environmental impact for each of the alternatives, so that they can be compared on a common basis.

Drobny and Smith identified ten criteria that represent basic requirements for an impact assessment methodology (4): an impact analysis method should be comprehensive, flexible, capable of detecting project-generated impacts, and objective, and should ensure input of expertise, utilize the state of the art, employ explicitly defined criteria, provide for assessment of impact magnitude, provide for overall assessment of total impacts, and detect environmentally sensitive areas. This list of criteria formed the basis for the Warner study (2), the Warner and Bromley study (5) and the Smith study (7).

Warner in 1973 and Warner and Bromley in 1974 divided impact methodologies into five main classes: *ad hoc* procedures, overlay techniques, checklists, matrices, and networks. *Ad hoc* procedures involve assembling a team of specialists to identify impacts in their areas of expertise, with minimal guidance beyond the requirements of NEPA. This approach was essentially utilized by all federal agencies in the period immediately following enactment of the NEPA.

Overlay techniques describe well-developed approaches used in planning and landscape architecture. These techniques are based on the use of a series of overlay maps depicting environmental factors or land features. The overlay approach is generally effective in selecting alternatives and identifying certain types of impacts; however, it cannot be used to quantify impacts or to identify secondary and tertiary interrelationships. Overlay techniques utilizing computerization for more effective data analysis have been developed.

Checklist approaches present types of impacts typically associated with particular categories of projects. From a master list of environmental factors and/or environmental impacts, impact statement preparers select and evaluate those impacts expected for the particular alternative under consideration. Checklists can be augmented by instructions on how to present and make use of data and by the inclusion of explicit criteria for impacts of certain magnitudes and importance.

Matrix methods are basically generalized checklists in which possible project activities are established along one axis, with potentially impacted environmental characteristics or conditions along the other axis. Usage differs from a checklist in that an attempt is made to identify various causal factors (project actions) producing specified impacts. Network approaches expand the concept of a matrix by introducing a cause-condition-effect network that allows identification of cumulative or indirect effects not adequately explained through simple cause-effect sequences represented by matrices.

Warner and Preston Study

In 1973 Warner and Preston (6) conducted a study of 17 methods of environmental impact assessment. In order to compare the methodologies selected, four components of an impact assessment were identified, namely, impact identification, measurement, interpretation, and communication. Tables 10-2–10-5 contain questions that represent the criteria associated with each of these four components. The 17 methods were also compared on resource requirements, replicability, and flexibility for various types of projects. Criteria questions associated with these three factors are shown in Table 10-6. Table 10-7 provides a summary of the evaluation of the 17 methods, whose characteristics are provided in Appendix D. The methods that achieve the greatest degree of compliance with the selected criteria include Dee et al. (10), Dee et al. (8), Stover, Georgia, Leopold, and Task Force.

Table 10-2 Criteria Questions for Impact Identification[a]

Criteria	Questions
Comprehensiveness	Does the methodology address a full range of impacts?
Specificity	Are specific environmental parameters identified?
Isolate project impacts	Does the method suggest ways of identifying project impacts?
Timing and duration	Does the method suggest construction-phase impacts vs. operational-phase impacts?
Data sources	Does the method require identification of data sources?

[a]From (6).

Table 10-3 Criteria Questions for Impact Measurement[a]

Criteria	Questions
Explicit indicators	Does the method suggest specific measurable indicators for impact quantification?
Magnitude	Does the method require determination of impact magnitude?
Objectivity	Does the method stress objective rather than subjective measurement?

[a]From (6).

Table 10-4 Criteria Questions for Impact Interpretation[a]

Criteria	Questions
Significance	Does the method require an assessment of significance on a local, regional, and national scale?
Explicit criteria	Does the method require that the criteria and assumptions in significance determination be stated?
Uncertainty	Does the method address uncertainty or the degree of confidence in impact projections?
Risk	Does the method focus on impacts of low probability of occurrence but high potential damage?
Alternatives comparison	Does the method provide a way of comparing alternatives?
Aggregation	Does the method provide a way for aggregation of information on impact measurement and interpretation?
Public involvement	Does the method provide a way for public input in the interpretation of impact significance?

[a]From (6).

Table 10-5 Criteria Questions for Impact Communication[a]

Criteria	Questions
Affected parties	Does the method link impacts to affected human groups?
Setting description	Does the method require a description of the environmental setting?
Summary format	Does the method contain a suggested summary format?
Key issues	Does the method suggest a way of highlighting key impacts or issues?
NEPA compliance	Does the method focus on NEPA/CEQ requirements?

[a]From (6).

Smith Study

In 1974 Smith (7) utilized 10 criteria for evaluation of 23 environmental impact assessment methodologies. The criteria are described as follows:

1 Be comprehensive—The environment contains intricate systems of living and nonliving elements bound together by complex interrelationships. An adequate methodology must consider impacts on these systems.

Table 10-6 Criteria Questions for Methodology Resource Requirements, Replicability, and Flexibility[a]

Criteria	Questions
	Resource requirements
Data requirements	Does the method use current data or are special studies required?
Manpower requirements	Are special skills required?
Time requirements	How much time is necessary to learn the method?
Costs	What are the costs of using the method?
Technologies	Are special technologies required?
	Replicability
Ambiguity	Is the method ambiguous?
Analyst bias	To what degree will different results occur depending on the analyst?
	Flexibility
Scale flexibility	Does the method apply to projects of different size or scale?
Range	Does the method apply to projects of different types?
Adaptability	Can the method be applied to different basic environmental settings?

[a]From (6).

2 Be flexible–Sufficient flexibility must be contained in the methodology, since projects of different size and scale result in different types of impacts.

3 Detect true impact–The actual impact is that change in environmental conditions resulting from a project, as opposed to the change that would naturally occur from existing conditions. Moreover, both short-term and long-term changes must be measured.

4 Be objective–The methodology must be objective, providing impersonal, unbiased, and constant measurements immune to outside tampering by political and other external forces. An objective and consistent procedure provides a firm foundation, which can be periodically updated, refined, and modified, thereby incorporating the experience gained through practical application. To be effective as a decision-making tool, environmental impact assessments also must be repeatable by different analysts and able to withstand scrutiny by various interest groups.

5 Ensure input of required expertise–Sound, experienced, professional judgment must be assured by a methodology, especially as subjectivity remains inherent in many aspects of environmental evaluation. Input of the necessary expertise can be achieved either through the design of the methodology itself or through the rules governing its use.

6 Utilize the state of the art–Maximum appropriate use of the state of the art must be made, drawing on the best available analytical techniques.

7 Employ explicitly defined criteria–Evaluation criteria, especially any quantified values, employed to assess the magnitude or importance of environmental

impacts should not be arbitrarily assigned. The methodology must provide explicitly defined criteria and explicitly stated procedures regarding the use of these criteria, with the rationale behind such criteria documented.

8 Assess actual magnitude of impacts–Means must be provided for an assessment based on specific levels of impact for each environmental concern, in the terms established for describing that concern (e.g., BOD, pH, and temperature for water quality). Assessment of magnitude based on generalities or relatives (qualitative comparisons between alternatives) is inadequate.

9 Provide for overall assessment of total impact–A means for aggregating multiple individual impacts is necessary to provide an evaluation of overall total environmental impact.

Table 10-7 Summary of Methodology Evaluations[a]

	Methods[b]																
	1	2	3	4	5	6	7	8	9	10	11	12	13	14	15	16	17
Type[c]	C	C	C–M	C	O	M	C	O	M	M	C	Nt	C	C	C	C	A
Comprehensiveness[d]	L	S	S	S	S	S	L	S	L	S	S	L	L	S	S	S	S
Specificity	L	L	L	L	L	L	L	L	L	L	L	S	L	S	L	L	N
Isolate projects impacts	N	L	L	S	N	N	N	N	N	L	N	N	L	L	S	N	N
Timing and duration	N	L	L	L	N	S	N	N	S	N	N	S	L	L	N	N	N
Data sources	N	N	S	L	N	N	N	N	N	N	N	S	N	N	N	N	N
Explicit indicators	N	L	L	L	L	N	S	S	L	N	L	N	S	S	S	S	N
Magnitude	N	L	L	L	L	L	L	N	L	S	S	N	L	L	S	L	N
Objectivity	N	L	S	L	S	L	L	L	N	N	S	N	N	S	N	L	N
Significance	N	S	N	S	N	S	S	N	S	S	N	N	S	S	S	N	N
Explicit criteria	N	L	L	L	N	L	N	L	N	N	N	S	N	S	S	L	N
Uncertainty	N	N	N	L	N	N	N	N	N	N	L	S	L	N	L	N	N
Risk	N	N	N	N	N	N	N	N	N	N	N	N	N	N	N	N	S
Alternatives comparison	L	L	L	S	L	L	N	L	L	N	L	N	L	L	L	L	N
Aggregation	L	L	L	L	S	*e*	S	L	*e*	*e*	S	*e*	L	*e*	L	L	*e*
Public involvement	N	N	S	N	N	N	S	N	N	N	N	N	N	S	N	L	S
Affected parties	N	N	S	N	N	N	N	N	S	N	N	N	S	N	N	S	N
Setting description	N	S	L	N	S	L	L	L	N	L	N	N	L	L	N	N	S
Summary format	L	L	L	S	L	L	N	L	L	S	L	N	L	L	L	L	N
Key issues	N	L	L	N	L	L	L	L	S	S	N	N	N	S	N	L	N
NEPA compliance	N	S	L	N	N	L	N	N	N	L	N	N	S	S	N	N	N
Resource requirements	L	N	S	N	N	L	L	L	L	L	N	L	N	L	L	N	L
Replicability	N	L	L	L	L	N	L	L	N	N	N	N	L	N	L	N	N
Flexibility	L	S	N	S	N	L	S	N	L	S	S	N	L	L	S	L	N

[a]From (6).

[b]1, Adkins; 2, Dee (1972); 3, Dee (1973); 4, Georgia; 5, Krauskopf; 6, Leopold; 7, Little; 8, McHarg; 9, Moore; 10, New York; 11, Smith; 12, Sorensen; 13, Stover; 14, Task Force; 15, Tulsa; 16, Walton; 17, WSCC.

[c]A, *ad hoc*; O, overlay; C, checklist; M, matrix; Nt, network.

[d]L, substantial compliance, low resource needs, or few replicability-flexibility limitations; S, partial compliance, moderate resource needs, or moderate limitations; N, no or minimal compliance, high resource needs, or major limitations.

[e]Aggregation not attempted.

10 Pinpoint critical impacts—The methodology must provide a warning system to pinpoint and emphasize particularly hazardous impacts. In some cases the sheer intensity or magnitude of impact may justify special attention in the planning process, regardless of how narrowly the impact might be felt.

Table 10-8 is a summary of the degree of compliance with the 10 criteria of each of the 23 methodologies. Based on this study, the methodologies that exhibit the greatest degree of compliance with the stated criteria include Dee (1972), McHarg, Baker and Gruendler, and Turner and Hausmanis. A summary of the features of the methodologies studied by Smith, and not previously evaluated by Warner and Preston, is included in Appendix E.

OTHER METHODOLOGIES

There are additional methodologies for impact analysis, which have been developed both prior to and subsequent to the comparative analysis studies discussed above. Table 10-9 contains a summary of various environmental impact assessment

Table 10-8 Methodology Summary Chart[a]

	Degree to which indicated requirement is fulfilled[b]									
Methodology	1	2	3	4	5	6	7	8	9	10
Eckenrode	S	S	S	N	N	N	N	N	N	N
Lamanna	S	S	S	N	N	N	N	N	N	N
McKenny	L	L	L	S	S	N	N	L	L	S
McHarg	L	L	N	L	L	L	L	N	N	S
Lacate	S	S	N	N	S	N	S	N	N	S
Baker and Gruendler	L	L	N	L	L	L	L	N	N	S
Turner and Hausmanis	L	L	N	L	L	L	L	N	N	S
Leopold	S	S	S	N	N	S	N	S	N	N
Manheim	N	L	S	N	N	S	S	S	S	S
Sorensen	S	L	L	S	N	N	N	N	N	S
Little	S	S	N	N	N	N	N	N	N	N
Adkins and Burke	S	S	S	N	N	S	N	S	L	N
Washington State	S	S	S	N	N	S	S	S	L	S
Hill	S	L	N	N	N	S	S	N	S	N
Klein	S	S	S	L	S	L	L	L	S	N
Oglesby	S	S	N	N	N	S	S	N	N	N
SE Wisconsin	S	S	N	N	N	S	N	N	N	N
Stover	L	S	L	S	S	S	S	S	L	N
Dearinger	S	S	S	L	S	L	L	L	S	N
Dee (1972)	L	S	L	L	L	L	L	L	L	L
Georgia	L	S	L	S	S	L	S	L	L	N
Orlob	S	S	N	L	S	L	L	N	N	N
Walton and Lewis	N	N	S	L	S	N	N	S	N	N

[a]From (7).
[b]N, little or no fulfillment; S, requirement fulfilled to some extent; L, requirement fulfilled to a large extent.

Table 10-9 Summary of Environmental Impact Assessment Methodologies

Environmental assessment method	Comparative study: Dickert, 1972 (3)	Warner, 1973 (2)	Warner and Preston, 1973 (6)	Smith, 1974 (7)	Method description (Appendix)
Leopold	M[a]	NS	M	C–A	D
Sorensen	M		N	C–A	D
Odum	M				D
Dee (1972)	NS	NS	C	C–C	D
Georgia	NS		C	C–C	D
Seattle	NS				
Task Force		NS	C		D
Adkins and Burke			C	C–B	D
Dee (1973)			C–M		D
Krauskopf			O		D
Little			C	C–B	D
McHarg			O	O	D
Moore			M		D
New York			M		D
Smith			C		D
Stover			C	C–B	D
Tulsa			C		D
Walton and Lewis			C	C–C	D
WSCC			AH		D
Eckenrode				AH	E
Lamanna				AH	E
McKenny				AH	E
Lacate				O	E
Baker and Gruendler				O	E
Turner and Hausmanis				O	E
Manheim				C–A	E
Washington State				C–B	E
Hill				C–B	E
Klein				C–B	E
Oglesby				C–B	E
SE Wisconsin				C–B	E
Dearinger				C–C	E
Orlob				C–C	E
Fischer and Davies					F
Pikul					F
BLM					F
Commonwealth					F
Alden					F
CERL					F
BOR-Wisconsin					F
Resource Planning					F
El Center					F
Kane					F
Tabors					F
Heuting					F
Harvard					F
Chen and Orlob					F
Schlesinger and Daetz					F

[a]M, matrix; NS, not specified; N, network; C, checklist; C–M, checklist matrix; O, overlay; AH, *ad hoc*; C–A, checklist type A; C–B, checklist type B; C–C, checklist type C.

methodologies that were evaluated in comparative studies, as well as additional ones that have been identified as impact methodologies. Appendix F provides a brief description of some of the features of those methodologies in Table 10-9 that represent methods not previously described (11).

Interaction Matrices

Matrices are methodologies that incorporate a list of project activities in addition to a checklist of potentially impacted environmental characteristics. The matrix approach for impact analysis was published by Leopold et al. in 1971 (12). The method involves the use of a matrix with 100 specified actions and 88 environmental items. An impact is identified at the interaction between an action and an environmental item. Figure 10-1 illustrates the concept of the Leopold matrix, and Table 10-10 contains a list of the 100 actions and 88 environmental items. In the use of the Leopold matrix, each action and its potential for creating an impact on each environmental item must be considered. Where an impact is anticipated, the matrix is marked with a diagonal line in the interaction box. The second step in using the Leopold matrix is to describe the interaction in terms of its magnitude and importance.

The magnitude of an interaction is the extensiveness or scale and is described by the assignment of a numerical value from one to ten, with ten representing a large magnitude and one, a small magnitude. Values near five on the magnitude scale represent impacts of intermediate extensiveness. Assignment of a numerical value for the magnitude of an interaction should be based on an objective evaluation of facts. The importance of an interaction is related to the significance,

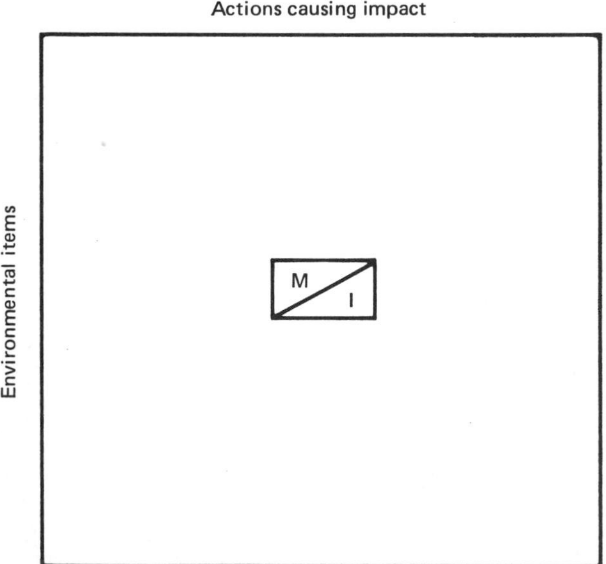

Figure 10-1 Leopold interaction matrix. M, magnitude; I, importance. From (12).

Table 10-10 Actions and Environmental Items in Leopold Interaction Matrix

Actions		Environmental items	
Category	**Description**	**Category**	**Description**
A Modification of regime	a Exotic fauna introduction	A Physical and chemical characteristics	
	b Biological controls	1 Earth	a Mineral resources
	c Modification of habitat		b Construction material
	d Alteration of ground cover		c Soils
			d Land form
	e Alteration of groundwater hydrology		e Force fields and background radiation
	f Alteration of drainage		f Unique physical features
	g River control and flow modification	2 Water	a Surface
	h Canalization		b Ocean
	i Irrigation		c Underground
	j Weather modification		d Quality
	k Burning		e Temperature
	l Surface or paving		f Recharge
	m Noise and vibration		g Snow, ice, and permafrost
B Land transformation and construction	a Urbanization		
	b Industrial sites and buildings	3 Atmosphere	a Quality (gases, particulates)
	c Airports		b Climate (micro, macro)
	d Highways and bridges		
	e Roads and trails		c Temperature
	f Railroads	4 Processes	a Floods
	g Cables and lifts		b Erosion
	h Transmission lines, pipelines, and corridors		c Deposition (sedimentation, precipitation)
			d Solution
	i Barriers including fencing		e Sorption (ion exchange, complexing)
	j Channel dredging and straightening		f Compaction and settling
	k Channel revertments		g Stability (slides, slumps)
	l Canals		
	m Dams and impoundments		h Stress-strain (earthquakes)
	n Piers, seawalls, marinas, and sea terminals		i Air movements
		B Biological conditions	
	o Offshore structures	1 Flora	a Trees
	p Recreational structures		b Shrubs
	q Blasting and drilling		c Grass
	r Cut and fill		d Crops
	s Tunnels and underground structures		e Microflora
			f Aquatic plants
C Resource extraction	a Blasting and drilling		g Endangered species
	b Surface excavation		h Barriers

Table 10-10 (*continued*) **Actions and Environmental Items in Leopold Interaction Matrix**

Actions		Environmental items	
Category	**Description**	**Category**	**Description**
	c Subsurface excavation and retorting		i Corridors
	d Well dredging and fluid removal	2 Fauna	a Birds
	e Dredging		b Land animals including reptiles
	f Clear cutting and other lumbering		c Fish and shellfish
	g Commercial fishing and hunting		d Benthic organisms
D Processing	a Farming		e Insects
	b Ranching and grazing		f Microfauna
	c Feed lots		g Endangered species
	d Dairying		h Barriers
	e Energy generation		i Corridors
	f Mineral processing	C Cultural factors	
	g Metallurgical industry	1 Land use	a Wilderness and open spaces
	h Chemical industry		b Wetlands
	i Textile industry		c Forestry
	j Automobile and aircraft		d Grazing
	k Oil refining		e Agriculture
	l Food		f Residential
	m Lumbering		g Commercial
	n Pulp and paper		h Industry
	o Product storage		i Mining and quarrying
E Land alteration	a Erosion control and terracing	2 Recreation	a Hunting
	b Mine sealing and waste control		b Fishing
	c Strip mining rehabilitation		c Boating
	d Landscaping		d Swimming
	e Harbor dredging		e Camping and hiking
	f Marsh fill and drainage		f Picnicking
F Resource renewal	a Reforestation		g Resorts
	b Wildlife stocking and management	3 Aesthetic and human interest	a Scenic views and vistas
	c Groundwater recharge		b Wilderness qualities
	d Fertilization application		c Open-space qualities
	e Waste recycling		d Landscape design
G Changes in traffic	a Railway		e Unique physical features
	b Automobile		f Parks and reserves
	c Trucking		g Monuments
	d Shipping		h Rare and unique species or ecosystems
			i Historical or archeological sites and objects
			j Presence of misfits
		4 Cultural status	a Cultural patterns (life-style)

Table 10-10 (*continued*) **Actions and Environmental Items in Leopold Interaction Matrix**

Actions		Environmental items	
Category	**Description**	**Category**	**Description**
	e Aircraft		b Health and safety
	f River and canal traffic		c Employment
			d Population density
	g Pleasure boating	5 Manufactured facilities and activities	a Structures
	h Trails		b Transportation network (movement, access)
	i Cables and lifts		
	j Communication		
	k Pipeline		c Utility networks
H Waste replacement and treatment	a Ocean dumping		d Waste disposal
	b Landfill		e Barriers
	c Emplacement of tailings, spoils, and overburden		f Corridors
		D Ecological relationships	a Salinization of water resources
	d Underground storage		b Eutrophication
	e Junk disposal		c Disease–insect vectors
	f Oil well flooding		d Food chains
	g Deep well emplacement		e Salinization of surficial material
	h Cooling water discharge		f Brush encroachment
			g Other
	i Municipal waste discharge including spray irrigation	E Others	
	j Liquid effluent discharge		
	k Stabilization and oxidation ponds		
	l Septic tanks, commercial and domestic		
	m Stack and exhaust emission		
	n Spent lubricants		
I Chemical treatment	a Fertilization		
	b Chemical deicing of highways, etc.		
	c Chemical stabilization of soil		
	d Weed control		
	e Insect control (pesticides)		
J Accidents	a Explosions		
	b Spills and leaks		
	c Operational failure		
K Others			

or assessment of the consequences, of the anticipated interaction. The scale of importance also ranges from one to ten, with ten representing a very important interaction and one, an interaction of relatively low importance. Assignment of an importance numerical value is based on the subjective judgment of the interdisciplinary team working on the environmental assessment study.

One of the attractive features of the Leopold matrix is that it can be expanded or contracted, that is, the number of actions can be increased or decreased from the total of 100, and the number of environmental factors can be increased or decreased from 88. The primary advantages of the Leopold matrix are that it is very useful as a gross screening technique for impact identification purposes and that it can provide a valuable means for impact communication in terms of a visual display of the impacted items and the major actions causing impact.

Summation of the number of rows and columns designated as having interactions can offer insight into impact assessment and interpretation. The Leopold matrix can also be utilized to identify beneficial as well as detrimental impacts through the use of an appropriate legend such as plus and minus signs. The Leopold matrix can also be employed to identify impacts for various temporal phases of a project, for example, construction, operation, and postoperation phases, and to describe impacts associated with various spatial boundaries, namely, at the site and in the region.

Many uses of the Leopold interaction matrix have involved the assignment of three levels of magnitude and importance. Major interactions would be assigned maximum numerical scores, with minor interactions being assigned minimal scores. Intermediate level interactions would be assigned numerical scores intermediate between the major and minor scores.

In Table 10-10 there are essentially no items in the list of environmental factors that are oriented to the socioeconomic environment. This does not mean that these items could not be added, but rather that in 1970 and 1971, the period of time in which the matrix concept was developed, less emphasis was given to these categories of environmental impact.

There have been several variations of the Leopold matrix utilized for impact analysis for various types of projects. The Federal Aviation Administration has used interaction matrices for aviation-type projects (13). The Oregon Highway Department has developed an interaction matrix for impact identification, and the various actions and environmental factors included in this matrix are shown in Table 10-11 (14). Shortened versions of the Leopold matrix have been employed in an environmental impact analysis for a coal mine, generation plant, county road and railroad project, water supply system, and transmission line (15).

Information other than numerical values for magnitude and importance can be included in the impact scales associated with identification of an interaction. In an earth-filled dam project the potential impact of various actions on environmental factors has been shown in 11 categories—neutral, five ranges of beneficial impact, and five ranges of detrimental impact (15). Scales have also been used to describe the probability of occurrence of an impact, with the scale ranging from low to intermediate to high probability of impact. Impact scales can also be developed to

Table 10-11 Highway Interaction Matrix

Actions that may cause impact: Category	Actions that may cause impact: Action	Environmental conditions: Category	Environmental conditions: Action
A Elements of design and location		A Physical and chemical characteristics	
1 Modification of regime	a Modification of habitat	1 Earth	a Mineral resources, precious
	b Alteration of groundwater hydrology		b Mineral resources, common
	c Canalization		c Soils
	d Irrigation		d Land form
	e Surfacing and paving	2 Water	a Surface
2 Land transformation and construction	a Highways and bridges		b Ocean–estuaries
	b Roads and trails		c Underground
	c Barriers including fencing		d Snow–ice
	d Channel dredging and straightening		e Recharge–percolation
	e Channel revetments		f Quality
	f Dams–impoundments		g Temperature
	g Piers–seawalls	3 Atmosphere	a Quality
	h Recreational structures		b Climate
	i Cut and fill		c Temperature
	j Tunnels and underground structures	4 Processes	a Floods
	k Erosion control		b Erosion (air or water)
	l Landscaping		c Deposition (air or water)
	m Harbor dredging		d Solution
	n Marsh fill and draining		e Compaction and settling
	o Scenic waysides		f Stability (slides and slumps)
	p Junkyard and billboard removal		g Air movements
3 Well drilling			h Fire
4 Resource renewal and protection	a Reforestation		i Evaporation
	b Scenic strip acquisition	B Biological conditions	
		1 Flora	a Trees
			b Shrubs
			c Grass

Table 10-11 *(continued)* **Highway Interaction Matrix**

Actions that may cause impact		Environmental conditions	
Category	**Action**	**Category**	**Action**
5 Changes in traffic	a Railway		d Crops
	b Automobile		e Microflora
	c Trucking		f Aquatic Plants
	d River and canal traffic		g Endangered species
	e Pleasure boating		h Barriers
	f Trails		i Corridors
	g Communication	2 Fauna	a Birds
	h Pipeline		b Land animals
B During construction			c Fish and shellfish
1 Modification of regime	a Exotic flora and fauna intro-		d Other aquatic organisms
	duction		e Insects
	b Biological controls		f Microfauna
	c Alteration of ground cover		g Endangered species
	d Alteration of drainage		h Barriers
	e River control and flow modifi-		i Corridors
	cation	C Cultural factors	
	f Burning	1 Land use	a Wilderness
2 Land transformation and	a Blasting and drilling		b Open space
construction	b Marsh fill and drainage		c Wetlands
	c Clearing and grubbing		d Forestry
	d Dams–impoundments		e Grazing
3 Resource extraction	a Blasting and drilling		f Agriculture
	b Surface excavation		g Residential
	c Subsurface excavation		h Commercial
	d Well drilling and fluid removal		i Industrial
	e Dredging		j Lakes and rivers
4 Changes in traffic	a Railway	2 Recreation	a Hunting
	b Automobile		b Fishing
	c Trucking		c Boating

	d River and canal traffic
	e Pleasure boating
	f Trails
	g Communication
	h Pipeline
5 Waste emplacement treatment	a Landfill
	b Emplacement of tailings, spoil and overburden
	c Liquid and exhaust discharge
	d Stack and exhaust emission
	e Spent lubricants
6 Chemical stabilization of soil	
7 Accidents	a Explosions
	b Spills and leaks
	c Operational failure
C Operation	
1 Waste emplacement and treatment	a Liquid effluent discharge
	b Septic tanks
	c Stack and exhaust emission
2 Chemical treatment	a Fertilization
	b Chemical de-icing
	c Weed control
	d Insect control
3 Accidents	a Explosions
	b Spills, leaks
	c Operational failures

	d Swimming
	e Camping
	f Hiking
	g Picnicking
	h Resorts
	i Winter sports
	j Rock hounding
3 Aesthetics and human interest	a Scenic views and vistas
	b Wilderness qualities
	c Open-space qualities
	d Landscape design
	e Unique physical features
	f Parks and reserves
	g Monuments
	h Rare or unique species or ecosystems
	i Historical or archeological sites and objects
	j Presence of incompatible features
4 Cultural status	a Cultural patterns
	b Health
	c Population density
	d Institutions
	e Minority groups
	f Economic groups
5 Manufactured facilities and activities	a Structures
	b Transportation
	c Utility networks
	d Waste disposal
	e Barriers
	f Corridors
	g Governmental activities

show the extent of potential reversibility associated with a beneficial or detrimental impact.

Moore Impact Matrix

Moore et al. (16) have developed an environmental impact matrix for describing the relationship between typical manufacturing activities and their potential ultimate impact on three regions of the Delaware coastal zone. The three regions include a heavily industrialized area in the coastal zone from the Chesapeake and Delaware Canal north to the Pennsylvania state line; an area consisting of several thousand acres of private, state, and federal wildlife refuges in a location from and including the Chesapeake and Delaware Canal, to but not including the city of Lewes; and a recreational area extending from and including the city of Lewes south to the Maryland state line. In all three regions historical/cultural sites and areas are important.

The basic philosophy of the Moore method is that a meaningful analysis of manufacturing environmental impacts must ultimately be based on determination of direct and indirect impacts on human uses. Figure 10-2 depicts the generalized flow of manufacturing-related events that ultimately impact on human uses of the environment. Interrelationships between various primary and secondary activities are also shown in Fig. 10-2. Figure 10-3 presents the generalized relationships contained in the Moore environmental impact matrix. The matrix is divided into four reasonably distinct categories: manufacturing and related activities, potential

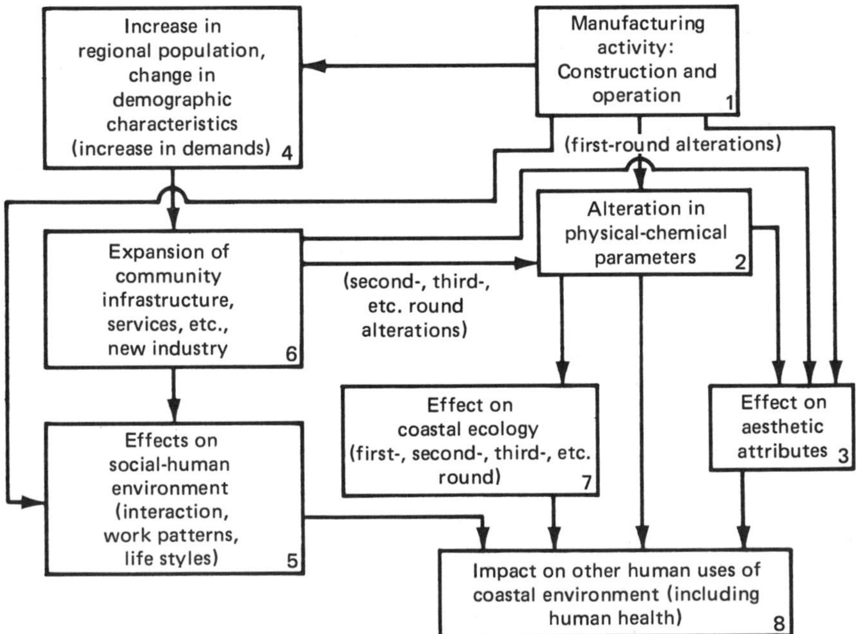

Figure 10-2 Schematic of manufacturing impact on other uses of the coastal environment. From (16).

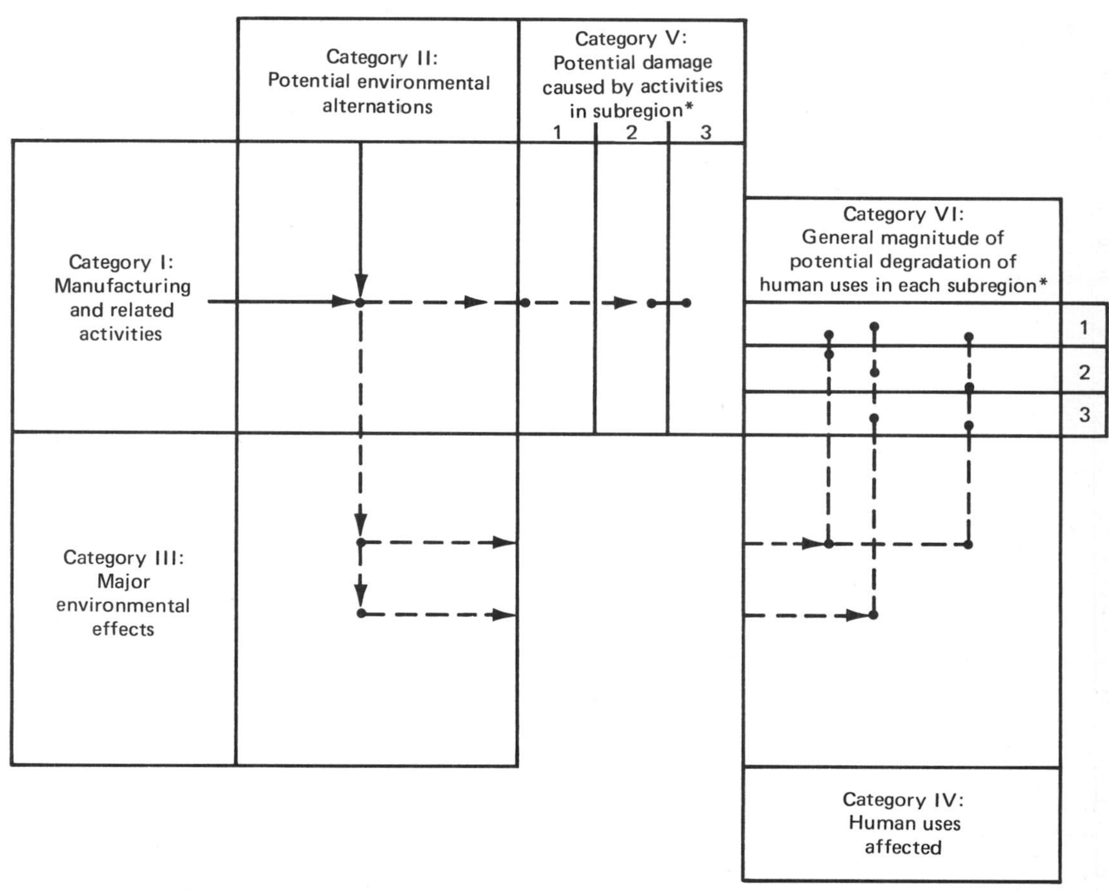

Figure 10-3 Moore interaction matrix. Each subregion has a four-level scale of measurement (negligible, low, moderate, high). From (16).

environmental alterations, major environmental effects, and human uses affected. Two categories of impact are also included: potential damage caused by activities and general magnitude of potential degradation on human uses. The potential for environmental damage is measured on a four-level scale (negligible, low, moderate, high); the general magnitude of potential degradation is measured on the same scale. The interaction category manufacturing and related activities (category I) includes the following (16):

A Construction.
- **1** Site preparation and building facilities (including connecting infrastructures, transmission facilities, etc.).
  - **a** Clearing.
  - **b** Filling.
  - **c** Dredging.
  - **d** Paving.
  - **e** Excavation.
  - **f** Erecting facilities.
  - **g** Erecting supporting structures.
  - **h** Transportation equipment and materials.
  - **i** Labor force commuting.

B Facilities and structures.
- **1** Production-related facilities.
  - **a** Contiguous property (fenced or unfenced).
  - **b** Plant and stacks, towers.
  - **c** Warehouse.
  - **d** Offices.
  - **e** Lighting systems.
  - **f** Parking lots and loading terminals.
  - **g** Open storage areas.

C Production activities and by-products.
- **1** Production residuals.
  - **a** Liquid.
    - **i** Biodegradable.
    - **ii** Nonbiodegradable.
  - **b** Gaseous.
    - **i** Particulate.
    - **ii** Nonvisible.
  - **c** Solid.
    - **i** Biodegradable.
    - **ii** Nonbiodegradable.
  - **d** Sound.
    - **i** Manufacturing.
- **2** Associated manufacturing activities.
  - **a** Work force commuting.
  - **b** Produce and raw materials shipping.
  - **c** Surface-water intake and discharge.
  - **d** Groundwater intake and discharge.
  - **e** Cooling water discharge.

3 Infrastructure (increased demand for and expansion of the community infrastructure and services due to increased population and manufacturing use).
   a Roads.
   b Rail.
   c Water transport.
   d Air transport.
   e Sewers and sewage treatment.
   f Electric power.
   g Utility transmission.
   h Schools.
   i Health facilities service.
   j Housing.
   k Recreation facilities.
      i Bars.
      ii Nightclubs.
      iii Bowling alleys.
      iv Tennis courts.
      v Theaters.
   l Cultural services.
      i Churches.
      ii Museums.
      iii Libraries.
   m Safety services.
   n Media.
      i Newspapers.
      ii Radio.
      iii TV.
4 Population.
   a Migration.
   b Spatial distribution.

Items in the potential environmental alterations category (16) (category II) are

A Aesthetics.
   1 Artificial coloration.
   2 Odor.
   3 Alteration of visual composition/profile.
B Land.
   1 Loss of open space.
   2 Loss of dunes.
   3 Loss of forest and vegetation.
   4 Loss of littorial shallow areas.
   5 Loss of marsh.
   6 Increase in erosion.
   7 Increase in frequency, intensity, and duration of noise.
   8 Increase in accumulation of glass, metals, plastics, cement, asphalt.
C Air.
   1 Increase in hazardous/toxic substances.

- 2 Increase in nitrogen oxide.
- 3 Increase in nonmethane hydrocarbons.
- 4 Increase in carbon monoxide.
- 5 Increase in sulfur dioxide.
- 6 Increase in particulate matter.
- 7 Increase in dust.

D Water.

- 1 Increase in phosphates and nitrates.
- 2 Increase in fecal coliform.
- 3 Change in frequency or volume of surface flow.
- 4 Groundwater balance.
- 5 Increase in suspended solids and turbidity.
- 6 Change in ambient temperature.
- 7 Change in salinity.
- 8 Change in pH.
- 9 Increase in hazardous/toxic and radioactive substances.
- 10 Decrease in dissolved oxygen.
- 11 Increase in gross solids.
- 12 Increase in dissolved solids.
- 13 Increase in oil accumulation.

E Community.

- 1 Change in per capita income level.
- 2 Change in marital category distribution.
- 3 Change in sex ratio.
- 4 Change in family size.
- 5 Change in age distribution.
- 6 Change in educational level.
- 7 Change in ethnic or racial composition.
- 8 Change in population per acre.

The major environmental effects category (16) (category III) includes the following items:

A Ecological.

- 1 Loss of stability in aquatic systems.
  - a Killing of aquatic organisms (reducing populations).
  - b Organic accumulation of lethal or sublethal substances.
  - c Alteration of areal base of aquatic food chains.
  - d Alteration of composition and abundance of aquatic micro flora and fauna.
  - e Elimination of species.
- 2 Loss of stability in terrestrial ecosystems.
  - a Alteration of areal base of terrestrial food chains.
  - b Killing of terrestrial organisms (reducing populations).
  - c Decrease of shelter and forage for terrestrial fauna.
  - d Elimination of species.

B Aesthetic.

- 1 Structural corrosion and deterioration (natural and constructed).
  - a Alteration of urban-rural composition.
  - b Alteration of community structural composition.

c Loss of integrity of historical/cultural areas.
d Loss of unique natural composition.
e Creation of odor.
f Decrease in water clarity.
g Decrease in air clarity.

C Physical/biological.
1 Alteration of chemical composition of air.
2 Alteration of chemical bacteriological characteristics of water (salinity, etc.).
3 Increase in radioactivity.
4 Alteration of natural sound patterns.

D Social/human environmental.
1 Disruption of established activity patterns.
2 Loss of community cohesion/traditions social interaction patterns.
3 Introduction of alien value systems.
4 Alteration of traditional visible economic patterns.
5 Alteration of work routines/patterns.
6 Alteration of political participation and existing power structure.
7 Alteration of established land-use patterns.
8 Increase in congestion.

The human uses affected category (16) (category IV) contains these items:

A Deterioration in biological human health.
1 Unrestricted natural/open areas.
2 Quietude.
3 Diversity.
4 Nonoffensive surroundings.

B Deterioration in real or perceived well-being (psychological).
1 Developing understanding of natural systems.
2 Visitation to natural/scenic areas.
3 Recreational driving.
4 Photography/painting.
5 Hiking/bicycling.
6 Sport fishing.
7 Hunting.
8 Clamming/crabbing.

C Deterioration in extensive recreation uses and visual access.
1 Swimming.
2 Boating.
3 Picnicking.
4 Photography/painting.
5 Camping.
6 Visitation to and perception of historical/cultural areas.
7 Beach/waterfront park activities.

D Deterioration in intensive recreational uses.
1 Water supply.
2 Commercial fishing.
3 Farming.
4 Recreational homes.

E Deterioration in other uses (including commercial, industrial, and residential).

The interactions between categories I and II can be described as temporary (T), operational-dependent (O), and permanent or long-lasting (P) environmental alterations. The interaction categories encompass the following boxes in Fig. 10-2: category I, 1, 6, and part of 4; category II, 2 and parts of 3, 4, and 7; category III, 5 and parts of 3 and 7; and category IV, 8.

Networks

Figure 10-4 illustrates a network analysis of dredging utilized by Sorenson (17). This particular network analysis identifies various interrelationships between the causal factors of dredging operations, such as removal of bottom material and production of material, and the impacted environmental items from these operations. Secondary and tertiary effects associated with dredging are identified in this network. Sorenson has also developed a network-type approach for consideration of the environmental impacts of various uses of the coastal zone. The Sorenson matrix approach for the coastal zone can be considered a stepped matrix in that one matrix interaction is depicted as leading to other interactions.

Example of Stepped Matrix

Impact matrices have been included in actual EISs. A 74-acre industrial park project in the southwestern portion of Fresno, California, made use of a stepped matrix (18). The project involved two separate and simultaneous actions. The first action was a loan to improve the processing facilities of two industries currently located within the project site. In addition to increased production, the loan also permitted greater control of the emission of objectionable odors. The second action was a grant to the city of Fresno, which facilitated the acquisition, renewal, and development of the site into improved sites for use by heavier industry. The environmental impact matrix that was developed is shown in Fig. 10-5. The steps involved in use of the matrix in Fig. 10-5 are as follows:

1 Enter the matrix at the upper left-hand corner under the heading Project Elements. In this example, the matrix is 2. Future Improvements.

2 Read to the right. A causal factor that may result in an impact is shown at Surfacing.

3 A dot (○) indicates that a relationship exists between 2. Future Improvements and Surfacing.

4 Read downward from the ○ until either a ☆, ⭑, □, ▫, or ∪ is encountered. If a ☆ appears, a major positive impact exists. A ⭑ indicates a minor positive impact exists. A □ indicates a major negative impact exists. A ▫ indicates a minor negative impact exists. A ∪ indicates an impact exists, but its magnitude or direction cannot be determined at present. Reading downward from Surfacing, a ▫ is shown.

5 Read to the left. A minor negative impact is a change in Subsurface Water. The 2 next to the ▫ indicates that the impact originates at 2. Future Improvements.

6 Read to the right.

7 In the column Initial Condition is the notation High Quality, indicating that the altered element is presently of high quality.

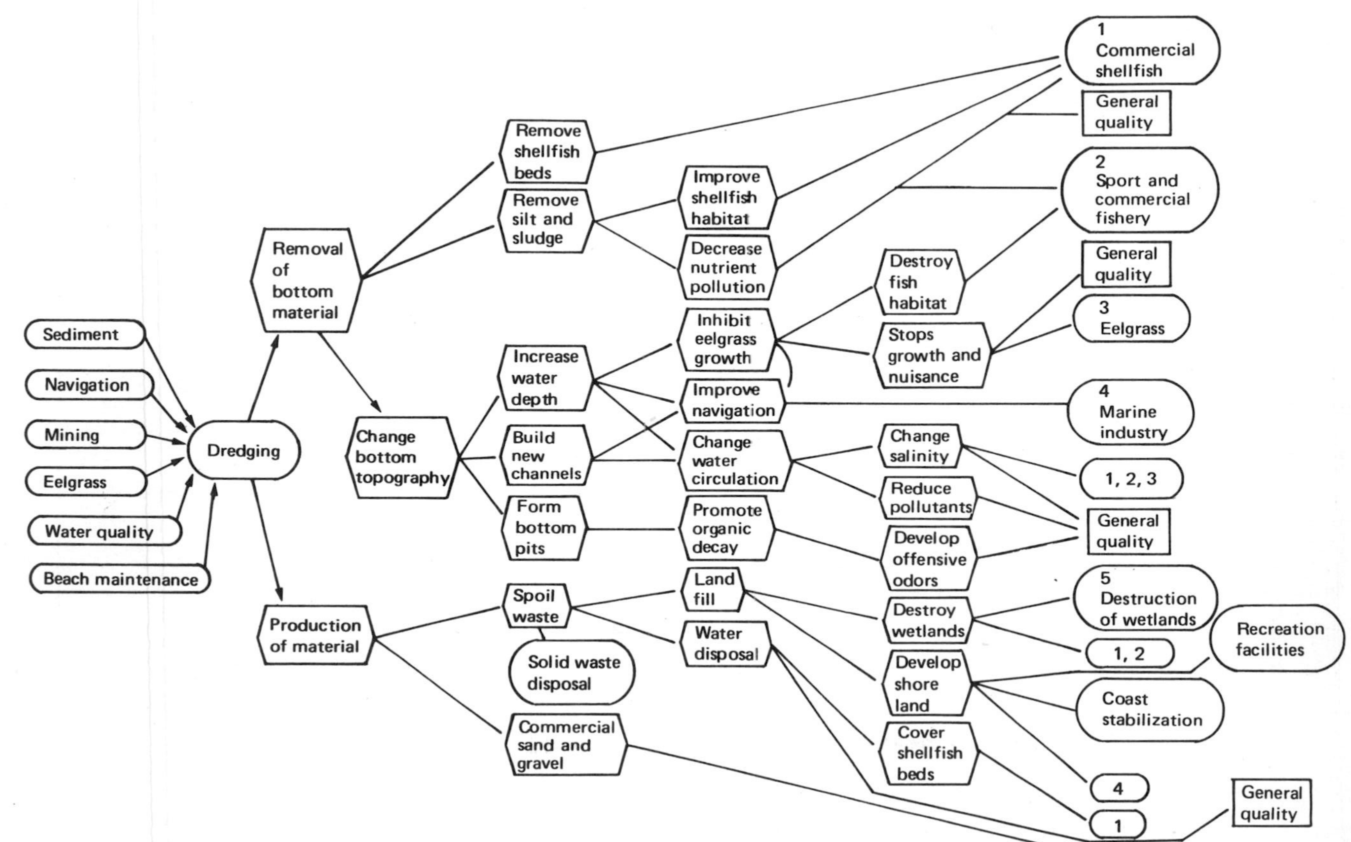

Figure 10-4 A network analysis of dredging. From (17).

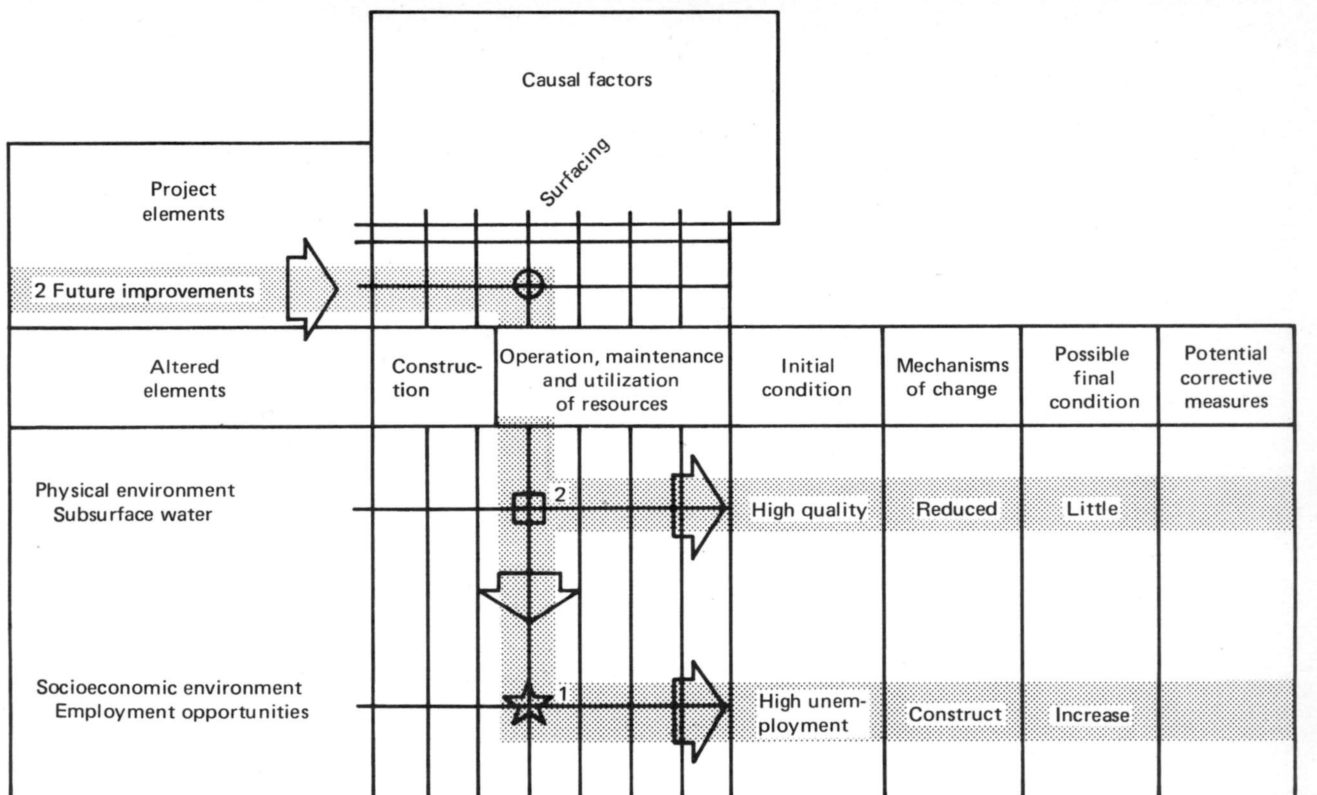

Figure 10-5 Guide to using impact matrix.

8 In the column Mechanisms of Change is a notation describing the way in which the element is altered.

9 In the column Possible Final Condition is a notation describing the altered element after the impact has taken place.

10 The Potential Corrective Measures column is reserved for those impacts against which some steps have been or could be taken to minimize the impacts.

Checklists

Checklists, four broad categories of which can be defined, represent one of the basic methodologies used in environmental impact assessment. Simple checklists are a list of parameters; however, no guidelines are provided on how environmental parameter data are to be measured and interpreted. Descriptive checklists include an identification of environmental parameters and guidelines on how parameter data are to be measured. Scaling checklists are similar to descriptive checklists, with the addition of information basic to subjective scaling of parameter values. Scaling-weighting checklists represent scaling checklists with information provided as to subjective evaluation of each parameter with respect to every other parameter.

Simple Checklists An example of a simple checklist is the methodology developed for the Department of Transportation (19). The method is basically a description of the impacts associated with transportation projects. Table 10-12 identifies potential environmental impacts of a transportation project, organized by category of impact as well as temporal phases. Impacts anticipated in Table 10-12 can be either beneficial or detrimental, depending upon specific circumstances. A similar list of potential impacts of a transportation project organized by phase of project development (19) is as follows:

I Planning and design phase.
- A Impact on land use through speculation in anticipation of development.
- B Impact of uncertainty on economic and social attributes of nearby areas.
- C Impact on other planning and provision of public services.
- D Acquisition and condemnation of property for project, with subsequent dislocation of families and businesses.

II Construction phase.
- A Displacement of people.
- B Noise.
- C Soil erosion and disturbance of natural drainage.
- D Interference with water table.
- E Water pollution.
- F Air pollution (including dust, dirt, and burning of debris).
- G Destruction of or damage to wildlife habitat.
- H Destruction of parks, recreation areas, historical sites.
- I Aesthetic impact of construction activity and destruction of or interference with scenic areas.
- J Impact of ancillary activities (e.g., disposal of earth, acquisition of gravel, and fill).

Table 10-12 Potential Environmental Impacts of a Transportation Project[a]

Category	Stage of project		
	Planning and design	Construction	Operation
I Noise impacts		x[b]	x
A Public health			
B Land use			
II Air quality impacts		x	x
A Public health			
B Land use			
III Water quality impacts		x	x
A Groundwater			
1 Flow and water-table alteration			
2 Interaction with surface drainage			
B Surface water			
1 Shoreline and bottom alteration			
2 Effects of filling and dredging			
3 Drainage and flood characteristics			
C Quality aspects			
1 Effect of effluent loadings			
2 Implication of other actions, such as			
a Disturbance of benthic layers			
b Alteration of currents			
c Changes in flow regime			
d Saline intrusion in groundwater			
3 Land use			
4 Public health			
IV Soil erosion impacts		x	x
A Economic and land use			
B Pollution and siltation			
V Ecological impacts		x	x
A Flora			
B Fauna (other than humans)			
IV Economic impacts	x	x	x
A Land use			
1 In immediate area of project			
2 In local jurisdiction served or traversed			
3 In region			
B Tax base	x	x	x
1 Loss through displacements			

Table 10-12 (*continued*) **Potential Environmental Impacts of a Transportation Project**[a]

Category	Stage of project		
	Planning and design	Construction	Operation
2 Gain through increased values			
C Employment			
1 Access to existing opportunities			
2 Creation of new jobs			
3 Displacement from jobs			
D Housing and public services			
1 Demand for new services			
2 Alteration in existing services			
E Income	x	x	x
F Damage to economically valuable natural resources		x	x
VII Sociopolitical impacts			
A Damage to, or use, of		x	x
1 Cultural resources			
2 Scientific resources			
3 Historical resources			
4 Recreation areas			
B Life-style and activities	x	x	x
1 Increased mobility			
2 Disruption of community			
C Perception of cost/benefit by different cohesive groups	x	x	x
1 Racial			
2 Ethnic			
3 Income class			
D Personal safety		x	x
VIII Aesthetic and visual impacts		x	x
A Scenic resources			
B Urban design			
C Noise			
D Air quality			
E Water quality			

[a]From (19).
[b]x denotes an impact that could be positive as well as negative, depending on circumstances.

K Commitment of resources to construction.
L Safety hazards.

III Operation of facility.
A Direct.
1 Noise.
2 Air pollution.

3 Water pollution.
4 Socioeconomic.
5 Aesthetic.
6 Effects on animal and plant life (ecology).
7 Demand for energy resources.

B Indirect.
1 Contiguous land use.
2 Regional development patterns.
3 Demand for housing and public facilities.
4 Impact on use of nearby environmental amenities (e.g., parks, woodlands, recreation areas).
5 Impact of additional and/or improved transportation into congested areas on those areas.
6 Differential usefulness for different economic and ethnic groups (and resulting problems and solutions).
7 Impact on life styles of increased mobility and other impacts.
8 Impact of improved facility on transportation and related technological development (and consequent impacts).

The following is information relative to potential impacts of transportation projects organized according to spatial boundaries (19):

I In immediate area of project.
A Displacement by project itself.
1 Residential.
2 Commercial.
3 Industrial.
4 Public facilities.
5 Recreational.
6 Natural resources.
7 Cultural resources.
8 Scenic resources.
9 Wildlife resources.

B Land-use choices affected by project.
1 Attracted by increased access.
a Ancillary uses (facility or user service).
b Users benefitting from access (certain industrial, commercial, residential, and public uses).
2 Disrupted by project.
a Incompatibility with noise, pollution, aesthetic, safety, and other effects of facility.
b Incompatibility with access-oriented uses.
c Incompatibility resulting from increased access of nonresident users (e.g., natural or wild areas).

C Neighborhood (or area) services, facilities, and living patterns affected by facility.
1 Disruption of service districts.
a Public facilities.

b Private nonprofit services.
c Retail establishments.
2 Effects on neighborhood cohesiveness and stability.

II In land jurisdiction served or traversed by facility.
A Effect on land-use planning and controls.
B Effect on planning and development of public facilities resulting from the project and land-use patterns generated or influenced by the project (including effect on tax base, cost of services).
C Effect on areas not directly contiguous with project of actions by those displaced, disrupted, attracted, or otherwise affected by the project.

III In region where project is located.
A Effect on regional development planning, inducement, and controls.
B Revenue effects, influencing other public projects.
C Economic effects, influencing private development in the region as a whole and differentially within it.

This methodology represents a simple checklist since various potential impacts are identified, but no detailed information is provided for measurement and interpretation. The only feature of the methodology dealing with impact interpretation is having an interdisciplinary team evaluate project impact significance on a scale of zero to ten.

The Federal Power Commission has developed detailed guidelines for preparing environmental reports (20). These guidelines provide a detailed description by project type of the categories of environmental impact that must be addressed. Specific areas to be included in the discussion of environmental impacts include impacts of the construction phase on land features and uses, species and aqueous systems, socioeconomic considerations, air and water environment, and waste disposal. Impacts associated with the operation and maintenance phase include those related to land features and uses, species and aqueous systems, socioeconomic considerations, air and water environment, solid waste, use of resources, maintenance, and accidents and catastrophes. As an example, the socioeconomic impacts to be addressed during the construction phase of the project are as follows (20):

Discuss the effect on local socioeconomic development in relationship to labor, housing, local industry, and public services. Discuss the need for relocations of families and businesses. Describe the beneficial effects, both direct and indirect, of the action on the human environment, such as benefits resulting from the services and products, and other results of the action (include tax benefits to local and state governments, growth of local tax base from new business and housing development in payrolls). Describe the impact on human elements, including the need for increased public services (schools, health facilities, police and fire protection, housing, waste disposal, markets, transportation, communication, energy supplies and recreational facilities).

Descriptive Checklists An environmental impact computer system (EICS) has been developed by the U.S. Army Construction Engineering Research Laboratory (21). This system uses computer techniques to identify potential environmental impacts from nine functional areas of army activities on eleven broad environmental categories (22). The nine functional areas are construction; operation, maintenance, and repair; training; mission change; real estate; procurement; army industrial activities; research, development, testing, and evaluation; and administration and support. Each of these functional areas has a number of additional basic activities. Examples of basic activities in the construction functional area include clearing trees, removing broken concrete, back-filling foundations, curing bituminous pavement, cleaning used concrete forms, installing insulation, and landscaping sites. A total of approximately 2,000 basic activities are identified in the nine functional areas. The environment in the EICS system is divided into 11 areas, including ecology, health science, air quality, surface water, groundwater, sociology, economics, earth science, land use, noise, and transportation. Within each of these categories additional parameters are defined. Approximately 1,000 specific environmental factors are defined for the 11 environmental categories. On this basis it is possible to have a checklist that addresses the impact of approximately 2,000 basic army activities on 1,000 environmental factors.

The computer system is used to identify potential impacts associated with various types of activities. In a sense this method is similar to a computerized interaction matrix. It is considered here as a descriptive checklist because each of the environmental factors is described in detail, with information given on actual measurement and data interpretation. The system codes each interaction into one of four categories: the first category indicates that the potential impact must be assessed every time the activity is carried out; the second, that the impact is usually present but may be absent depending upon individual circumstances; the third, that the impact arises in a small but predictable number of cases and its presence should be considered in the individual circumstance; and finally, if there is no indication of potential impact, then the particular activity under consideration has no impact upon this environmental factor.

The Atomic Energy Commission has also developed specific guidelines associated with environmental impact assessment (23). Appendix G contains a detailed description of various environmental factors that should be considered when evaluating environmental impacts of nuclear power plant operation. Detailed information is provided for data analysis and interpretation, as well as data presentation.

Scaling Checklists Scaling checklists use scaling techniques for assessing the environmental impact of a project. Adkins and Burke developed a method for transportation projects that involves scaling impacts of alternatives on a relative basis from minus five to plus five (24). The environmental factors employed in the Adkins-Burke approach are transportation, environmental, sociological, and economic parameters. Table 10-13 is an example of the application of the Adkins-Burke methodology in environmental impact assessment. Two alternatives are considered, with the evaluations ranging from minus five to plus five in accordance with the Adkins-Burke scale. The summary rating in Table 10-13 is a means of displaying relative impacts by the ratio of the plus ratings to the minus ratings, as well as the algebraic average of all ratings. Table 10-14 provides an overall

Table 10-13 Example of Adkins-Burke Method in Environmental Category[a]

Factor	Definition or explanation	Rating alternative 1	2	3	Comments
A Community (local area)					
1 Noise pollution	Relation to present levels, Policy and Procedures Memorandum 20-8 (PPM 20-8)				
a Adjacent to freeway		−2	−1		Relief of street traffic helps offset
b General area		+3	+1		Improves due to relief of street traffic
2 Air pollution	PPM 20-8				
a Adjacent to freeway		+2	+1		Relief of street traffic
b General area		+5	+2		Relief of street traffic
3 Drainage	Effects on chances of flooding, etc.				
a Adjacent to freeway		+1	0		Route 1 will help slightly
b General area		0	0		
4 Water supply					
a Water pollution	PPM 20-8, permanent or serious temporary	0	0		Little, if any, effect
b Water quantity	Interference with movement or level of groundwater	0	0		Little, if any, effect
5 Waste disposal	PPM 20-8, access to, interference, etc.	0	0		Little, if any, effect
6 Flora effects	NEPA and PPM 20-8, irreplaceable losses, etc.	0	0		Little, if any, effect
7 Fauna effects	NEPA and PPM 20-8, breeding or nesting, etc.	0	0		Little, if any, effect
8 Parks	PPM 20-8, improvement or damage to	+5	+2		Improves access to
9 Playgrounds	PPM 20-8, improvement or damage to	+5	0		Route 1 improves access to
10 Archeological sites	NEPA and PPM 20-8, loss of or access to, etc.	0	0		None affected
11 Historical sites	PPM 20-8, loss of or access to, etc.	+2	+1		Improves access to
12 Open space		+3	+1		Opens area by removing structures, some undesirable
13 Visual aspects	PPM 20-8, community view of freeway				
a Adjacent to freeway		+3	+1		Through proper treatment areas improved
b General area		+2	0		Route 1 would help, route 2 not likely to help

Table 10-13 (*continued*) Example of Adkins-Burke Method in Environmental Category[a]

Factor	Definition or explanation	Rating alternative 1	2	3	Comments
14 Safety	PPM 20-8, any change in hazards				
a Traffic		+3	+1		Route 1 gives more relief to streets and removes railroads
b Pedestrian		+5	+1		Route 1 gives more relief to streets and removes railroads, route 1 more persons involved
c Other		—	—		
15 Other	PPM 20-8, e.g., other resources				
B Freeway motorist experience	PPM 20-8				
1 View of freeway	Appearance and security	+3	+1		Route 1 clearer and nicer view
2 View of adjacent area	Aesthetics or special sights	0	+1		Route 2 could give special views on curves
3 Panoramic views	Vistas	+1	+3		Route 2 good, route 1 downtown area
4 Area hazards	Hazards to freeway users and vehicles	+3	−1		Route 1 would displace hazards, route 2 would expose motorists to industrial smog, etc.

Summary rating

	Alternative 1	2	3		Alternative 1	2	3
No. of plus ratings	15	12		Algebraic sum of ratings	44	14	
No. of minus ratings	1	2		Average of ratings	2.75	1.00	
Ratio of plus ratings	0.94	0.86					

[a] From (24).

Table 10-14 Adkins-Burke Method: Overall Comparison of Ratings[a]

Parameters	No. of plus ratings	No. of minus ratings	Total no. of ratings	Algebraic sum of ratings	Ratio of plus ratings	Average rating
Transportation						
Local area						
Alt. 1	7	6	13	18	0.54	1.38
Alt. 2	4	2	6	1	0.67	0.17
Alt. 3						
Metropolitan area						
Alt. 1	8	0	8	34	1.00	4.25
Alt. 2	6	1	7	7	0.86	1.00
Alt. 3						
Environmental						
Alt. 1	15	1	16	44	0.94	2.75
Alt. 2	12	2	14	14	0.86	1.00
Alt. 3						
Sociological						
Community						
Alt. 1	9	2	11	27	0.82	2.46
Alt. 2	6	3	9	−1	0.67	−0.11
Alt. 3						
Metropolitan						
Alt. 1	9	0	9	31	1.00	3.44
Alt. 2	6	1	7	7	0.86	1.00
Alt. 3						
Economic						
Alt. 1	15	14	29	27	0.52	0.93
Alt. 2	14	14	28	−11	0.50	−0.39
Alt. 3						
All ratings						
Alt. 1	63	23	86	181	0.73	2.10
Alt. 2	48	23	71	17	0.68	0.24
Alt. 3						

[a]From (24).

comparison of the alternatives relative to the four environmental categories used in the Adkins-Burke method. Again, the summary of the overall evaluation is based on the number of plus and minus ratings, as well as the algebraic average rating.

Another scaling checklist methodology is the technique developed by the Soil Conservation Service (25). Basically it involves consideration of a number of environmental factors, with the use of a scaling system for impact evaluation.

Scaling-Weighting Checklists An environmental evaluation system was developed in 1972 at Battelle Laboratories for the Bureau of Reclamation (8). This system consists of a description of the environmental factors included in the checklist as well as instructions for scaling the values of each parameter and assigning importance units. Figure 10-6 shows the individual environmental

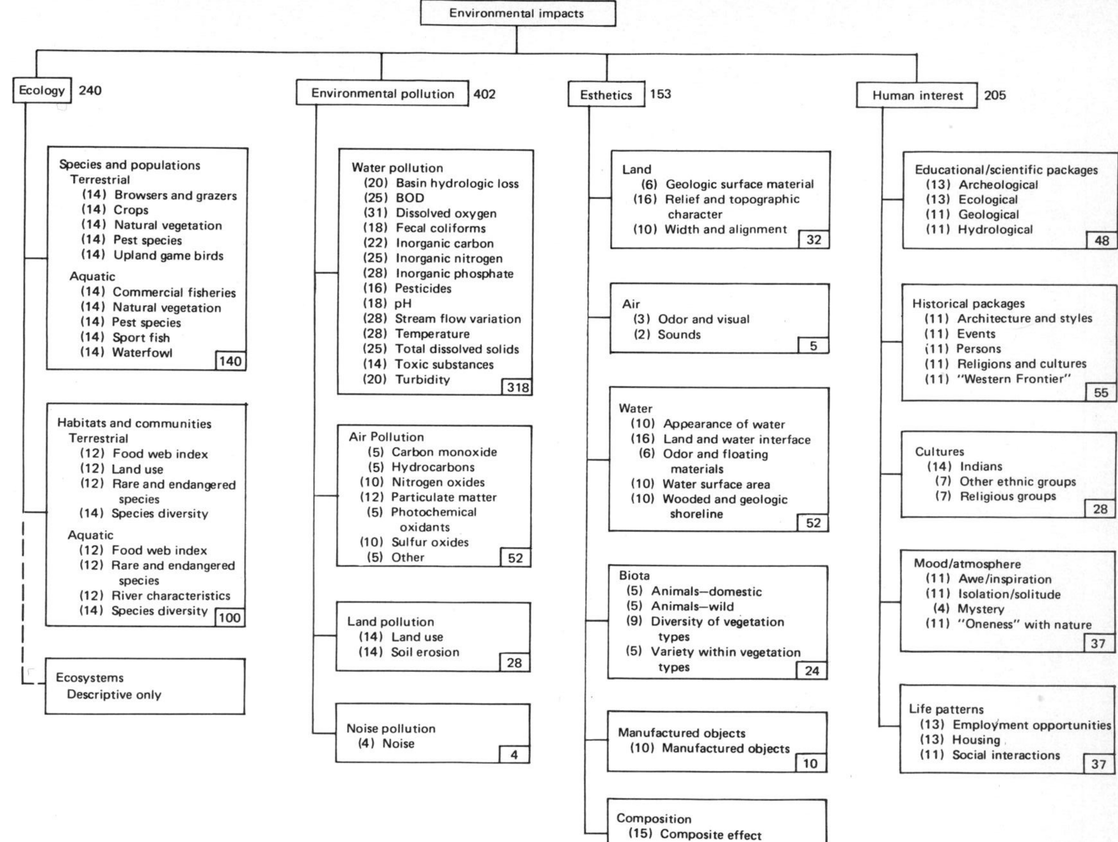

Figure 10-6 Battelle environmental evaluation system. Numbers in parentheses are parameter importance units. Numbers enclosed in boxes represent the total.

parameters organized into 4 categories, 17 components, and 78 environmental parameters.

The Battelle environmental evaluation system was developed for water resources projects. The major feature of this system is that environmental impact is expressed in commensurate units. In impact analysis there is need for development of common units of comparison, since various environmental factors are measured in differing units. The steps involved in the development of commensurate units include transformation of parameter estimates into an environmental quality (EQ) scale, assignment of importance weights (PIU) to the individual parameters, and the multiplication of scale values by importance values to obtain environmental impact units (EIU).

Transformation of parameter estimates into an EQ scale is based on the fact that there is a certain range of anticipated values for a given parameter, with the range dependent upon the units of measurement of the parameter. For example, dissolved oxygen in water will typically range between 0 and 10 mg/liter, while total suspended particulates in the atmosphere may range from 20 to several thousand $\mu g/m^3$. To help transform these parameter estimates into an environmental quality scale, value function graphs are presented for each of the 78 parameters used in the system. Figures 10-7–10-10 show four value function graphs employed in the Battelle system. Parameter values are shown on the abscissa, while the EQ scale is shown on the ordinate. Environmental quality can range from 0 to 1.0, with 0 representing poor quality and 1.0, very good quality. The specific rationale associated with these four graphs is described in the methodology report (8).

Assignment of importance units to each of the individual parameters is based on the ranked pairwise comparison technique, in which subjective judgment determines the relative importance or significance of individual parameters. As an example of this technique, consider the distribution of 100 PIUs among three environmental factors. After some discussion on the part of the interdisciplinary

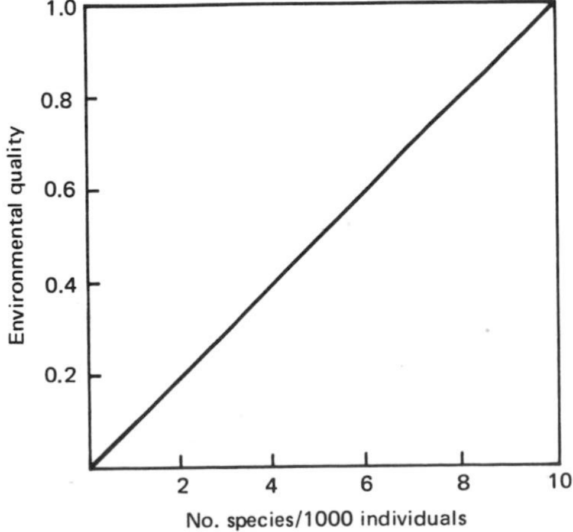

Figure 10-7 Species diversity.

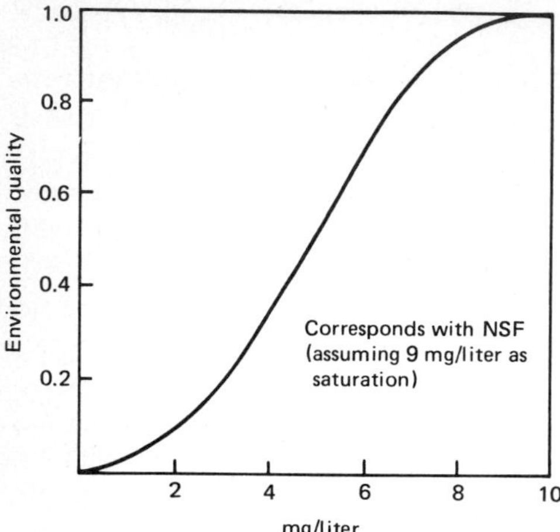

Figure 10-8 Dissolved oxygen.

team, factor B is considered to be more important than factor C, and both B and C are considered to be more important than factor A. This is the ranking step, and the rationale used to reach this decision should be documented. Next, factor B is assigned a value of 1.0. Factor C is considered relative to factor B and assigned an importance on a scale from 0 to 1. For purposes of this example, factor C is considered to be one-half as important as factor B. Then consider factor A relative to factor C, and in this example a value of one-fifth is applied. The assignment of the 100 PIUs would then be on the basis of the following proportionalities:

$$\text{Factor B} = \frac{1.0}{1.6}(100) = 63$$

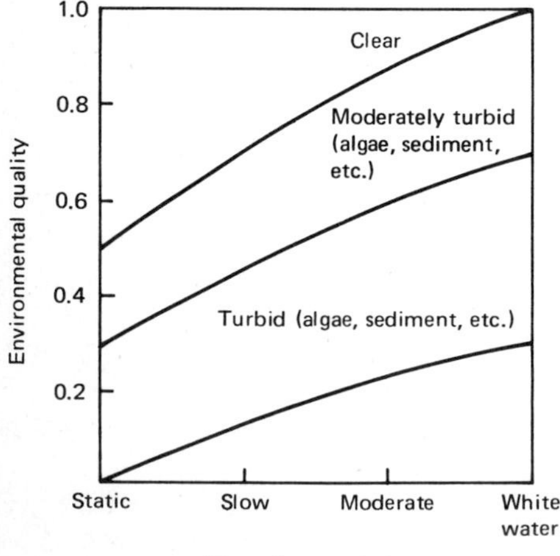

Figure 10-9 Appearance of water.

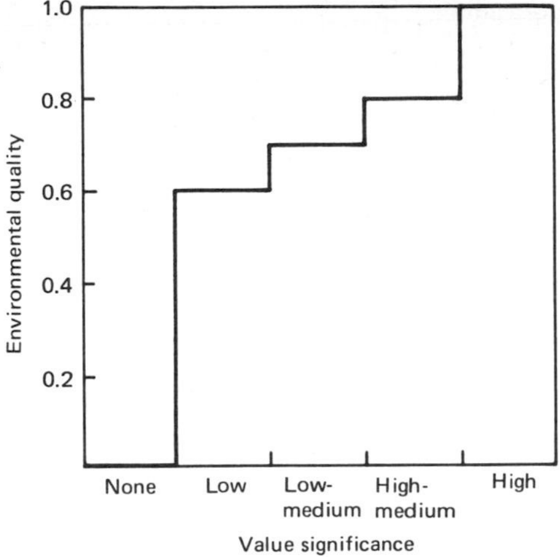

Figure 10-10 Historical external package.

$$\text{Factor C} = \frac{0.5}{1.6}(100) = 31$$

$$\text{Factor A} = \frac{0.1}{1.6}(100) = 6$$

This approach was utilized in the Battelle environmental evaluation system to arrive at a distribution of 1,000 PIUs. Figure 10-6 shows the results of this allocation, with the environmental pollution category containing the greatest number of points.

Using the Battelle system involves the following steps:

1 Obtain parameter data without the project for each of the 78 environmental factors. Convert these parameter data into EQ scale values for each of the 78 parameters. Multiply these scale values by the PIU for each of the individual parameters to develop a composite score for the environment without the project.

2 For each alternative predict the change in the environmental parameters.

3 Utilizing predicted changes in the parameter values, determine the environmental quality scale for each parameter and each alternative.

4 Multiply the environmental quality values for each alternative by each PIU, and aggregate the information for a total composite score.

The numerical evaluation system provides a tool that serves to guide environmental impact analysis. The Battelle environmental evaluation system is a very highly organized methodology, and as such, it helps to ensure systematic, all-inclusive approaches and to identify critical changes. As is the case with other methodologies, very little emphasis is given to socioeconomic factors in this method.

One of the key points to note is that there is no passing or failing score in the Battelle system, since the resultant numerical evaluations must be subjected to professional interpretation. The methodology is valuable for an analysis of trade offs within a component, within a category, or between categories. Table 10-15 shows

Table 10-15 Trade-Off Analysis Using the Battelle Environmental Evaluation System

Category	Alt. A[a]	Alt. B	Alt. C
(240)[b] Ecology	215	200	220
(402) Environmental pollution	340	350	310
(153) Aesthetics	110	120	135
(205) Human interest	180	175	175

[a]No action.
[b]The number in parentheses is the maximum for a category.

an example of a trade-off presentation using the Battelle environmental evaluation system.

One of the key criticisms of the Battelle system is that it is an inflexible methodology in terms of application to projects of different types. However, the concept of this methodology has been converted and applied to a rapid transit system project in Atlanta (7). It has also been applied to various water resources projects within the Bureau of Reclamation as well as to a multipurpose reservoir project of the U.S. Army Corps of Engineers (26).

Another example of a scaling-weighting checklist is the optimum pathway matrix developed by Odum et al. (9). This methodology was addressed to an evaluation of eight alternatives for completion of a section of I-75 north of Atlanta. The analysis consisted of first defining factors to be included in the method. A total of 56 factors were identified and sorted into four general groups, as follows:

Group E—economic and highway engineering factors
Group L—environmental and land-use considerations
Group R—recreation considerations
Group S—social and human considerations

Data were then developed for each of the 56 factors from existing engineering reports and other sources of information. A summary of the information developed for each of the eight alternatives is shown in Table 10-16. The next step consisted of unitizing the information for each factor and for each alternative. The unitization (scaling) was accomplished by the use of the following mathematical formulation:

$$S_i = \frac{1}{\text{max. of } X_1, X_2, \ldots, X_8}$$

where S_i = scaling factor for ith factor
$X_1, X_2, \ldots, X_8$ = value of eight alternatives for ith factor

$$\text{UV}_{ij} = S_i X_{ij}$$

where UV_{ij} = unitized value for ith factor and jth alternative
X_{ij} = value for ith factor and jth alternative

Table 10-16 Components in Odum Optimum Pathway Matrix

Factor	Component	Relative weights		Classi-fication[a]	Alternatives[b]							
		Initial	Long term		G	G-1	T	T-1	F	F-1	P	O
	Land removed for right-of-way, acres											
1	Pine	− 3	−10	L	1,407	1,411	1,438	1,442	1,352	1,356	811	826
2	Mixed	− 4	−10	L	1,619	1,619	1,406	1,406	1,092	1,092	1,263	1,445
3	Hardwood	− 7	−10	L	215	226	277	288	344	355	152	164
4	Agricultural	− 3	−10	L	313	282	305	275	325	294	257	279
5	Idle	+ 5	+ 8	L	227	237	175	185	147	157	61	51
6	Water	− 7	− 4	L	17	27	23	33	22	32	52	19
7	Swamp	− 4	− 4	L	0	0	0	0	0	0	0	2
8	Mined land	+ 8	+10	L	69	69	68	68	0	0	0	0
9	Urban	− 6	+10	L-S	212	199	175	156	68	40	22	30
10	Water supplies affected	− 2	0	L-S	1	1	1	1	0	0	0	0
11	Unique areas	− 2	− 2	L-S	0	0	0	0	0	0	1	1
12	Streams crossed	− 2	− 2	L	29	27	28	36	36	24	23	24
13	Small abridgements	− 2	− 2	L	6	4	4	2	2	0	0	0
	Major bridges											
	Across main body of lake											
14	Number	− 5	− 5	L-R	0	0	0	0	1	1	1	0
15	Length, ft	− 5	− 5	L-R	0	0	0	0	1,400	1,400	3,100	0
	Across other than main body											
16	Number	− 3	− 3	L-R	2	3	2	2	1	2	0	2
17	Length, ft	− 3	− 3	L-R	1,600	2,100	1,700	2,300	500	1,000	0	1,500
18	Composite soil limitations	− 4	− 3	L-E	293	293	300	300	320	320	328	303
19	Max. sediment possible	− 6	− 3	L-E	315	313	295	292	350	397	270	290
20	Min. sediment expected	− 8	− 3	L-E	44	44	40	40	48	48	37	40
21	Area to be paved	− 2	− 2	L-E	75	666	649	640	553	644	519	609
	Area affected by noise, mi^2											
22	Some	− 7	− 4	S-L	103	103	90	90	63	63	64	73
23	Great	− 7	− 7	S-L	33	33	50	50	84	84	95	66

Table 10-16 (*continued*) **Components in Odum Optimum Pathway Matrix**

Factor	Component	Relative weights: Initial	Relative weights: Long term	Classi-fication[a]	Alternatives[b]: G	G-1	T	T-1	F	F-1	P	O
24	Total system cost, $ × 10^6	–20	0	E	108	101	103	95	106	98	89	82
25	Annual costs, $ × 10^6	0	–20	E	8.7	8.3	8.5	7.8	8.6	8.0	7.2	6.8
26	Total excavation/system, yd × 10^6	– 2	0	E–L	25.3	25.0	25.5	25.3	20.6	20.1	20.9	25.5
27	Annual road user costs, $ × 10^6	–10	–10	E–S	90.8	91.9	90.0	91.1	88.5	89.1	94.2	103.7
28	Benefit/cost ratio	–10	–10	E	–17.1	– 9.6	– 9.8	– 3.2	0.0	– 0.9	– 4.1	– 8.2
29	Interstate highway mileage	– 2	– 2	S–E	27.1	27.1	27.1	27.1	24.5	24.5	27.4	31.9
30	Taxable land removed, acres	– 8	– 8	S–E	3,955	3,995	3,737	3,736	3,039	3,084	2,557	2,803
31	Public land removed, acres	– 8	– 8	S–E	242	197	200	204	401	338	151	92
32	Total family displacemements	–20	0	S–E	177	99	177	97	146	66	101	53
	Nearby residences affected (noise)											
	Daytime activity disturbance											
33	Some	– 2	– 2	S	3,072	3,015	3,222	3,432	1,555	1,667	1,956	2,521
34	Great	– 5	– 5	S	789	797	2,088	1,901	1,368	1,076	998	975
	Nighttime sleep disturbance											
35	Some	– 4	– 4	S	1,431	1,072	1,185	826	958	599	637	598
36	Great	–10	–10	S	174	215	470	513	263	306	193	268
	Churches affected by noise											
37	Some	– 2	– 2	S	10	8	6	7	6	7	5	7
38	Great	– 5	– 5	S	2	2	2	3	4	4	3	3
	Schools affected by noise											
39	Some	– 3	– 3	S	2	1	0	0	0	0	0	3
40	Great	–10	–10	S	0	0	1	6	1	1	1	0
	Lives saved/route											
41	Short term (1981)	+50	0	S–E	16	16	42	42	82	82	5	5
42	Long term (1993)	0	+50	S–E	377	377	389	389	385	385	301	305
43	Number interchanges	+ 2	+ 6	S–E	16	16	14	14	13	13	10	10

	Secondary growth impact											
44	Potential for development	0	+25	S–E	10	10	10	10	8	8	4	6
45	Suitability for development	0	+25	S–E	+ 8	+ 9	+ 8	+ 9	− 7	− 6	− 9	− 5
46	Water quality	5	2	R–S	− 1	− 3	− 1	− 2	− 2	− 3	− 2	− 2
47	Visual disturbance	2	2	R	− 2	− 3	− 2	− 3	− 5	− 6	− 5	− 2
48	Hunting and game	4	1	R	0	− 1	0	− 1	− 6	− 7	− 6	− 4
49	Natural character loss	4	4	R–S	− 1	− 2	− 1	− 2	− 4	− 5	− 3	− 2
50	Safe access	6	6	R–S–E	+ 6	+ 7	+ 6	+ 7	+ 6	+ 7	+ 4	+ 2
51	Impact on planned reservoir use pattern	1	2	R–S–L	+ 2	+ 3	+ 2	+ 3	+ 3	+ 4	− 2	− 3
52	Driving for pleasure	2	1	R–S	+ 3	+ 4	+ 3	+ 4	+ 8	+ 9	+ 8	+ 7
	Composite noise effect											
53	Camping	− 3	− 3	R	44	44	48	48	89	89	51	23
54	Picnic	− 2	− 2	R	40	40	87	87	26	26	17	9
	Recreational land loss											
55	Present	5	0	R	3	− 4	− 1	− 2	− 6	− 7	− 4	− 3
56	Potential	0	3	R	1	− 2	− 1	− 2	− 4	− 5	− 5	− 3

[a]L, environmental and land-use considerations; S, social and human considerations; R, recreation considerations; E, economic and highway engineering factors.

[b]G = southwestern highway route, G-1 = similar to route G except in connection to existing highway in the northwest, T = western route, T-1 = similar to route T except in connection to existing highway in the northwest, F = centermost route, F-1 = similar to route F except in connection to existing highway in the northwest, P = eastern route, O = northeastern route.

The next step consisted of an interdisciplinary decision on the relative weights of initial and long-term effects of project construction on individual factors. The range of values selected was from minus 20 to plus 50, as shown in Table 10-16. After the relative weights were decided, composite weighting values were developed for each of the 56 factors. These values were calculated by assuming that the long-term effects were 10 times greater than the initial effects. The primary reason for this assumption is that the operational period for a highway project is at least 10 times longer than the construction period. Then the composite weighting values were unitized by dividing each weight by the sum of all weights in accordance with the following mathematical relationship:

$$N_i = \frac{W_i}{\sum_{1}^{56} W_i}$$

where N_i = unitized weighting value
W_i = composite weighting factor

The final step consisted of the calculation of environmental indices for each route in accordance with the following mathematical equation:

$$I_j = \sum_{1}^{56} N_i S_i X_{ij} \pm e(N_i S_i X_{ij})$$

where I_j = environmental index
e = error term to allow for misjudgment on relative weights by ±50%, stochastically selected

The key feature of the Odum method is that an error term is included to allow for misjudgment on relative weights. This error term is handled by computational analysis using a package stochastic computer program. An environmental index was calculated 20 times for each route with the computer program. Table 10-17 provides information on the indices for 20 iterations for route G.

Figure 10-11 shows the relative environmental indices for each of the eight routes in the study. Additional computational analysis considered the effects of various subgroups or combinations of subgroups on the optimum route, as well as the assignment of equal weight to initial and long-term effects.

The Odum methodology, in conjunction with the Battelle environmental evaluation system, has been employed in the evaluation of the relative environmental impact of eight alternative navigation routes (27).

SUMMARY

More than 50 impact analysis methodologies have been developed in response to the requirements of the NEPA, but no universal methodology has been developed to

Table 10-17 Computational Analysis for Odum Optimum Pathway Matrix for Route G[a]

Pass no.	Resultant value
1	–20.815
2	– 6.738
3	– 1.986
4	–15.550
5	– 7.423
6	– 8.737
7	–12.596
8	– 5.359
9	–12.800
10	– 8.908
11	– 4.012
12	– 8.064
13	–12.650
14	– 8.472
15	– 4.082
16	– 8.584
17	– 3.563
18	– 3.817
19	–13.681
20	–11.035
Mean	– 8.944
Standard deviation	4.736
95% Confidence interval	–11.160 to –6.727

[a]From (9).

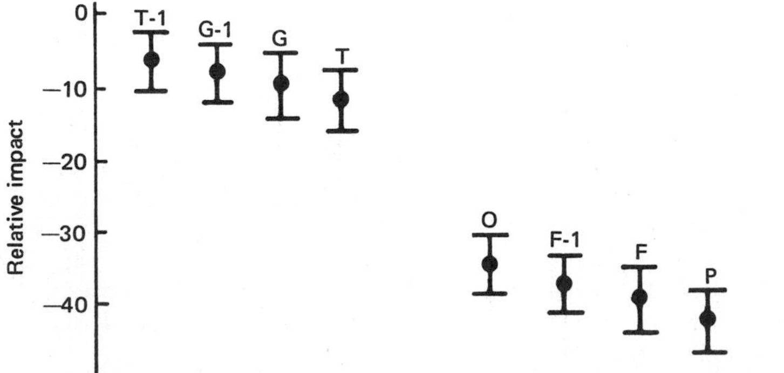

Figure 10-11 Graphical presentation of optimum pathway matrix results. T-1, G-1, G, T, O, F-1, F, and P are the routes.

date. This chapter has examined several methodologies and their comparative features. Two primary methods employed in impact analysis include checklists and matrices. Checklists can be further subdivided into simple, descriptive, scaling, and scaling-weighting checklists. Matrices can be subdivided into simple matrices and stepped matrices.

The key point with regard to all environmental impact analysis methodologies is that they are useful tools for examining relative environmental impacts of alternatives. They represent a tool that must be applied with professional judgment, and their results must also be interpreted using professional judgment.

SELECTED REFERENCES

1 *National Environmental Policy Act of 1969*, PL 91-190, 91st Cong., S. 1075, Jan. 1, 1970.

2 Warner, M. L.: "Environmental Impact Analysis: An Examination of Three Methodologies," p. 28, Ph.D. dissertation, University of Wisconsin, Madison, 1973.

3 Dickert, T. G.: Methods For Environmental Impact Assessment: A Comparison, in Thomas G. Dickert and Katherine R. Domeny (eds.), "Environmental Impact Assessment: Guidelines and Commentary," pp. 127–143, University of California, Berkeley, 1974.

4 Drobny, N. L., and M. A. Smith: Review of Environmental Impact Assessment Methodologies, internal working paper, Battelle-Columbus, Columbus, Ohio, 1973.

5 Warner, M. L., and D. W. Bromley: "Environmental Impact Analyses: A Review of Three Methodologies," Tech. Rept., Wisconsin Water Resources Center, University of Wisconsin, Madison, 1974.

6 Warner, M. L., and E. H. Preston: "A Review of Environmental Impact Assessment Methodologies," report prepared by Battelle-Columbus for U.S. Environmental Protection Agency, Washington, D.C., Oct. 1973.

7 Smith, M. A.: "Field Test of an Environmental Impact Assessment Methodology," Rept. ERC-1574, Environmental Resources Center, Georgia Institute of Technology, Atlanta, Aug. 1974.

8 Dee, Norbert, et al.: "Environmental Evaluation System for Water Resource Planning," final report prepared by Battelle-Columbus for Bureau of Reclamation, Jan. 31, 1972.

9 Odum, E. P., et al.: "Optimum Pathway Matrix Analysis Approach to the Environmental Decision-Making Process," Institute of Ecology, University of Georgia, Athens, 1971.

10 Dee, Norbert, et al.: "Planning Methodology for Water Quality Management: Environmental Evaluation System," Battelle-Columbus, Columbus, Ohio, July 1973.

11 Viohl, R. C., Jr., and K. G. M. Mason: Environmental Impact Assessment Methodologies: An Annotated Bibliography, *Counc. Planning Libr. Exchange Biblio.* 691, Nov. 1974.

12 Leopold, L. B., et al.: A Procedure for Evaluating Environmental Impact, *Geolog. Surv. Circ.* 645, 1971.

13 Federal Aviation Administration: Procedures for Environmental Impact Statement Preparation, FAA Order 1050.1A, Washington, D.C., June 19, 1973.
14 Oregon Highway Department, personal communication, July 1973.
15 Chase, G. B.: Matrix Analyses in Environmental Impact Assessment, paper presented at the Engineering Foundation Conference on Preparing Environmental Impact Statements, Henniker, N.H., July 29–Aug. 3, 1973.
16 Moore, J. L., et al.: A Methodology for Evaluating Manufacturing Environmental Impact Statements for Delaware's Coastal Zone, app. D, report prepared by Battelle-Columbus, for the State of Delaware, June 15, 1973.
17 Sorensen, J. C.: "A Framework for Identification and Control of Resource Degradation and Conflict in the Multiple Use of the Coastal Zone," University of California, Berkeley, June 1971.
18 Economic Development Administration: "Final Environmental Statement, Fruit/Church Industrial Park, Fresno, California," Washington, D.C., Feb. 1973.
19 A. D. Little, Inc.: "Transportation and Environment: Synthesis for Action: Impact of National Environmental Policy Act of 1969 on the Department of Transportation," 3 vols., prepared for Office of the Secretary, Department of Transportation, July 1971.
20 Federal Power Commission: Implementation of the National Environmental Policy Act of 1969, Order 485, Order Amending Part 2 of the General Rules to Provide Guidelines for the Preparation of Applicants' Environmental Reports Pursuant to Order 415-C, Washington, D.C., June 7, 1973.
21 Lee, E. Y. S., et al.: "Environmental Impact Computer System," Tech. Rept. E-37, Construction Engineering Research Laboratory, U.S. Army, Champaign, Ill., Sept. 1974.
22 Jain, R. K., et al.: "Environmental Impact Assessment Study for Army Military Programs," Tech. Rept. D-13, Construction Engineering Research Laboratory, U.S. Army, Champaign, Ill., Dec. 1973.
23 U.S. Atomic Energy Commission: "Preparation of Environmental Reports for Nuclear Power Plants," pp. 4.2-15–4.2-57, Regulatory Guide 4.2, Washington, D.C., Mar. 2, 1973.
24 Adkins, W. G., and D. Burke, Jr.: "Social, Economic and Environmental Factors in Highway Decision Making," Res. Rept. 148-4, Texas Transportation Institute, Texas A&M University, College Station, Nov. 1974.
25 Soil Conservation Service: "Environmental Assessment Procedure," U.S. Department of Agriculture, Washington, D.C., May 1974.
26 Corps of Engineers: Preparation and Coordination of Environmental Statements, app. C, Regulation 1105-2-507, Department of the Army, Washington, D.C., Apr. 15, 1974.
27 Canter, L. W., P. G. Risser, and L. G. Hill: "Effects Assessment of Alternate Navigation Routes from Tulsa, Oklahoma to Vicinity of Wichita, Kansas," University of Oklahoma, Norman, June 1974.

Chapter 11

Public Participation in Environmental Decison Making

The basic purpose of public participation is to promote productive use of inputs and perceptions from private citizens and public interest groups in order to improve the quality of environmental decision making. These citizen-oriented activities are variously referred to as citizen participation, public participation, public involvement, and citizen involvement. Interest groups include those representative of industry, development, conservation, and preservation. Public participation is required in environmental decision making by the CEQ (1). This chapter focuses on basic definitions and concepts regarding public participation, identification of "publics," conceptual models for public participation in environmental decision making, conduction of a public participation program, and incorporation of public participation findings in EISs.

BASIC DEFINITIONS AND CONCEPTS

Public participation can be defined as a continuous, two-way communication process, which involves promoting full public understanding of the processes and mechanisms through which environmental problems and needs are investigated and solved by the responsible agency; keeping the public fully informed about the status and progress of studies and findings and implications of plan formulation and evaluation activities; and actively soliciting from all concerned citizens their opinions and perceptions of objectives and needs and their preferences regarding resource use and alternative development or management strategies and any other information and assistance relative to plan formulation and evaluation (2).

In essence, public participation involves both information feedforward and feedback (3, 4). Feedforward is the process whereby information is communicated from public officials to citizens concerning public policy. Feedback is the communication of information from citizens to public officials regarding public policy. Feedback information should be useful to decision makers in making timing and content decisions.

One of the essential features of a public participation program in environmental assessment is that it should provide information in an appropriate form and timely manner. The program should be planned with clearly defined objectives and should incorporate several public participation techniques. There should be no restrictions regarding the individuals and organizations participating in hearings or other public programs. A good public participation program does not occur by accident. It requires an analysis of who will receive the information to be presented, careful preparation of informational materials, and examination of the responses received.

ADVANTAGES AND DISADVANTAGES OF PUBLIC PARTICIPATION

There are advantages and disadvantages associated with public participation in environmental decision making. Benefits accrue when affected persons likely to be unrepresented in environmental assessment processes are provided an opportunity to present their views. Members of the public may provide useful information to decision makers, especially when some values or factors cannot be easily quantified. This process helps to enhance public confidence in the agency and the decision making because citizens can clearly see that all issues have been fully and carefully considered. Public participation can also serve as a safety valve by providing a forum for pent-up feelings.

An additional benefit of public participation is that the agency, by constructing a record of decision making (draft statement, review, and final statement), provides for both judicial and public examination of the factors and considerations in the decision-making process. Thus an added accountability is placed on political and administrative decision makers since the process is open to public view. Openness exerts pressure on administrators to adhere to required procedures in decision making. Finally, through public participation the agency is forced to be responsive to issues beyond those of the immediate project.

Disadvantages, or costs, of public participation include the potential for confusion of the issues since many new perspectives may be introduced. It is possible to receive erroneous information that results from the lack of knowledge on the part of the participants. Additional disadvantages include uncertainty of the results of the process of public participation, as well as potential project delay and increased project costs. A properly planned public participation program need not represent a major funding item, and it need not cause an extensive period of delay in the process of decision making.

One of the problems relating to citizen participation, and to properly interpreting and responding to it, is the tendency for citizens to lose interest during the long development period for the project. Citizens other than those with an intense personal interest are not likely to maintain active participation for long periods.

OBJECTIVES OF PUBLIC PARTICIPATION

There are several objectives of public participation in the environmental impact assessment process, with one or more objectives being relevant to each time phase in the project development. Six general objectives are as follows:

1 Information, education, and liaison.
2 Identification of problems, needs, and important values.
3 Idea generation and problem solving.
4 Reaction and feedback on proposals.
5 Evaluation of alternatives.
6 Conflict resolution consensus.

The first of these objectives is directed toward education of the citizenry on EISs, their purpose, and the process of citizen participation. In addition, this objective includes dissemination of information on the study progress and findings, as well as data on potential environmental impacts.

Identification of problems, needs, and important values is related to the determination of the environmental resources important to various segments of the public in an area. In addition, this objective is focused on defining areas of environmental problems and needs and the relation of potential solutions being addressed in the project study.

The third objective is directed toward identification of alternatives that may not have been considered in normal planning processes. In addition to specific alternatives for identified needs, it is possible also to enumerate mitigating measures for various alternatives so as to minimize adverse environmental effects.

The fourth objective attempts to probe public perceptions of the actions and resource interrelations. In addition, this objective can be used to assess significance of various types of impacts. The alternatives evaluation objective is closely related to reaction and feedback on proposals. In the process of evaluation of alternatives, valuable information can be received about the significance of unquantified and quantified environmental amenities. Public reaction to value trade offs in the process of selection can also be assessed.

The final objective is related to resolving conflicts that exist over the proposed action. This objective may involve mediation of differences among various interest groups, development of mechanisms for environmental costs compensation, and effort directed toward arriving at a consensus opinion on a preferred action.

Successful accomplishment of this objective can avoid unnecessary and costly litigation.

An indication of how these objectives fit with the overall environmental impact assessment process is shown in Fig. 11-1 and, in a more general sense, in Fig. 11-2. One of the key aspects, which is noted in Fig. 11-1, is that the publics are actually involved in all of the basic planning activities associated with the development of a recommended plan of action. This suggests that public participation should be an iterative process in order to be most effective. Figure 11-2 identifies the broad activities in impact assessment relative to the six objectives enumerated above. An X denotes that the public participation objective should be a functional part of the impact assessment activity.

IDENTIFICATION OF VARIOUS PUBLICS

There are several ways of categorizing various publics that might be involved in a public participation for environmental impact assessment. One group of publics consists of four separate categories:

1 Persons who are immediately affected by the project and live in the vicinity of the project.

2 Ecologists ranging from preservationists to those who want to ensure that development is as effectively integrated into the needs of the environment as possible. Persons in this group are willing to incur substantial financial costs for environmental protection.

3 Business and commercial developers who would benefit from initiation of the proposed action.

4 The part of the general public who enjoy a high standard of living and who do not want to sacrifice this standard in order to preserve wilderness or scenic areas or have pollution-free air and water.

The Corps of Engineers has defined the following publics in conjunction with water resources development projects (2):

1 Individual citizens, including the general public and key individuals who do not express their preferences through, or participate in, any groups or organizations.

2 Sporting groups.

3 Conservation/environmental groups.

4 Farm organizations.

5 Property owners and users, representing those persons who will be or might be displaced by any alternative under study.

6 Business and industrial groups, including Chambers of Commerce and selected trade and industrial associations.

7 Professional groups and organizations, such as the American Institute of Planners, American Society of Civil Engineers, and others.

8 Educational institutions, including universities, high schools, and vocational schools. General participation is by a few key faculty members and students or student groups and organizations.

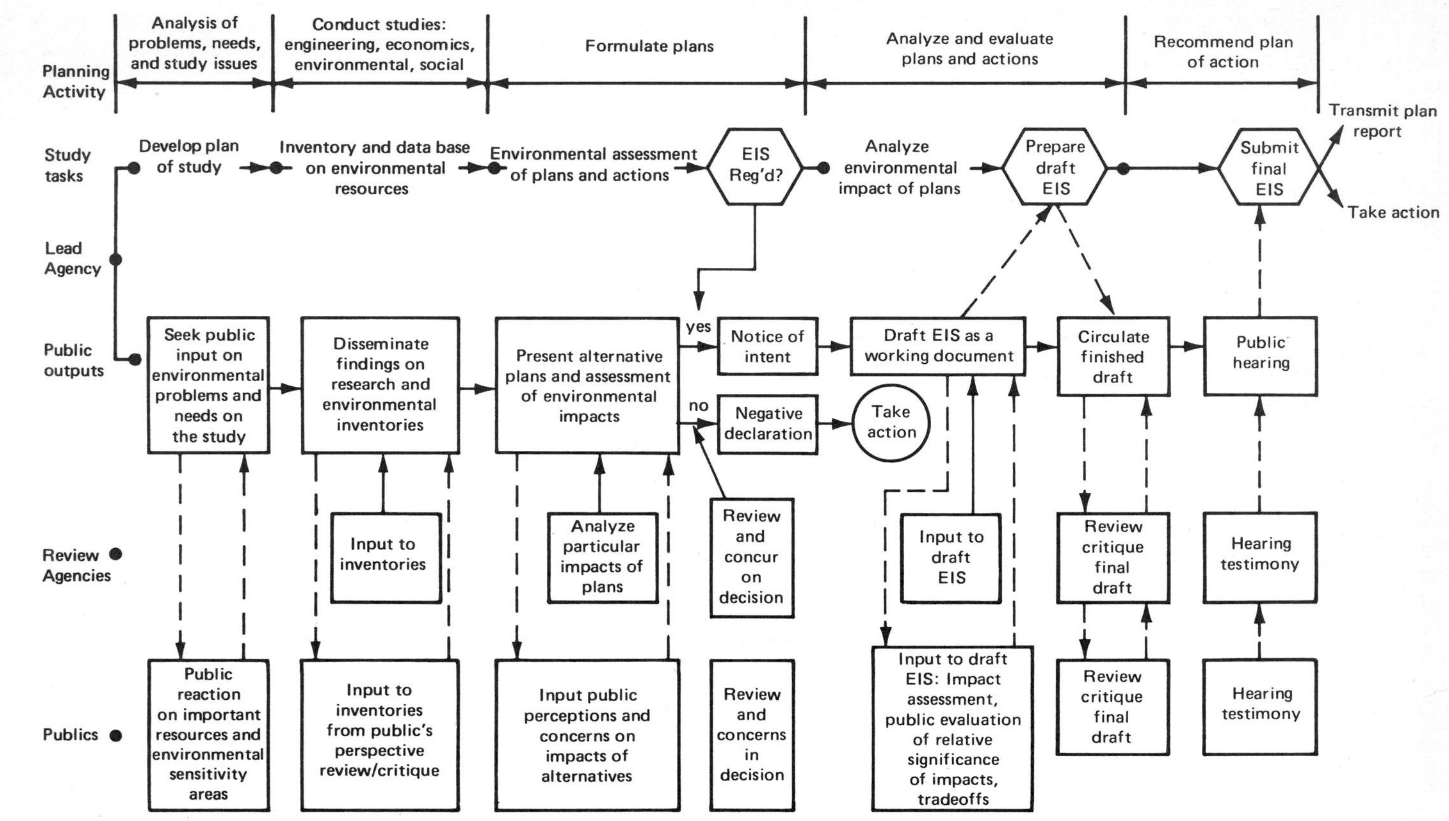

Figure 11-1 Public participation in environmental impact assessment.

Objectives	Impact assessment activities				
	Plan of study for environmental problems and issues	Environmental resource inventory (environmental objectives)	Formulate plans and assess environmental impacts	Analysis and evaluation of plans and impacts (draft EIS)	Select course of action and prepare final EIS
Information, education, liaison	X	X	X	X	X
Identification of problems, needs, and values	X	X	X	X	
Idea generation and problem solving			X		X
Reaction and feedback		X	X	X	X
Evaluation of alternatives				X	X
Conflict resolution consensus				X	X

Figure 11-2 Public participation objectives for impact assessment activities.

Communication characteristics				Impact assessment objectives					
Level of public contact achieved	Ability to handle specific interest	Degree of two-way commu-nication	Public participation techniques	Inform/ educate	Identify problems/ values	Get ideas/ solve problems	Feedback	Evaluate	Resolve conflict/ consensus
M	L	L	Public hearings		X		X		
M	L	M	Public meetings	X	X		X		
L	M	H	Informal small group meetings	X	X	X	X	X	X
M	L	M	General public information meetings	X					
L	M	M	Presentations to community organization	X	X		X		
L	H	H	Information coordination seminars	X			X		
L	M	L	Operating field offices		X	X	X	X	
L	H	H	Local planning visits		X		X	X	
L	H	L	Planning brochures and workbooks	X		X	X	X	
M	M	L	Information brochures and pamphlets	X					
L	H	H	Field trips and site visits	X	X				
H	L	M	Public displays	X		X	X		
M	L	M	Model demonstration projects	X			X	X	X
H	L	L	Material for mass media	X					
L	H	M	Response to public inquiries	X					
H	L	L	Press releases inviting comments	X			X		
L	H	L	Letter requests for comments			X	X		
L	H	H	Workshops		X	X	X	X	X
L	H	H	Charettes			X		X	X
L	H	H	Advisory committees		X	X	X	X	
L	H	H	Task forces		X	X		X	
L	H	H	Employment of community residents		X	X			X
L	H	H	Community interest advocates			X		X	X
L	H	H	Ombudsman or representative		X	X	X	X	X

Figure 11-3 Capabilities of public participation techniques. L, low; M, medium; H, high.

9 Service clubs and civic organizations, including service clubs in a community such as Rotary Club, Lions Club, League of Women Voters, and others.

10 Labor unions.

11 State and local governmental agencies, including planning commissions, councils of government, and individual agencies.

12 State and local elected officials.

13 Federal agencies.

14 Other groups and organizations, possibly including various urban groups, economic opportunity groups, political clubs and associations, minority groups, religious groups and organizations, and many others.

15 Media, including the staff of newspapers, radio, television, and various trade media.

The primary purpose of considering the various publics that may be involved in the environmental impact assessment process is to aid in focusing objectives for a program, as well as in interpreting the results of the application of various public participation techniques.

TECHNIQUES OF PUBLIC PARTICIPATION

There are many techniques that can be utilized to satisfy the feedforward and feedback features of a public participation program. The most traditional public participation technique is the public hearing, which is a formal meeting for which written statements are received and a transcript is kept. The public hearing is generally not an appropriate forum for public participation in the environmental impact assessment process. Other available techniques include informal public information sessions as well as project workshops. Figure 11-3 summarizes the capabilities of a variety of public participation techniques. These capabilities are expressed in terms of communication characteristics and of the various objectives for public participation in environmental impact assessment.

A public participation program was recently held in conjunction with a water resources study in the Mid-Arkansas River Basin located in northern Oklahoma and south central Kansas (5). This program consisted of 7 public information sessions and 13 specific project workshops. These 20 different meetings were held in various cities and towns within the basin. The meetings were considered to be successful based on the following considerations:

1 Approximately 1,600 persons attended the 20 meetings.

2 Over 40 percent of the attendees came from distances of more than 10 mi from the meeting cities; thus persons were attracted from all portions of the basin.

3 Question and answer periods at each meeting were characterized by active participation by many attendees.

4 Opinion questionnaires on various aspects of the study were received from over 70 percent of the attendees.

5 Analysis of the questionnaire results indicated that the majority of the attendees considered the meetings to be very beneficial.

One of the key aspects of these 20 meetings was that local sponsors were solicited for each meeting, with the sponsors efforts directed toward coordinating meeting time and place and promoting attendance. Each meeting consisted of a slide presentation and an informal question and answer session in which every effort was made to provide information of interest to the public. In addition, materials were handed out to each attendee.

PRACTICAL SUGGESTIONS FOR PUBLIC PARTICIPATION

The following list of items represents some very practical ideas and suggestions that can be useful in organizing a public participation program:

1 Coordinate with various federal, state, and local agencies that have interests and responsibilities in the same geographic or technical areas of the study. Develop formal agreements or informal relationships.

2 Develop list of groups and citizens in the geographic area who have previously expressed interests or potential interests in the study.

3 Assemble a newspaper clipping file on project needs and previous history of the project or study.

4 Try to convey the attitude—"What can we do in this study to assist you in your local planning problems? How can we coordinate with other local planning efforts and projects?" This is in contrast to the attitude—"We are here to solve your problems and prepare plans and studies for you."

5 Disseminate study information through the news media (newspaper, radio, and television) and through regular publication of a planning newsletter. The mailing list should encompass all state and federal interests as well as local groups and individuals who have participated in previous meetings or shown interest in the study.

6 Every third or fourth issue of the newsletter should contain a mail-in coupon for persons wanting to continue to receive the newsletter. Each issue should contain a coupon for suggestions of other persons or groups that should receive the newsletter.

Some practical suggestions for actually conducting a public meeting are as follows:

1 Keep data presentations simple. The purpose of data presentations is to inform, and not to confuse or disillusion.

2 An outline of any project or alternative should include a discussion of location, features, benefits and costs, and beneficial and detrimental environmental impacts.

3 Use simple visual aids. Slides of the actual area prove very beneficial.

4 Discuss project timing (prior authorizations and resolutions) previous to the meeting and the anticipated timing of future required steps.

5 Discuss general concept, features and requirements of benefit–cost ratio, or other relevant economic analysis.

6 Only those persons who can speak on general matters, comingled with engineering expertise, should be considered for meeting with the public. It is well known that not everybody has the ability to speak, answer questions, and perhaps debate while holding a specific image in front of the public. The ability to speak well is not the only trait on which the selection should be based. A person who can answer questions quickly and confidently from an audience in which the sentiment is mixed or opposed can establish a profound positive relation with the audience.

7 Avoid use of technical jargon or words that may be hard to understand, especially with local groups unaccustomed to engineering and ecological terminology.

8 Be familiar with the area.

9 Be earnest, sincere, and willing to work on problems with individual groups.

INCORPORATION OF RESULTS

The feedback loop from public participation must be used in environmental impact assessment, or the purpose of public participation will not have been fully satisfied. Public participation results can be useful in defining project need, describing unique features of the environmental setting, and identifying environmental impacts. Results can also be utilized by the proposing agency in assigning significance (importance) values to both environmental items and impacts. Finally, selection of the most desirable alternative for meeting project need can be aided by public participation.

Two levels are suggested for incorporation of public participation information in an environmental impact assessment report. First, all public meetings and the entirety of the planned and accomplished public participation program should be summarized. Any information relative to the objectives outlined earlier and obtained through questionnaire surveys or other public participation techniques should be summarized. Second, public preference can be used in selection of the alternative to become the proposed action for meeting a project need. One means of preference utilization is through use of a weighted-ranking technique developed for marketing purposes (6). This technique will be illustrated by the following sample problem.

Selecting among X Alternatives, Including the No-Action Alternative; Steps in Weighted-Ranking Technique

The first step is to select factors for consideration in the decision process and assemble data for each alternative. Public participation can be used to select the factors that are of importance in the decision process. As an example, the following factors will be employed: deterioration of water quality, loss of habitat, public preference, decrease in flooding, and increase in recreational potential. The data utilized for making the decision can be either absolute or relative, and the method is still valid. Table 11-1 contains the data that are used in the example.

The next step is to develop factor importance coefficients for each factor. Again, public participation could be incorporated at this point. The basic approach involves considering every factor versus every other factor and assigning a value of one to the factor considered to be the most important and zero to the one considered to be the least important. If a decision cannot be made regarding relative importance, it is possible to assign to both factors under consideration an equal

Table 11-1 Data for Weighted-Ranking Technique Sample Problem

Factor	Alternative		
	A[a]	B	C
Deterioration of water quality	None	Most	Some
Loss of habitat	None	5,000 acres	3,000 acres
Public preference	Oppose	Favor	Accept
Decrease in flooding	None	95%	95%
Increase in recreational potential	None	500 acres	100 acres

[a]No-action alternative.

Table 11-2 Factor Importance Coefficients for Sample Problem

Factor	Assigned values										Sum	FIC[a]
Deterioration of water quality	1	0	0.5	1							2.5	0.25
Loss of habitat	0				0	1	0				1.0	0.10
Public preference		1			1			1	1		4.0	0.40
Decrease in flooding			0.5			0		0		1	1.5	0.15
Increase in recreational potential				0			1		0	0	1	0.10
											10	1.00

[a]Factor importance coefficient.

Table 11-3 Alternative Choice Coefficients for Sample Problem

Factor	Alternative	Assigned values			Sum	ACC[a]
Deterioration of water quality	A	1	1		2	0.67
	B	0		0	0	0
	C		0	1	1	0.33
Loss of habitat	A	1	1		2	0.67
	B	0		0	0	0
	C		0	1	1	0.33
Public preference	A	0	0		0	0
	B	1		1	2	0.67
	C		1	0	1	0.33
Decrease in flooding	A	0	0		0	0
	B	1		0.5	1.5	0.50
	C		1	0.5	1.5	0.50
Increase in recreational potential	A	0	0		0	0
	B	1		1	2	0.67
	C		1	0	1	0.33

[a]Alternative choice coefficient.

Table 11-4 Summary of FIC and ACC Values from Sample Problem

		ACC		
Factor	FIC	A	B	C
Deterioration of water quality	0.25	0.67	0	0.33
Loss of habitat	0.10	0.67	0	0.33
Public preference	0.40	0	0.67	0.33
Decrease in flooding	0.15	0	0.50	0.50
Increase in recreational potential	0.10	0	0.67	0.33

value of 0.5. Table 11-2 contains the factor importance coefficients developed for the sample program. The sum of the values in Table 11-2 should equal $N(N-1)/2$, where N is the number of factors incorporated in the decision process. The sum of the factor importance coefficients should equal 1.00. It is noted that this step is the key step involving judgment in selecting among alternatives.

The third step is to develop alternative choice coefficients for each alternative relative to each factor. Again, this involves considering each alternative for each factor, and, by considering two alternatives at a time, assigning a value of one for the most desirable and zero for the least desirable. Alternatives that have equal characteristics for a given factor are assigned values of 0.5 each. Table 11-3 summarizes the alternative choice coefficients for each alternative and each factor.

The final step in the process is to multiply the factor importance coefficient for each factor by the alternative choice coefficient for each factor. A summary of the values from Tables 11-2 and 11-3 is presented in Table 11-4. Table 11-5 shows the final coefficient matrix. The alternative of choice in this process is the one with the greatest sum value, and in the case of this sample problem this is alternative B. This example incorporates public opinion or preference for the alternatives under consideration for meeting a given need.

SUMMARY

Public participation is becoming an increasingly important component of environmental impact assessment. At a minimum public participation should consist of

Table 11-5 Final Coefficient Matrix for Sample Problem

	FIC × ACC		
Factor	A	B	C
Deterioration of water quality	0.17	0	0.08
Loss of habitat	0.07	0	0.03
Public preference	0	0.27	0.13
Decrease in flooding	0	0.07	0.08
Increase in recreational potential	0	0.07	0.03
	0.24	0.41	0.35

solicitation of basic information and reaction to various alternatives under consideration. There are a variety of techniques for accomplishing public participation, and the results of a well-planned program can be used to select the alternative to become the proposed action for meeting a project need.

SELECTED REFERENCES

1 Council on Environmental Quality: Preparation of Environmental Impact Statements: Guidelines, *Fed. Reg.*, vol. 38, no. 147, pp. 20550–20562, Aug. 1, 1973.
2 Corps of Engineers: Public Participation in Water Resources Planning, EC-1165-2-100, Washington, D.C., May 1971.
3 Bishop, A. B.: "Public Participation in Water Resources Planning," pp. 60–85, Publ. IWR 70-7, Institute for Water Resources, Washington, D.C., Dec. 1970.
4 Bishop, A. B.: Public Participation in Environmental Impact Assessment, paper presented at Engineering Foundation Conference on Preparation of Environmental Impact Statements, New England College, Henniker, N.H., July 29–Aug. 3, 1973.
5 Canter, L. W.: "Public Participation in the Mid-Arkansas River Basin Study," 2 vols., report to Tulsa District of Corps of Engineers, Tulsa, Okla., Nov. 1973.
6 Dean, B. V., and M. J. Nishry: Scoring and Profitability Models for Evaluating and Selecting Engineering Projects, *J. Operations Res. Soc. Am.*, vol. 13, no. 4, pp. 550–569, July–Aug. 1965.

Chapter 12

Practical Considerations in Writing Impact Statements

Several comparative analyses of EISs have focused on deficiencies in writing style. Common criticisms include lack of continuity in writing style, inadequate description of the environmental setting, inadequate documentation and referencing, minimal use of visual display techniques for the environmental setting as well as the impacts of alternatives, and apparent bias by the writers and proposing agency. This chapter is oriented to some practical considerations and suggestions for writing EISs and alleviating common criticisms.

USE OF VISUAL DISPLAY MATERIALS

Liberal usage of maps, photographs, and drawings in EISs can be of tremendous value to reviewers as well as to agency decision makers. Several types of maps are of general value in most EISs. Topographic maps allow for presentation of area features and terrain as well as surface-water drainage characteristics. Land-use and master plan maps are becoming increasingly important, particularly because of secondary effects resulting from implementation of proposed actions. These maps are useful in describing the relationship between the proposed action and existing land-use plans, policies, and controls. Maps can be used to display environmental characteristics such as air and water quality levels, noise levels, plant and animal species distributions, and historical and archeological sites. In addition, maps can be useful for depicting socioeconomic characteristics such as population distribution, housing, income levels, and transportation systems.

Incorporation of photographs illustrating the environmental setting for the proposed action is beneficial. It must be realized by preparers of EISs that many agency decision makers will be unfamiliar with the project area. In addition, statement reviewers from other governmental agencies as well as the private sector will probably have little prior knowledge of the setting. The adage that one picture is worth a thousand words is definitely true. Photographs are also useful in depicting conditions of the environment that need improvement. For example, if a reservoir project is proposed for flood control, photographs depicting recent flood damages are useful for establishing project need. Need-related photographs can also be useful for many other impact statements, including those associated with community and industrial development.

The use of simplified architectural/engineering drawings or conceptual drawings to depict the proposed action upon its completion can aid the reader in conceiving visual impacts of the proposed action on the environment. Drawings of typical units such as housing units or various processes in a waste-water treatment plant are examples of specific visual materials that enable the nonexpert to develop an understanding of the proposed action and its potential impact on the environment.

STATEMENT DOCUMENTATION

Many EISs are characterized by lack of adequate documentation for sources of information as well as for the scientific approaches utilized. It is important to indicate the various environmental factors that were considered and deemed to be not relevant for the impact statement. For example, if it is determined that an area has no rare or endangered plant or animal species, then this should be stated. In like manner, the absence of historic and archeological sites from a project area needs to be verified and stated. Lack of indication that attention has been given to items such as rare and endangered plant or animal species, as well as historic and archeological sites, generally leads the reviewer to conclude that these items were not considered.

Proper scientific referencing should be included in EISs. If many remarks contained in EISs are not supported by scientific referencing, even though these remarks may be true, the reviewer tends to conclude that adequate study has not been conducted. Therefore, lack of referencing often forms the basis for criticisms of EISs. There are many referencing techniques that can be used, including footnotes, a numbered list of selected references, or a system such as the one advocated by the American Institute of Biological Sciences. It may be desirable to identify general documents as well as specific sources of information. In addition, a master file of reference materials used in conjunction with a given EIS writing effort may be assembled and maintained.

Detailed information on the environmental setting should appropriately be included in appendices, with only summary tables included in the text. The CEQ guidelines encourage liberal use of appendices in EISs (1). Detailed calculations associated with prediction and assessment of the impact of alternatives should also be included in appendices, with summary tables provided in the text.

Finally, many EISs could be improved through use of a consistent numbering system for sections, tables, and figures, which aids in continuity and review. It also provides a means for cross-checking materials.

GENERAL WRITING SUGGESTIONS

Many writing suggestions for EISs are appropriate for any technical report. Some relevant suggestions are as follows (2):

1 Do not use cliches or catch words. They may be understandable to the writers but unfamiliar to reviewers and agency decision makers.

2 Every effort should be made to make EISs succinct and clear, with minimal use of written texts and liberal use of visual display methods.

3 Do not use vague generalities in the EIS. Examples include "developer will exercise supervision and control to prevent siltation," "special consideration will be given to providing in-plant controls," and "construction noise will be minimized." While statements such as these may be true and applicable, it is better to identify exact methods of implementation and enforcement if they are known.

4 Try to avoid creating a credibility gap as a result of too many technical errors and mistakes in the EIS. In any scientific writing, and particularly for documents prepared under very stringent time deadlines, many errors and omissions may occur. There is no foolproof way of eliminating this problem. One approach is to subject the draft statement to internal review by persons unfamiliar with details of the project.

5 Both pro and con information with regard to a proposed action should be presented. Value judgments indicating that the writer or writers are attempting to justify the project are easily identifiable in an EIS.

6 Efforts should be made to provide as complete a document as possible within the time frame and monetary constraints associated with a given EIS writing effort. Even though all potential impacts of a project may not be analyzed in the same detail, it is useful to identify all known items of potential impact even though they may be dealt with only in a cursory manner.

7 Care should be taken to prevent plagiarism from existing documents. This particular concern can be met if a proper referencing and documentation system is used.

8 Attempts should be made to provide a document that has continuity from one section to another. Poor organization of the contents of an EIS will lead to review criticisms, even though appropriate substantive materials are present.

9 Since EISs are to be written by interdisciplinary teams, it is not uncommon to find differences in writing style within an impact statement. This can result in contradictory information and statements in various sections of an impact statement. One of the most important steps in EIS preparation is the internal review of the completed draft report. Conflicts of professional opinion or scientific information in substantive areas should not be omitted, but rather they should be included to provide decision makers and reviewers more complete information and

to enable better realization of the consequences of initiating a particular action. Conflicting statements based on misinformation or errors in interpretation should be eliminated from EISs.

10 One of the criticisms of EISs is that the information is so general it is of little relevance to an analysis of environmental impacts. It is desirable to be as specific as possible, within the bounds of data availability. If general information or information from other locales is used, it should be clearly identified, with appropriate explanations as to why this information is considered to be relevant.

SUMMARY

This chapter has identified selected writing problems associated with impact statement preparation. There is nothing unique about an EIS that would tend to exempt it from the principles of technical report writing. This chapter contains practical suggestions and ideas for improving the technical and nontechnical content of EISs.

SELECTED REFERENCES

1 Council on Environmental Quality: Preparation of Environmental Impact Statements: Guidelines, *Fed. Reg.*, vol. 38, no. 147, pp. 20550–20562, Aug. 1, 1973.
2 Garing, Taylor, and Associates: "A Handbook Approach to the Environmental Impact Report," 2d ed., pp. 19–46, Arroyo Grande, Calif., 1974.

Chapter 13

Future of Environmental Impact Assessment

The environmental impact assessment process is less than a decade old and, as such, represents a developing technology. The future of this process can be described according to the five steps defined in Chapter 2. These steps include basics, description of the environmental setting, impact prediction and assessment, selection of the proposed action, and preparation of the EIS. This chapter is addressed to future trends associated with these five steps and also to research needs in environmental impact analysis.

BASICS

Basic components of the environmental impact assessment process include guidelines, knowledge of the study, and an interdisciplinary team. Guidelines are being constantly updated by the CEQ and all federal agencies. Legal actions regarding agency compliance with the spirit and intent of the NEPA will continue, with court decisions eventually becoming components of guidelines. As more specific guidelines are issued, more description will be provided of actions needing EISs and actions requiring negative declarations.

The continuing evolvement of the environmental impact assessment process, when considered along with passage of numerous state and local environmental policy acts, leads to the conclusion that the number of EISs will continue to increase. Policy decisions by some agencies have caused even greater emphasis on EIS preparation. For example, the 1974 Housing and Community Development Act of the Department of Housing and Urban Development requires local preparation of EISs for block grant programs. State and local environmental legislation helps to direct attention to the preparation of EISs for private developments as well as for public projects.

Knowledge of project need and all possible alternatives for meeting that need is increasingly important as agency planners and project developers become more responsive to including environmental concerns in decision making. Persons at every level of agency employment are gaining knowledge of the environmental assessment process, and thus they can provide more usable information to interdisciplinary teams working on EISs. In response to requirements for an interdisciplinary approach to environmental decision making, many agencies and organizations have employed persons whose disciplines have previously been excluded from agency employment practices. An interdisciplinary approach for project decision making will be increasingly used throughout the United States.

DESCRIPTION OF ENVIRONMENTAL SETTING

One of the first steps in producing an EIS is to prepare an adequate description of the physical-chemical, biological, cultural, and socioeconomic environments. Base-line information for describing each of the necessary components is not always available. Table 13-1 summarizes some types of base-line information that are often found to be inadequate with regard to environmental statement preparation (1). As time progresses, data inadequacies will be overcome as a result of manifold information gathering by public and private organizations.

An area of future importance in conjunction with describing the environmental setting is associated with information storage and retrieval systems. Modest efforts are currently being made to describe the environmental setting through the use of information storage and retrieval techniques, and in some instances, computer-

Table 13-1 Impact Areas with Often Inadequate Base-line Data[a]

Type of data	Remarks
Geology and soils	Data gathered for specific project types
Groundwater	Interpretation often difficult
Natural plant communities	Maps often inadequate
Topoclimate and microclimate	Data often inadequate or conflicting
Census data for low-density or rapidly growing areas	Often out of date or misleading
Property records	Not kept in a form for geographical reference
Standard land-use classification	Has little relation to environmental carrying capacity

[a]After (1).

mapping approaches are utilized. Another potential future area in environmental impact assessment is associated with remote sensing for the purpose of describing large-scale environmental features.

IMPACT PREDICTION AND ASSESSMENT

Impact prediction techniques have been developed for describing many project consequences on the physical-chemical environment. Less quantitative techniques are available for the other broad categories of environmental concern. Technology is being developed to provide more scientific and quantitative approaches for impact prediction for all environmental categories. Most of the predictive techniques employed at the current time are directed toward the primary, or direct, effects of alternatives. An area of increasing importance in environmental impact assessment is the prediction of secondary effects of project development and implementation. Technology is also being developed for predictions of changes resulting from these occurrences.

With regard to assessment or interpretation of predicted impacts, standards are currently available for many parameters within the physical-chemical and socio-economic environments. Less quantitative standards are available for other areas of concern. As time progresses, more quantitative and qualitative standards will be developed, thus enabling interdisciplinary teams to interpret more properly the consequences of project implementation.

SELECTION OF PROPOSED ACTION

Most agencies are currently incorporating environmental factors early in the planning process for meeting identified needs, implying that environmental impacts are considered for each alternative along with technical and economic factors. In excess of 50 environmental impact assessment methodologies have been developed, with these methodologies being characterized primarily as either checklists or matrices. Numerous research efforts are in progress to develop additional methodologies that will provide more complete information on impact identification, prediction, and interpretation.

One of the key aspects of current guidelines is the incorporation of public participation throughout the process leading to a decision regarding a proposed action. Public participation techniques will continue to evolve, with the primary thrust being away from formalized public hearings and toward the use of methods for accomplishing a process of information feedforward and feedback.

PREPARATION OF ENVIRONMENTAL IMPACT STATEMENT

EISs have varied in length from fairly short versions in the first few years of preparation to rather extensive documents at the current time. Future EISs will become more simplified, with supporting information and calculations provided as separate appendices or attachments. Simplified statements will be possible as acceptable methodologies for impact analysis become better developed. Another area of current and future focus in impact statement preparation is more appropriate documentation of information. Many of the questions generated from the review of draft EISs have been associated with lack of referencing of information and inadequate quantification of identified impacts. These two areas should be improved in future statements.

RESEARCH NEEDS

The process of environmental impact assessment has led to the delineation of research needs associated with the environment in project decision making. Research needs related to individual impacts are as follows (1):

1 Compilation of a complete and accurate environmental basic data bank—Physiographically unique areas should be completely characterized and data stored so that they are readily available. This applies particularly to the following types of information that are often deficient: geology and soils, groundwater, natural plant communities, climate, current census and school data for low-density and rapidly growing areas, property records, and realistic land-use classifications.

2 Refinement of techniques for measuring impacts on biological communities—Sophisticated systems models for describing ecosystems should be researched in order to accurately predict future conditions, as well as present conditions.

3 Refinement of techniques for measuring impacts on social well-being—Techniques already existing in sociology and psychology should be modified and new techniques researched and developed to assess comprehensively the impacts of specific projects on the surrounding community.

4 More complete consideration of irreversible impacts—In addition to measuring the magnitude of an environmental change, research should develop improved characterization of changes in terms of both reversibility and probability of occurrence. Extension of risk analysis often used in investment decisions would seem to be a particularly promising approach.

5 Better methods for predicting cumulative impacts—Quantification techniques are needed for predicting long-term changes induced by a project and for assessing the cumulative impacts of new land uses that may result.

Research needs for overall assessment of comprehensive single project impacts include the following (1):

1 Objective assessment of the relative desirability of impacts—To avoid the controversy often generated by EISs, methods are needed for objectively determining whether particular impacts are beneficial or detrimental to a community by an explicit understanding and consideration of the community's value structure.

2 Refinement of materials balance and input-output techniques—These and other sophisticated economic models should be refined to be more applicable to the impacts of smaller-scale single projects.

3 Refinement of environmental impact matrices—Further research should be

performed to systematically assign impact weighting factors, to meaningfully identify areas of impact and relevant project components, and to separate secondary from primary impacts.

Research needs for formulation and comparison of alternative projects (1) include

1 More systematic techniques for determining the best use of project sites—The environmental carrying capacity concept should be refined for individual projects, as well as for multiple projects. One promising approach is the map-overlay technique popularized by McHarg; further work is needed to accurately identify and weigh the various environmental indicators to establish air-, water-, and land-carrying capacities.

2 Refinement of cost-benefit analysis—Continued effort is needed to quantify the intangible and long-term costs and benefits of a particular project and to estimate meaningfully discount rates and opportunity costs.

3 Better evaluation of economic-environmental trade offs—Better methods are needed for computing scores for alternative projects to realistically weight both the economic and environmental consequences of particular projects.

It has been proposed that the current EIS preparation system should be considered as an interim measure, to be eventually replaced by a more streamlined procedure closely tied to integrated land-use planning (2). Although this proposal may never be adopted, the concept of jointly considering land use and environmental quality is valid. Research needs associated with land use-environmental quality relationships are as follows (1):

1 Environmental sciences priorities for land-use research.
- **a** Air and water pollution impact of various land-use problems.
- **b** Determination of biological and physical factors that make certain ecological areas inherently "fragile."
- **c** Alternative rates of energy consumption resulting from various settlement patterns.
- **d** Ecological aspects of the preservation of inland and coastal wetlands.
- **e** Environmental impacts of specific large-scale interbasin transfers of water.
- **f** Maintenance costs of synthetic biological communities.
- **g** Criteria for the location of land uses to minimize human stress caused by noise.

2 Data requirements for land-use research.
- **a** Comprehensive assessment of data deficiencies specifically applied to research needs in environmental and land-use planning.

3 Settlement patterns and their relationship to land use.
- **a** Settlement trends in the United States as determinants of land-use effects of alternative settlement patterns and rates on the supply and demand for environmental resources.
- **b** Effects of changes in the supply, demand, and pricing of environmental resources on settlement patterns at national, regional, and local levels.
- **c** Impact of changing recreational and leisure-oriented living patterns on settlement and activity patterns, including the impact of second homes and recreational vehicles.

   d Policy implications of the relationships among existing central cities, regional hinterlands, new communities, and growth centers in logging regions.
   e Policy implications of the development of central cities and of outlying settlements on the lives of minority groups and the poor.

4 Institutional policies and mechanisms for land-use allocation.
   a Alternative incentive and penalty systems to achieve land-use goals.
   b Public participation in land-use planning.
   c Responsiveness of policies to weak and strong social interest groups.
   d Control procedures related to the location of new communities.
   e Public-private arrangements in land ownership, regulations, and use.

SUMMARY

The future of environmental impact assessment will be characterized by manifold activities. More federal guidelines and the passage of additional state and local environmental policy acts will result in the preparation of more EISs. Although more statements will be prepared, it is anticipated that their average length will be reduced because of developing technologies for impact prediction and assessment and methodologies for evaluation of alternatives. Public participation in all phases of environmental decision making will increasingly occur in the future. Finally, current and future research efforts will provide information relevant to many of today's unanswered questions.

SELECTED REFERENCES

1 Stone, R. and A. Herson: Research Needs for Environmental Impact Analysis, *ASCE Ann. Natl. Environ. Eng. Conv., Kansas City, Mo.*, Meeting Preprint 2374, Oct. 21–25, 1974.

2 Warner, M. L., and D. W. Bromley: "Environmental Impact Analyses: A Review of Three Methodologies," Tech. Rept., Wisconsin Water Resources Center, University of Wisconsin, Madison, 1974.

Appendix A

National Environmental Policy Act

Public Law 91-190
91st Congress, S. 1075
January 1, 1970

An Act

To establish a national policy for the environment, to provide for the establishment of a Council on Environmental Quality, and for other purposes.

Be it enacted by the Senate and House of Representatives of the United States of America in Congress assembled, That this Act may be cited as the "National Environmental Policy Act of 1969".

Purpose

Sec. 2. The purposes of this Act are: To declare a national policy which will encourage productive and enjoyable harmony between man and his environment; to promote efforts which will prevent or eliminate damage to the environment and biosphere and stimulate the health and welfare of man; to enrich the understanding of the ecological systems and natural resources important to the Nation; and to establish a Council on Environmental Quality.

TITLE I

Declaration of National Environmental Policy

Sec. 101. (a) The Congress, recognizing the profound impact of man's activity on the interrelations of all components of the natural environment, particularly the profound influences of population growth, high-density urbanization, industrial expansion, resource exploitation, and new and expanding technological advances and recognizing further the critical importance of restoring and maintaining environmental quality to the overall welfare and development of man, declares that it is the continuing policy of the Federal Government, in cooperation with State and local governments, and other concerned public and private organizations, to use all practicable means and measures, including financial and technical assistance, in a manner calculated to foster and promote the general welfare, to create and maintain conditions under which man and nature can exist in productive harmony, and fulfill the social, economic, and other requirements of present and future generations of Americans.

(b) In order to carry out the policy set forth in this Act, it is the continuing responsibility of the Federal Government to use all practicable means, consistent with other essential considerations of national policy, to improve and coordinate Federal plans, functions, programs, and resources to the end that the Nation may—

(1) fulfill the responsibilities of each generation as trustee of the environment for succeeding generations;

(2) assure for all Americans safe, healthful, productive, and esthetically and culturally pleasing surroundings;

(3) attain the widest range of beneficial uses of the environment without degradation, risk to health or safety, or other undesirable and unintended consequences;

(4) preserve important historic, cultural, and natural aspects of our national heritage, and maintain, wherever possible, an environment which supports diversity and variety of individual choice;

(5) achieve a balance between population and resource use which will permit high standards of living and a wide sharing of life's amenities; and

(6) enhance the quality of renewable resources and approach the maximum attainable recycling of depletable resources.

(c) The Congress recognizes that each person should enjoy a healthful environment and that each person has a responsibility to contribute to the preservation and enhancement of the environment.

Sec. 102. The Congress authorizes and directs that, to the fullest extent possible: (1) the policies, regulations, and public laws of the United States shall be interpreted and administered in accordance with the policies set forth in this Act, and (2) all agencies of the Federal Government shall—

(A) utilize a systematic, interdisciplinary approach which will insure the integrated use of the natural and social sciences and the environmental design arts in planning and in decisionmaking which may have an impact on man's environment;

(B) identify and develop methods and procedures, in consultation with the Council on Environmental Quality established by title II of this Act, which will insure that presently unquantified environmental amenities and values may be given appropriate consideration in decisionmaking along with economic and technical considerations;

(C) include in every recommendation or report on proposals for legislation and other major Federal actions significantly affecting the quality of the human environment, a detailed statement by the responsible official on—

(i) the environmental impact of the proposed action,

(ii) any adverse environmental effects which cannot be avoided should the proposal be implemented,

(iii) alternatives to the proposed action,

(iv) the relationship between local short-term uses of man's environment and the maintenance and enhancement of long-term productivity, and

(v) any irreversible and irretrievable commitments of resources which would be involved in the proposed action should it be implemented.

Prior to making any detailed statement, the responsible Federal official shall consult with and obtain the comments of any Federal agency which has jurisdiction by law or special expertise with respect to any environmental impact involved. Copies of such statement and the comments and views of the appropriate Federal, State, and local agencies, which are authorized to develop and enforce environmental standards, shall be made available to the President, the Council on Environmental Quality and to the public as provided by section 552 of title 5. United States Code, and shall accompany the proposal through the existing agency review processes;

(D) study, develop, and describe appropriate alternatives to recommended courses of action in any proposal which involves unresolved conflicts concerning alternative uses of available resources;

(E) recognize the worldwide and long-range character of environmental problems and, where consistent with the foreign policy of the United States, lend appropriate support to initiatives, resolutions, and programs designed to maximize international cooperation in anticipating and preventing a decline in the quality of mankind's world environment;

(F) make available to States, counties, municipalities, institutions, and individuals, advice and information useful in restoring, maintaining, and enhancing the quality of the environment;

(G) initiate and utilize ecological information in the planning and development of resource-oriented projects; and

(H) assist the Council on Environmental Quality established by title II of this Act.

Sec. 103. All agencies of the Federal Government shall review their present statutory authority, administrative regulations, and current policies and procedures for the purpose of determining whether there are any deficiencies or inconsistencies therein which prohibit full compliance with the purposes and provisions of this Act and shall propose to the President not later than July 1, 1971, such measures as may be necessary to bring their authority and policies into conformity with the intent, purposes, and procedures set forth in this Act.

Sec. 104. Nothing in Section 102 or 103 shall in any way affect the specific statutory obligations of any Federal agency (1) to comply with criteria or standards of environmental quality, (2) to coordinate or consult with any other Federal or State agency, or (3) to act, or refrain from acting contingent upon the recommendations or certification of any other Federal or State agency.

Sec. 105. The policies and goals set forth in this Act are supplementary to those set forth in existing authorizations of Federal agencies.

TITLE II

Council on Environmental Quality

Sec. 201. The President shall transmit to the Congress annually beginning July 1, 1970, an Environmental Quality Report (hereinafter referred to as the "report") which shall set forth (1) the status and condition of the major natural, manmade, or altered environmental classes of the Nation, including, but not limited to, the air, the aquatic, including marine, estuarine, and fresh water, and the terrestrial environment, including, but not limited to, the forest, dryland, wetland, range, urban, suburban, and rural environment; (2) current and foreseeable trends in the quality, management and utilization of such environments and the effects of those trends on the social, economic, and other requirements of the Nation; (3) the adequacy of available natural resources for fulfilling human and economic requirements of the Nation in the light of expected population pressures; (4) a review of the programs and activities (including regulatory activities) of the Federal Government, the State and local governments, and nongovernmental entities or individuals, with particular reference to their effect on the environment and on the conservation, development and utilization of natural resources; and (5) a program for remedying the deficiencies of existing programs and activities, together with recommendations for legislation.

Sec. 202. There is created in the Executive Office of the President a Council on Environmental Quality (hereinafter referred to as the "Council"). The Council shall be composed of three members who shall be appointed by the President to serve at his pleasure, by and with the advice and consent of the Senate. The President shall designate one of the members of the Council to serve as Chairman. Each member shall be a person who, as a result of his training, experience, and attainments, is exceptionally well qualified to analyze and interpret environmental trends and information of all kinds; to appraise programs and activities of the Federal Government in the light of the policy set forth in title I of this Act; to be conscious of and responsive to the scientific, economic, social, esthetic, and cultural needs and interests of the Nation; and to formulate and recommend national policies to promote the improvement of the quality of the environment.

Sec. 203. The Council may employ such officers and employees as may be necessary to carry out its functions under this Act. In addition, the Council may employ and fix the compensation of such experts and consultants as may be necessary for the carrying out of its functions under this Act, in accordance with section 3109 of title 5, United States Code (but without regard to the last sentence thereof).

Sec. 204. It shall be the duty and function of the Council–

(1) to assist and advise the President in the preparation of the Environmental Quality Report required by section 201;

(2) to gather timely and authoritative information concerning the conditions and trends in the quality of the environment both current and prospective, to analyze and interpret such information for the purpose of determining whether such conditions and trends are interfering, or are likely to interfere, with the achievement of the policy set forth in title I of this Act, and to compile and submit to the President studies relating to such conditions and trends;

(3) to review and appraise the various programs and activities of the Federal Government in the light of the policy set forth in title I of this Act for the purpose of determining the extent to which such programs and activities are contributing to the achievement of such policy, and to make recommendations to the President with respect thereto;

(4) to develop and recommend to the President national policies to foster and promote the improvement of environmental quality to meet the conservation, social, economic, health, and other requirements and goals of the Nation;

(5) to conduct investigations, studies, surveys, research, and analyses relating to ecological systems and environmental quality;

(6) to document and define changes in the natural environment, including the plant and animal systems, and to accumulate necessary data and other information for a continuing analysis of these changes or trends and an interpretation of their underlying causes;

(7) to report at least once each year to the President on the state and condition of the environment; and

(8) to make and furnish such studies, reports thereon, and recommendations with respect to matters of policy and legislation as the President may request.

Sec. 205. In exercising its powers, functions, and duties under this Act, the Council shall–

(1) consult with the Citizens' Advisory Committee on Environmental Quality established by Executive Order numbered 11472, dated May 29, 1969, and with such representatives of science, industry, agriculture, labor, conservation organizations, State and local governments and other groups, as it deems advisable; and

(2) utilize, to the fullest extent possible, the services, facilities, and information (including statistical information) of public and private

agencies and organizations, and individuals, in order that duplication of effort and expense may be avoided, thus assuring that the Council's activities will not unnecessarily overlap or conflict with similar activities authorized by law and performed by established agencies.

Sec. 206. Members of the Council shall serve full time and the Chairman of the Council shall be compensated at the rate provided for Level II of the Executive Schedule Pay Rates (5 U.S.C. 5313). The other members of the Council shall be compensated at the rate provided for Level IV or the Executive Schedule Pay Rates (5 U.S.C. 5315).

Sec. 207. There are authorized to be appropriated to carry out the provisions of this Act not to exceed $300,000 for fiscal year 1970, $700,000 for fiscal year 1971, and $1,000,000 for each fiscal year thereafter.

Approved January 1, 1970.

LEGISLATIVE HISTORY

HOUSE REPORTS: No. 91-378, 91-378, pt. 2, accompanying H.R. 12549 (Comm. on Merchant Marine & Fisheries) and 91-765 (Comm. of Conference).
SENATE REPORT No. 91-296 (Comm. on Interior & Insular Affairs).
CONGRESSIONAL RECORD, Vol. 115 (1969): July 10: Considered and passed Senate. Sept. 23: Considered and passed House, amended, in lieu of H.R. 12549. Oct. 8: Senate disagreed to House amendments; agreed to conference. Dec. 20: Senate agreed to conference report. Dec. 22: House agreed to conference report.

Appendix B

Council on Environmental Quality Guidelines

Title 40—Protection of the Environment

CHAPTER V—COUNCIL ON ENVIRONMENTAL QUALITY

PART 1500—PREPARATION OF ENVIRONMENTAL IMPACT STATEMENTS: GUIDELINES

On May 2, 1973, the Council on Environmental Quality published in the *Federal Register*, for public comment, a proposed revision of its guidelines for the preparation of environmental impact statements. Pursuant to the National Environmental Policy Act (P.L. 91-190, 42 U.S.C. 4321 et seq.) and Executive Order 11514 (35 FR 4247) all Federal departments, agencies, and establishments are required to prepare such statements in connection with their proposals for legislation and other major Federal actions significantly affecting the quality of the human environment. The authority for the Council's guidelines is set forth below in §1500.1. The specific policies to be implemented by the guidelines is set forth below in §1500.2.

The Council received numerous comments on its proposed guidelines from environmental groups, Federal, State, and local agencies, industry, and private individuals. Two general themes were presented in the majority of the comments.

From *Federal Register*, vol. 38, no. 147, pp. 20550–20562, Aug. 1, 1973.

First, the Council should increase the opportunity for public involvement in the impact statement process. Second, the Council should provide more detailed guidance on the responsibilities of Federal agencies in light of recent court decisions interpreting the Act. The proposed guidelines have been revised in light of the specific comments relating to these general themes, as well as other comments received, and are now being issued in final form.

The guidelines will appear in the Code of Federal Regulations in Title 40, Chapter V, at Part 1500. They are being codified, in part, because they affect State and local governmental agencies, environmental groups, industry, and private individuals, in addition to Federal agencies, to which they are specifically directed, and the resultant need to make them widely and readily available.

Authority: National Environmental Policy Act (P.L. 91-190, 42 U.S.C. 4321 et seq.) and Executive Order 11514.

§1500.1 Purpose and Authority

(a) This directive provides guidelines to Federal departments, agencies, and establishments for preparing detailed environmental statements on proposals for legislation and other major Federal actions significantly affecting the quality of the human environment at required by section 102(2)(C) of the National Environmental Policy Act (P.L. 91-190, 42 U.S.C. 4321 et. seq.) (hereafter "the Act"). Underlying the preparation of such environmental statements is the mandate of both the Act and Executive Order 11514 (35 FR 4247) of March 5, 1970, that all Federal agencies, to the fullest extent possible, direct their policies, plans and programs to protect and enhance environmental quality. Agencies are required to view their actions in a manner calculated to encourage productive and enjoyable harmony between man and his environment, to promote efforts preventing or eliminating damage to the environment and biosphere and stimulating the health and welfare of man, and to enrich the understanding of the ecological systems and natural resources important to the Nation. The objective of section 102(2)(C) of the Act and of these guidelines is to assist agencies in implementing these policies. This requires agencies to build into their decisionmaking process, beginning at the earliest possible point, an appropriate and careful consideration of the environmental aspects of proposed action in order that adverse environmental effects may be avoided or minimized and environmental quality previously lost may be restored. This directive also provides guidance to Federal, State, and local agencies and the public in commenting on statements prepared under these guidelines.

(b) Pursuant to section 204(3) of the Act the Council on Environmental Quality (hereafter "the Council") is assigned the duty and function of reviewing and appraising the programs and activities of the Federal Government, in the light of the Act's policy, for the purpose of determining the extent to which such programs and activitics arc contributing to the achievement of such policy, and to make recommendations to the President with respect thereto. Section 102(2)(B) of the Act directs all Federal agencies to identify and develop methods and procedures, in consultation with the Council, to insure that unquantified environmental values be given appropriate consideration in decisionmaking along with economic and technical considerations; section 102(2)(C) of the Act directs that copies of all environmental impact statements be filed with the Council; and section 102(2)(H) directs all Federal agencies to assist the Council in the performance of its functions. These provisions have been supplemented in sections 3(h) and (i) of Executive Order 11514 by directions that the Council issue guidelines to Federal agencies for preparation of environmental impact statements and such other instructions to agencies and requests for reports and information as may be required to carry out the Council's responsibilities under the Act.

§1500.2 Policy

(a) As early as possible and in all cases prior to agency decision concerning recommendations or favorable reports on proposals for (1) legislation significantly affecting the quality of the human environment (see §§1500.5(i) and 1500.12) (hereafter "legislative actions") and (2) all other major Federal actions significantly affecting the quality of the human environment (hereafter "administrative actions"), Federal agencies will, in consultation with other appropriate Federal, State and local agencies and the public assess in detail the potential environmental impact.

(b) Initial assessments of the environmental impacts of proposed action should be undertaken concurrently with initial technical and economic studies and, where required, a draft environmental impact statement prepared and circulated for comment in time to accompany the proposal through the existing agency review processes for such action. In this process, Federal agencies shall: (1) Provide for circulation of draft environmental statements to other Federal, State, and local agencies and for their availability to the public in accordance with the provisions of these guidelines; (2) consider the comments of the agencies and the public; and (3) issue final environmental impact statements responsive to the comments received. The purpose of this assessment and consultation process is to provide agencies and other decisionmakers as well as members of the public with an understanding of the potential environmental effects of proposed actions, to avoid or minimize adverse effects wherever possible, and to restore or enhance environmental quality to the fullest extent practicable. In particular, agencies should use the environmental impact statement process to explore alternative actions that will avoid or minimize adverse impacts and to evaluate both the long- and short-range implications of proposed actions to man, his physical and social surroundings, and to nature. Agencies should consider the results of their environmental assessments along with their assessments of the net economic, technical and other benefits of proposed actions and use all practicable means, consistent with other essential considerations of national policy, to restore environmental quality as well as to avoid or minimize undesirable consequences for the environment.

§1500.3 Agency and OMB Procedures

(a) Pursuant to section 2(f) of Executive Order 11514, the heads of Federal agencies have been directed to proceed with measures required by section 102(2)(C) of the Act. Previous guidelines of the Council directed each agency to establish its own formal procedures for (1) identifying those agency actions requiring environmental statements, the appropriate time prior to decision for the consultations required by section 102(2)(C) and the agency review process for which the environmental statements are to be available, (2) obtaining information required in their preparation, (3) designating the officials who are to be responsible for the statements, (4) consulting with and taking account of the comments of appropriate Federal, State and local agencies and the public, including obtaining the comment of the Administrator of the Environmental Protection Agency when required under section 309 of the Clean Air Act, as amended, and (5) meeting the requirements of section 2(b) of Executive Order 11514 for providing timely public information on Federal plans and programs with environmental impact. Each agency, including both departmental and sub-departmental components having such procedures, shall review its procedures and shall revise them, in consultation with the Council, as may be necessary in order to respond to requirements imposed by these revised guidelines as well as by such previous directives. After such consultation, proposed revisions of such agency procedures shall be published in the *Federal Register* no later than October 30, 1973. A minimum 45-day period for public comment shall be provided, followed by publication of final procedures no later than forty-five (45) days after the conclusion of the comment period. Each agency shall submit seven (7) copies of all such procedures to the Council. Any future revision of such agency procedures shall similarly be proposed and adopted only after prior consultation with the Council and, in the case of substantial revision, opportunity for public comment. All revisions shall be published in the *Federal Register*.

(b) Each Federal agency should consult, with the assistance of the Council and the Office of Management and Budget if desired, with other appropriate Federal agencies in the development and revision of the above procedures so as to achieve consistency in dealing with similar activities and to assure effective coordination among agencies in their review of proposed activities. Where applicable, State and local review of such agency procedures should be conducted pursuant to procedures established by Office of Management and Budget Circular No. A-85.

(c) Existing mechanisms for obtaining the views of Federal, State, and local agencies on proposed Federal actions should be utilized to the maximum extent practicable in dealing with environmental matters. The Office of Management and Budget will issue instructions, as necessary, to take full advantage of such existing mechanisms.

§1500.4 Federal Agencies Included; Effect of the Act on Existing Agency Mandates

(a) Section 102(2)(C) of the Act applies to all agencies of the Federal Government. Section 102 of the Act provides that "to the fullest extent possible: (1) The policies, regulations, and public laws of the United States shall be interpreted and administered in accordance with the policies set forth in this Act," and section 105 of the Act provides that "the policies and goals set forth in this Act are supplementary to those set forth in existing authorizations of Federal agencies." This means that each agency shall interpret the provisions of the Act as a supplement to its existing authority and as a mandate to view traditional policies and missions in the light of the Act's national environmental objectives. In accordance with this purpose, agencies should continue to review their policies, procedures, and regulations and to revise them as necessary to ensure full compliance with the purposes and provisions of the Act. The phrase "to the fullest extent possible" in section 102 is meant to make clear that each agency of the Federal Government shall comply with that section unless existing law applicable to the agency's operations expressly prohibits or makes compliance impossible.

§1500.5 Types of Actions Covered by the Act

(a) "Actions" include but are not limited to:

(1) Recommendations or favorable reports relating to legislation including requests for appropriations. The requirement for following the section 102(2)(C) procedure as elaborated in these guidelines applies to both (i) agency recommendations on their own proposals for legislation (see §1500.12); and (ii) agency reports on legislation initiated elsewhere. In the latter case only the agency which has primary responsibility for the subject matter involved will prepare an environmental statement.

(2) New and continuing projects and program activities: directly undertaken by Federal agencies; or supported in whole or in part through Federal contracts, grants, subsidies, loans, or other forms of funding assistance (except where such assistance is solely in the form of general revenue sharing funds, distributed under the State and Local Fiscal Assistance Act of 1972, 31 U.S.C. 1221 et. seq. with no Federal agency control over the subsequent use of such funds); or involving a Federal lease, permit, license certificate or other entitlement for use.

(3) The making, modification, or establishment of regulations, rules, procedures, and policy.

§1500.6 Identifying Major Actions Significantly Affecting the Environment

(a) The statutory clause "major Federal actions significantly affecting the quality of the human environment" is to be construed by agencies with a view to the overall, cumulative impact of the action proposed, related Federal actions and projects in the area, and further actions contemplated. Such actions may be localized in their impact, but if there is potential that the environment may be significantly affected, the statement is to be prepared. Proposed major actions, the environmental impact of which is likely to be highly controversial, should be covered in all cases. In considering what constitutes major action significantly affecting the environment, agencies should bear in mind that the effect of many Federal decisions about a project or complex of projects can be individually limited but cumulatively considerable. This can occur when one or more agencies over a period of years puts into a project individually minor but collectively major resources, when one decision involving a limited amount of money is a precedent for action in much larger cases or represents a decision in principle about a future major course of action, or when several Government agencies individually make decisions about partial aspects of a major action. In all such cases, an environmental statement should be prepared if it is reasonable to anticipate a cumulatively significant impact on the environment from Federal action. The Council, on the basis of a written assessment of the impacts involved, is available to assist agencies in determining whether specific actions require impact statements.

(b) Section 101(b) of the Act indicates the broad range of aspects of the environment to be surveyed in any assessment of significant effect. The Act also indicates that adverse significant effects include those that degrade the quality of the environment, curtail the range of beneficial uses of the environment, and serve short-term, to the disadvantage of long-term, environmental goals. Significant effects can also include actions which may have both beneficial and detrimental effects, even if on balance the agency believes that the effect will be beneficial. Significant effects also include secondary effects, as described more fully, for example, in §1500.8(a)(iii)(B). The significance of a proposed action may also vary with the setting, with the result that an action that would have little impact in a rural setting or vice versa. While a precise definition of environmental "significance," valid in all contexts, is not possible, effects to be considered in assessing significance include, but are not limited to, those outlined in Appendix II of these guidelines.

(c) Each of the provisions of the Act, except section 102(2)(C), applies to all Federal agency actions. Section 102(2)(C) requires the preparation of a detailed environmental impact statement in the case of "major Federal actions significantly affecting the quality of the human environment." The identification of major actions significantly affecting the environment is the responsibility of each Federal agency, to be carried out against the background of its own particular operations. The action must be a (1) "major" action, (2) which is a "Federal action," (3) which has a "significant" effect, and (4) which involves the "quality of the human environment." The words "major" and "significantly" are intended to imply thresholds of importance and impact that must be met before a statement is required. The action causing the impact must also be one where there is sufficient Federal control and responsibility to constitute "Federal action" in contrast to cases where such Federal control and responsibility are not present as, for example, when Federal funds are distributed in the form of general revenue sharing to be used by State and local governments (see §1500.5(ii)). Finally, the action must be one that significantly affects the quality of the human environment either by directly affecting human beings or by indirectly affecting human beings through adverse effects on the environment. Each agency should review the typical classes of actions that it undertakes and, in consultation with the Council, should develop specific criteria and methods for identifying those actions likely to require environmental statements and those actions likely not to require environmental statements. Normally this will involve:

(i) Making an initial assessment of the environmental impacts typically associated with principal types of agency action.

(ii) Identifying on the basis of this assessment, types of actions which normally do, and types of actions which normally do not require statements.

(iii) With respect to remaining actions that may require statements depending on the circumstances, and those actions determined under the preceding paragraph (C)(4)(ii) of this section as likely to require statements, identifying: (*a*) what basic information needs to be gathered; (*b*) how and when such information is to be assembled and analyzed; and (*c*) on what bases environmental assessments and decisions to prepare impact statements will be made. Agencies may either include this substantive guidance in the procedures issued pursuant to §1500.3(a) of these guidelines, or issue such guidance as supplemental instructions to aid relevant agency personnel in implementing the impact statement process. Pursuant to §1500.14 of these guidelines, agencies shall report to the Council by June 30, 1974, on the progress made in developing such substantive guidance.

(d)(1) Agencies should give careful attention to identifying and defining the purpose and scope of the action which would most appropriately serve as the subject of the statement. In many cases, broad program statements will be required in order to assess the environmental effects of a number of individual actions on a given geographical area (e.g., coal leases), or environmental impacts that are generic or common to a series of agency actions (e.g., maintenance or waste handling practices), or the overall impact of a large-scale program or chain of contemplated projects (e.g., major lengths of highway as opposed to small segments). Subsequent statements on major individual actions will be necessary where such actions have significant environmental impacts not adequately evaluated in the program statement.

(2) Agencies engaging in major technology research and development programs should develop procedures for periodic evaluation to determine when a program statement is required for such programs. Factors to be considered in making this determination include the magnitude of Federal investment in the program, the likelihood of widespread application of the technology, the degree of environmental impact which would occur if the technology were widely applied, and the extent to which continued investment in the new technology is likely to restrict future alternatives. Statements must be written late enough in the development process to contain meaningful information, but early enough so that this information can practically serve as an input in the decision-making process. Where it is anticipated that a statement may ultimately be required but that its preparation is still premature, the agency should prepare an evaluation briefly setting forth the reasons for its determination that a statement is not yet necessary. This evaluation should be periodically updated, particularly when significant new information becomes available concerning the potential environmental impact of the program. In any case, a statement must be prepared before research activities have reached a state of investment or commitment to implementation likely to determine subsequent development or restrict later alternatives. Statements on technology research and development programs should include an analysis not only of alternative forms of

the same technology that might reduce any adverse environmental impacts but also of alternative technologies that would serve the same function as the technology under consideration. Efforts should be made to involve other Federal agencies and interested groups with relevant expertise in the preparation of such statements because the impacts and alternatives to be considered are likely to be less well defined than in other types of statements.

(e) In accordance with the policy of the Act and Executive Order 11514 agencies have a responsibility to develop procedures to insure the fullest practicable provision of timely public information and understanding of Federal plans and programs with environmental impact in order to obtain the views of interested parties. In furtherance of this policy, agency procedures should include an appropriate early notice system for informing the public of the decision to prepare a draft environmental statement on proposed administrative actions (and for soliciting comments that may be helpful in preparing the statement) as soon as is practicable after the decision to prepare the statement is made. In this connection, agencies should: (1) maintain a list of administrative actions for which environmental statements are being prepared; (2) revise the list at regular intervals specified in the agency's procedures developed pursuant to §1500.3(a) of these guidelines (but not less than quarterly) and transmit each such revision to the Council; and (3) make the list available for public inspection on request. The Council will periodically publish such lists in the *Federal Register*. If an agency decides that an environmental statement is not necessary for a proposed action (i) which the agency has identified pursuant to §1500.6(c)(4)(ii) as normally requiring preparation of a statement, (ii) which is similar to actions for which the agency has prepared a significant number of statements, (iii) which the agency has previously announced would be the subject of a statement, or (iv) for which the agency has made a negative determination in response to a request from the Council pursuant to §1500.11(f), the agency shall prepare a publicly available record briefly setting forth the agency's decision and the reasons for that determination. Lists of such negative determinations, and any evaluations made pursuant to §1500.6 which conclude that preparation of a statement is not yet timely, shall be prepared and made available in the same manner as provided in this subsection for lists of statements under preparation.

§1500.7 Preparing Draft Environmental Statements: Public Hearings

(a) Each environmental impact statement shall be prepared and circulated in draft form for comment in accordance with the provisions of these guidelines. The draft statement must fulfill and satisfy to the fullest extent possible at the time the draft is prepared the requirements established for final statements by section 102(2)(C). (Where an agency has an established practice of declining to favor an alternative until public comments on a proposed action have been received, the draft environmental statement may indicate that two or more alternatives are under consideration.) Comments received shall be carefully evaluated and considered in the decision process. A final statement with substantive comments attached shall then be issued and circulated in accordance with applicable provisions of §§1500.10, 1500.11, or 1500.12. It is important that draft environmental statements be prepared and circulated for comment and furnished to the Council as early as possible in the agency review process in order to permit agency decisionmakers and outside reviewers to give meaningful consideration to the environmental issues involved. In particular, agencies should keep in mind that such statements are to serve as the means of assessing the environmental impact of proposed agency actions, rather than as a justification for decisions already made. This means that draft statements on administrative actions should be prepared and circulated for comment prior to the first significant point of decision in the agency review process. For major categories of agency action, this point should be identified in the procedures issued pursuant to §1500.3(a). For major categories of projects involving an applicant and identified pursuant to §1500.6(c)(c)(ii) as normally requiring the preparation of a statement, agencies should include in their procedures provisions limiting actions which an applicant is permitted to take prior to completion and review of the final statement with respect to his application.

(b) Where more than one agency (1) directly sponsors an action, or is directly involved in an action through funding, licenses, or permits, or (2) is involved in a group of actions directly related to each other because of their functional interdependence and geographical proximity, consideration should be given to preparing one statement for all the Federal actions involved (see §1500.6(d)(1)). Agencies in such cases should consider the possibility of joint preparation of a statement by all agencies concerned, or designation of a single "lead agency" to assume supervisory responsibility for preparation of the statement. Where a lead agency prepares the statement, the other agencies involved should provide assistance with respect to their areas of jurisdiction and expertise. In either case, the statement should contain an environmental assessment of the full range of Federal actions involved, should reflect the views of all participating agencies, and should be prepared before major or irreversible actions have been taken by any of the participating agencies. Factors relevant in determining an appropriate lead agency include the time sequence in which the agencies become involved, the magnitude of their respective involvement, and their relative expertise with respect to the project's environmental effects. As necessary, the Council will assist in resolving questions of responsibility for statement preparation in the case of multi-agency actions. Federal Regional Councils, agencies and the public are encouraged to bring to the attention of the Council and other relevant agencies appropriate situations where a geographic or regionally focused statement would be desirable because of the cumulative environmental effects likely to result from multi-agency actions in the area.

(c) Where an agency relies on an applicant to submit initial environmental information, the agency should assist the applicant by outlining the types of information required. In all cases, the agency should make its own evaluation of the environmental issues and take responsibility for the scope and content of draft and final environmental statements.

(d) Agency procedures developed pursuant to §1500.3(a) of these guidelines should indicate as explicitly as possible those types of agency decisions or actions which utilize hearings as part of the normal agency review process, either as a result of statutory requirement or agency practice. To the fullest extent possible, all such hearings shall include consideration of the environmental aspects of the proposed action. Agency procedures shall also specifically include provision for public hearings on major actions with environmental impact, whenever appropriate, and for providing the public with relevant information, including information on alternative courses of action. In deciding whether a public hearing is appropriate, an agency should consider: (1) The magnitude of the proposal in terms of economic costs, the geographic area involved, and the uniqueness or size of commitment of the resources involved; (2) the degree of interest in the proposal, as evidenced by requests from the public and from Federal, State and local authorities that a hearing be held; (3) the complexity of the issue and the likelihood that information will be presented at the hearing which will be of assistance to the agency in fulfilling its responsibilities under the Act; and (4) the extent to which public involvement already has been achieved through other means, such as earlier public hearings, meetings with citizen representatives, and/or written comments on the proposed action. Agencies should make any draft environmental statements to be issued available to the public at least fifteen (15) days prior to the time of such hearings.

§1500.8 Content of Environmental Statements

(a) The following points are to be covered:

(1) A description of the proposed action, a statement of its purposes, and a description of the environment affected, including information, summary technical data, and maps and diagrams where relevant, adequate to permit an assessment of potential environmental impact by commenting agencies and the public. Highly technical and specialized analyses and data should be avoided in the body of the draft impact statement. Such materials should be attached as appendices or footnoted with adequate bibliographic references. The statement should also succinctly describe the environment of the area affected as it exists prior to a proposed action, including other Federal activities in the area affected by the proposed action which are related to the proposed action. The interrelationships and cumulative environmental impacts of the proposed action and other related Federal projects shall be presented in the statement. The amount of detail provided in such descriptions should be commensurate with the extent and expected impact of the action and with the amount of information required at the particular level of decisionmaking (planning, feasibility, design, etc.). In order to ensure accurate descriptions and environmental assessments, site visits should be made where feasible. Agencies should also take care to identify, as appropriate, population and growth characteristics of the affected area and any population and growth assumptions used to justify the project or program or to determine secondary population and growth impacts resulting from the proposed action and its alternatives (see paragraph (a)(1)(3)(ii), of this section). In discussing these population aspects, agencies should give consideration to using the rates of growth in the region of the project contained in the projection compiled for the Water Resources Council by the Bureau of Economic Analysis of the Department of Commerce and the Economic Research Service of the Department of Agriculture (the "OBERS" projection). In any event it is essential that the sources of data used to identify, quantify or evaluate any and all environmental consequences be expressly noted.

(2) The relationship of the proposed action to land use plans, policies, and controls for the affected area. This requires a discussion of how the proposed action may conform or conflict with the objectives and specific terms of approved or proposed Federal, State, and local land use plans, policies, and controls, if any, for the area affected including those developed in response to the Clean Air Act or the Federal Water Pollution Control Act Amendments of 1972. Where a conflict or inconsistency exists, the statement should describe the extent to which the agency has reconciled its proposed action with the plan, policy or control, and the reasons why the agency has decided to proceed notwithstanding the absence of full reconciliation.

(3) The probable impact of the proposed action on the environment.

(i) This requires agencies to assess the positive and negative effects of the proposed action as it affects both the national and international environment. The attention given to different environmental factors will vary according to the nature, scale, and location of proposed actions. Among factors to consider should be the potential effect of the action on such aspects of the environment as those listed in Appendix II of these guidelines. Primary attention should be given in the statement to discussing those factors most evidently impacted by the proposed action.

(ii) Secondary or indirect, as well as primary or direct, consequences for the environment should be included in the analysis. Many major Federal actions, in particular those that involve the construction or licensing of infrastructure investments (e.g., highways, airports, sewer systems, water resource projects, etc.), stimulate or induce secondary effects in the form of associated investments and changed patterns of social and economic activities. Such secondary effects, through their impacts on existing community facilities and activities, through inducing new facilities and activities, or through changes in natural conditions, may often be even more substantial than the primary effects of the original action itself. For example, the effects of the proposed action on population and growth may be among the more significant secondary effects. Such population and growth impacts should be estimated if expected to be significant (using data identified as indicated in §1500.8(a)(1)) and an assessment made of the effect of any possible change in population patterns or growth upon the resource base, including land use, water, and public services, of the area in question.

(4) Alternatives to the proposed action, including, where relevant, those not within the existing authority of the responsible agency. (Section 102(2)(D) of the Act requires the responsible agency to "study, develop, and describe appropriate alternatives to recommended courses of action in any proposal which involves unresolved conflicts concerning alternative uses of available resources"). A rigorous exploration and objective evaluation of the environmental impacts of all reasonable alternative actions, particularly those that might enhance environmental quality or avoid some or all of the adverse environmental effects, is essential. Sufficient analysis of such alternatives and their environmental benefits, costs and risks should accompany the proposed action through the agency review process in order not to foreclose prematurely options which might enhance environmental quality or have less detrimental effects. Examples of such alternatives include: the alternative of taking no action or of postponing action pending further study; alternatives requiring actions of a significantly different nature which would provide similar benefits with different environmental impacts (e.g., nonstructural alternatives to flood control programs, or mass transit alternatives to highway construction); alternatives related to different designs or details of the proposed action which would present different environmental impacts (e.g., cooling ponds vs. cooling towers for a power plant or alternatives that will significantly conserve energy); alternative measures to provide for compensation of fish and wildlife losses, including the acquisition of land, waters, and interests therein. In each case, the analysis should be sufficiently detailed to reveal the agency's comparative evaluation of the environmental benefits, costs and risks of the proposed action and each reasonable alternative. Where an existing impact statement already contains such an analysis, its treatment of alternatives may be incorporated provided that such treatment is current and relevant to the precise purpose of the proposed action.

(5) Any probable adverse environmental effects which cannot be avoided (such as water or air pollution, undesirable land use patterns, damage to life systems, urban congestion, threats to health or other consequences adverse to the environmental goals set out in section 101(b) of the Act). This should be a brief section summarizing in one place those effects discussed in paragraph (a)(3) of this section that are adverse and unavoidable under the proposed action. Included for purposes of contrast should be a clear statement of how other avoidable adverse effects discussed in paragraph (a)(2) of this section will be mitigated.

(6) The relationship between local short-term uses of man's environment and the maintenance and enhancement of long-term productivity. This section should contain a brief discussion of the extent to which the proposed action involves tradeoffs between short-term environmental gains at the expense of long-term losses, or vice versa, and a discussion of the extent to which the proposed action forecloses future options. In this context short-term and long-term do not refer to any fixed time periods, but should be viewed in terms of the environmentally significant consequences of the proposed action.

(7) Any irreversible and irretrievable commitments of resources that would be involved in the proposed action should it be implemented. This requires the agency to identify from its survey of unavoidable impacts in paragraph (a)(5) of this section the extent to which the action irreversibly curtails the range of potential uses of the environment. Agencies should avoid construing the term "resources" to mean only the labor and materials devoted to an action. "Resources" also means the natural and cultural resources committed to loss or destruction by the action.

(8) An indication of what other interests and considerations of Federal policy are thought to offset the adverse environmental effects of the proposed action identified pursuant to paragraphs (a)(3) and (5) of this section. The statement should also indicate the extent to which these stated countervailing benefits could be realized by following reasonable alternatives to the proposed action (as identified in paragraph (a)(4) of this section) that would avoid some or all of the adverse environmental effects. In this connection, agencies that prepare cost-benefit analyses of proposed actions should attach such analyses, or summaries thereof, to the environmental impact statement, and should clearly indicate the extent to which environmental costs have not been reflected in such analyses.

(b) In developing the above points agencies should make every effort to convey the required information succinctly in a form easily understood, both by members of the public and by public decisionmakers, giving attention to the substance of the information conveyed rather than to the particular form, or length, or detail of the statement. Each of the above points, for example, need not always occupy a distinct section of the statement if it is otherwise adequately covered in discussing the impact of the proposed action and its alternatives—which items should normally be the focus of the statement. Draft statements should indicate at appropriate points in the text any underlying studies, reports, and other information obtained and considered by the agency in preparing the statement including any cost-benefit analyses prepared by the agency, and reports of consulting agencies under the Fish and Wildlife Coordination Act, 16 U.S.C. 661 et seq., and the National Historic Preservation Act of 1966, 16 U.S.C. 470 et seq., where such consultation has taken place. In the case of documents not likely to be easily accessible (such as internal studies or reports), the agency should indicate how such information may be obtained. If such information is attached to the statement, care should be taken to ensure that the statement remains an essentially self-contained instrument, capable of being understood by the reader without the need for undue cross reference.

(c) Each environmental statement should be prepared in accordance with the precept in section 102(2)(A) of the Act that all agencies of the Federal Government "utilize a systematic, interdisciplinary approach which will insure the integrated use of the natural and social sciences and the environmental design arts in planning and decisionmaking which may have an impact on man's environment." Agencies should attempt to have relevant disciplines represented on their own staffs; where this is not feasible they should make appropriate use of relevant Federal, State, and local agencies or the professional services of universities and outside consultants. The interdisciplinary approach should not be limited to the preparation of the environmental impact statement, but should also be used in the early planning stages of the proposed action. Early application of such an approach should help assure a systematic evaluation of reasonable alternative courses of action and their potential social, economic, and environmental consequences.

(d) Appendix I prescribes the form of the summary sheet which should accompany each draft and final environmental statement.

§1500.9 Review of Draft Environmental Statements by Federal, Federal-State, State, and Local Agencies and by the Public

(a) *Federal agency review.* (1) *In general.* A Federal agency considering an action requiring an environmental statement should consult with, and (on the basis of a draft environmental statement for which the agency takes responsibility) obtain the comment on the environmental impact of the action of Federal and Federal-State agencies with jurisdiction by law or special expertise with respect to any environmental impact involved. These Federal and Federal-State agencies and their relevant areas of expertise include those identified in Appendices II and III to these guidelines. It is recommended that the listed departments and agencies establish contact points, which may be regional offices, for providing comments on the environmental statements. The requirement in section 102(2)(C) to obtain comment from Federal agencies having jurisdiction or special expertise is in addition to any specific statutory obligation of any Federal agency to coordinate or consult with any other Federal or State agency. Agencies should, for example, be alert to consultation requirements of the Fish and Wildlife Coordination Act, 16 U.S.C. 661 et seq., and the National Historic Preservation Act of 1966, 16 U.S.C. 470 et seq. To the extent possible, statements or findings concerning environmental impact required by other statutes, such as section 4(f) of the Department of Transportation Act of 1966, 49 U.S.C. 1653(f), or section 106 of the National Historic Preservation Act of 1966, should be combined with compliance with the environmental impact statement requirements of section 102(2)(C) of the Act to yield a single document which meets all applicable requirements. The Advisory Council on Historic Preservation, the Department of Transportation, and the Department of the Interior, in consultation with the Council, will issue any necessary supplementing instructions for furnishing information or findings not forthcoming under the environmental impact statement process.

(b) *EPA review.* Section 309 of the Clean Air Act, as amended (42 U.S.C. §1857h-7), provides that the Administrator of the Environmental Protection Agency shall comment in writing on the environmental impact of any matter relating to his duties and responsibilities, and shall refer to the Council any matter that the Administrator determines is unsatisfactory from the standpoint of public health or welfare or environmental quality. Accordingly, wherever an agency action related to air or water quality, noise abatement and control, pesticide regulation, solid waste disposal, generally applicable environmental radiation criteria and standards, or other provision of the authority of the Administrator is involved, Federal agencies are required to submit such proposed actions and their environmental impact statements, if such have been prepared, to the Administrator for review and comment in writing. In all cases where EPA determines that proposed agency action is environmentally unsatisfactory, or where EPA determines that an environmental statement is so inadequate that such a determination cannot be made, EPA shall publish its determination and notify the Council as soon as practicable. The Administrator's comments shall constitute his comments for the purposes of both section 309 of the Clean Air Act and section 102(2)(C) of the National Environmental Policy Act.

(c) *State and local review.* Office of Management and Budget Circular No. A-95 (Revised) through its system of State and areawide clearinghouses provides a means for securing the views of State and local environmental agencies, which can assist in the preparation and review of environmental impact statements. Current instructions for obtaining the views of such agencies are contained in the joint OMB-CEQ memorandum attached to these guidelines as Appendix IV. A current listing of clearinghouses is issued periodically by the Office of Management and Budget.

(d) *Public review.* The procedures established by these guidelines are designed to encourage public participation in the impact statement process at the earliest possible time. Agency procedures should make provision for facilitating the comment of public and private organizations and individuals by announcing the availability of draft environmental statements and by making copies available to organizations and individuals that request an opportunity to comment. Agencies should devise methods for publicizing the existence of draft statements, for example, by publication of notices in local newspapers or by maintaining a list of groups, including relevant conservation commissions, known to be interested in the agency's activities and directly notifying such groups of the existence of a draft statement, or sending them a copy, as soon as it has been prepared. A copy of the draft statement should in all cases be sent to any applicant whose project is the subject of the statement. Materials to be made available to the public shall be provided without charge to the extent practicable, or at a fee which is not more than the actual cost of reproducing copies required to be sent to other Federal agencies, including the Council.

(e) *Responsibilities of commenting entities.* (1) Agencies and members of the public submitting comments on proposed actions on the basis of draft environmental statements should endeavor to make their comments as specific, substantive, and factual as possible without undue attention to matters of form in the impact statement. Although the comments need not conform to any particular format, it would assist agencies reviewing comments if the comments were organized in a manner consistent with the structure of the draft statement. Emphasis should be placed on the assessment of the environmental impacts of the proposed action, and the acceptability of those impacts on the quality of the environment, particularly as contrasted with the impacts of reasonable alternatives to the action. Commenting entities may recommend modifications to the proposed action and/or new alternatives that will enhance environmental quality and avoid or minimize adverse environmental impacts.

(2) Commenting agencies should indicate whether any of their projects not identified in the draft statement are sufficiently advanced in planning and related environmentally to the proposed action so that a discussion of the environmental interrelationships should be included in the final statement (see §1500.8(a)(1)). The Council is available to assist agencies in making such determinations.

(3) Agencies and members of the public should indicate in their comments the nature of any monitoring of the environmental effects of the proposed project that appears particularly appropriate. Such monitoring may be necessary during the construction, startup, or operation phases of the project. Agencies with special expertise with respect to the environmental impacts involved are encouraged to assist the sponsoring agency in the establishment and operation of appropriate environmental monitoring.

(f) Agencies seeking comment shall establish time limits of not less than forty-five (45) days for reply, after which it may be presumed, unless the agency or party consulted requests a specified extension of time, that the agency or party consulted has no comment to make. Agencies seeking comment should endeavor to comply with requests for extensions of time of up to fifteen (15) days. In determining an appropriate period for comment, agencies should consider the magnitude and complexity of the statement and the extent of citizen interest in the proposed action.

§1500.10 Preparation and Circulation of Final Environmental Statements

(a) Agencies should make every effort to discover and discuss all major points of view on the environmental effects of the proposed action and its alternatives in the draft statement itself. However, where opposing professional views and responsible opinion have been overlooked in the draft statement and are brought to the agency's attention through the commenting process, the agency should review the environmental effects of the action in light of those views and should make a meaningful reference in the final statement to the existence of any responsible opposing view not adequately discussed in the draft statement, indicating the agency's response to the issues raised. All substantive comments received on the draft (or summaries thereof where response has been exceptionally voluminous) should be attached to the final statement, whether or not each such comment is thought to merit individual discussion by the agency in the text of the statement.

(b) Copies of final statements, with comments attached, shall be sent to all Federal, State, and local agencies and private organizations that made substantive comments on the draft statement and to individuals who requested a copy of the final statement, as well as any applicant whose project is the subject of the statement. Copies of final statements shall in all cases be sent to the Environmental Protection Agency to assist it in carrying out its responsibilities under section 309 of the Clean Air Act. Where the number of comments on a draft statement is such that distribution of the final statement to all commenting entities appears impracticable, the agency shall consult with the Council concerning alternative arrangements for distribution of the statement.

§1500.11 Transmittal of Statements to the Council; Minimum Periods for Review; Requests by the Council

(a) As soon as they have been prepared, ten (10) copies of draft environmental statements, five (5) copies of all comments made thereon (to be forwarded to the Council by the entity making comment at the time comment is forwarded to the responsible agency), and ten (10) copies of the final text of environmental statements (together with the substance of all comments received by the responsible agency from Federal, State, and local agencies and from private organizations and individuals) shall be supplied to the Council. This will serve to meet the statutory requirement to make environmental statements available to the President. At the same time that copies of draft and final statements are sent to the Council, copies should also be sent to relevant commenting entities as set forth in §§1500.9 and 1500.10(b) of these guidelines.

(b) To the maximum extent practicable no administrative action subject to section 102(2)(C) is to be taken sooner than ninety (90) days after a draft environmental statement has been circulated for comment, furnished to the Council and, except where advance public disclosure will result in significantly increased costs of procurement to the Government, made available to the public pursuant to these guidelines; neither should such administrative action be taken sooner than thirty (30) days after the final text of an environmental statement (together with comments) has been made available to the Council, commenting agencies, and the public. In all cases, agencies should allot a sufficient review period for the final statement so as to comply with the statutory requirement that the "statement and the comments and views of appropriate Federal, State, and local agencies * * * accompany the proposal through the existing agency review processes." If the final text of an environmental statement is filed within ninety (90) days after a draft statement has been circulated for comment, furnished to the Council and made public pursuant to this section of these guidelines, the minimum thirty (30) day period and the ninety (90) day period may run concurrently to the extent that they overlap. An agency may at any time supplement or amend a draft or final environmental statement, particularly when substantial changes are made in the proposed action, or significant new information becomes available concerning its environmental aspects. In such cases the agency should consult with the Council with respect to the possible need for or desirability of recirculation of the statement for the appropriate period.

(c) The Council will publish weekly in the *Federal Register* lists of environmental statements received during the preceding week that are available for public comment. The date of publication of such lists shall be the date from which the minimum periods for review and advance availability of statements shall be calculated.

(d) The Council's publication of notice of the availability of statements is in addition to the agency's responsibility, as described in §1500.9(d) of these guidelines, to insure the fullest practicable provision of timely public information concerning the existence and availability of environmental statements. The agency responsible for the environmental statement is also responsible for making the statement, the comments received, and any underlying documents available to the public pursuant to the provisions of the Freedom of Information Act (5 U.S.C., 552), without regard to the exclusion of intra- or interagency memoranda when such memoranda transmit comments of Federal agencies on the environmental impact of the proposed action pursuant to §1500.9 of these guidelines. Agency procedures prepared pursuant to §1500.3(a) of these guidelines shall implement these public information requirements and shall include arrangements for availability of environmental statements and comments at the head and appropriate regional offices of the responsible agency and at appropriate State and areawide clearinghouses unless the Governor of the State involved designates to the Council some other point for receipt of this information. Notice of such designation of an alternate point for receipt of this information will be included in the Office of Management and Budget listing of clearinghouses referred to in §1500.9(c).

(e) Where emergency circumstances make it necessary to take an action with significant environmental impact without observing the provisions of these guidelines concerning minimum periods for agency review and advance availability of environmental statements, the Federal agency proposing to take the action should consult with the Council about alternative arrangements. Similarly where there are overriding considerations of expense to the Government or impaired program effectiveness, the responsible agency should consult with the Council concerning appropriate modifications of the minimum periods.

(f) In order to assist the Council in fulfilling its responsibilities under the Act and under Executive Order 11514, all agencies shall (as required by section 102(2)(H) of the Act and section 3(i) of Executive Order 11514) be responsive to requests by the Council for reports and other information dealing with issues arising in connection with the implementation of the Act. In particular, agencies shall be responsive to a request by the Council for the preparation and circulation of an environmental statement, unless the agency determines that such a statement is not required, in which case the agency shall prepare an environmental assessment and a publicly available record briefly setting forth the reasons for its determination. In no case, however, shall the Council's silence or failure to comment or request preparation, modification, or recirculation of an environmental statement or to take other action with respect to an environmental statement be construed as bearing in any way on the question of the legal requirement for or the adequacy of such statement under the Act.

§1500.12 Legislative Actions

(a) The Council and the Office of Management and Budget will cooperate in giving guidance as needed to assist agencies in identifying legislative items believed to have environmental significance. Agencies should prepare impact statements prior to submission of their legislative proposals to the Office of Management and Budget. In this regard, agencies should identify types of repetitive legislation requiring environmental impact statements (such as certain types of bills affecting transportation policy or annual construction authorizations).

(b) With respect to recommendations or reports on proposals for legislation to which section 102(2)(C) applies the final text of the environmental statement and comments thereon should be available to the Congress and to the public for consideration in connection with the proposed legislation or report in cases where the scheduling of congressional hearings on recommendations or reports on proposals for legislation which the Federal agency has forwarded to the Congress does not allow adequate time for the completion of a final text of an environmental statement (together with comments), a draft environmental statement may be furnished to the Congress and made available to the public pending transmittal of the comments as received and the final text.

§1500.13 Application of Section 102(2)(C) Procedure to Existing Projects and Programs

Agencies have an obligation to reassess ongoing projects and programs in order to avoid or minimize adverse environmental effects. The section 102(2)(C) procedure shall be applied to further major Federal actions having a significant effect on the environment even though they arise from projects or programs initiated prior to enactment of the Act on January 1, 1970. While the status of the work and degree of completion may be considered in determining whether to proceed with the project, it is essential that the environmental impacts of proceeding are reassessed pursuant to the Act's policies and procedures and, if the project or program is continued, that further incremental major actions be shaped so as to enhance and restore environmental quality as well as to avoid or minimize adverse environmental consequences. It is also important in further action that account be taken of environmental consequences not fully evaluated at the outset of the project or program.

§1500.14 Supplementary Guidelines: Evaluation of Procedures

(a) The Council after examining environmental statements and agency procedures with respect to such statements will issue such supplements to these guidelines as are necessary.

(b) Agencies will continue to assess their experience in the implementation of the section 102(2)(C) provisions of the Act and in conforming with these guidelines and report thereon to the Council by June 30, 1974. Such reports should include an identification of the problem areas and suggestions for revision or clarification of these guidelines to achieve effective coordination of views on environmental aspects (and alternatives, where appropriate) of proposed actions without imposing unproductive administrative procedures. Such reports shall also indicate what progress the agency has made in developing substantive criteria and guidance for making environmental assessments as required by §1500.6(c) of this directive and by section 102(2)(B) of the Act.

Effective date. The revisions of these guidelines shall apply to all draft and final impact statements filed with the Council after January 28, 1973.

Russell E. Train,
Chairman.

APPENDIX I—SUMMARY TO ACCOMPANY DRAFT AND FINAL STATEMENTS

(Check one) ( ) Draft. ( ) Final Environmental Statement.

Name of responsible Federal agency (with name of operating division where appropriate) Name, address, and telephone number of individual at the agency who can be contacted for additional information about the proposed action or the statement.

1. Name of action (Check one) ( ) Administrative Action. ( ) Legislative Action.

2. Brief description of action and its purpose. Indicate what States (and counties) particularly affected, and what other proposed Federal actions in the area, if any, are discussed in the statement.

3. Summary of environmental impacts and adverse environmental effects.

4. Summary of major alternatives considered.

5. (For draft statements) List all Federal, State, and local agencies and other parties from which comments have been requested. (For final statements) List all Federal, State, and local agencies and other parties from which written comments have been received.

6. Date draft statement (and final environmental statement, if one has been issued) made available to the Council and the public.

APPENDIX II–AREAS OF ENVIRONMENTAL IMPACT AND FEDERAL AGENCIES AND FEDERAL STATE AGENCIES[1] WITH JURISDICTION BY LAW OR SPECIAL EXPERTISE TO COMMENT THEREON[2]

AIR

Air Quality

Department of Agriculture–
    Forest Service (effects on vegetation)
Atomic Energy Commission (radioactive substances)
Department of Health, Education, and Welfare
Environmental Protection Agency
Department of the Interior–
    Bureau of Mines (fossil and gaseous fuel combustion)
    Bureau of Sport Fisheries and Wildlife (effect on wildlife)
    Bureau of Outdoor Recreation (effects on recreation)
    Bureau of Land Management (public lands)
    Bureau of Indian Affairs (Indian lands)
National Aeronautics and Space Administration (remote sensing, aircraft emissions)
Department of Transportation–
    Assistant Secretary for Systems Development and Technology (auto emissions)
    Coast Guard (vessel emissions)
    Federal Aviation Administration (aircraft emissions)

Weather Modification

Department of Agriculture–
    Forest Service
Department of Commerce–

[1] River Basin Commissions (Delaware, Great Lakes, Missouri, New England, Ohio, Pacific Northwest, Souris-Red-Rainy, Susquehanna, Upper Mississippi) and similar Federal-State agencies should be consulted on actions affecting the environment of their specific geographic jurisdictions.

[2] In all cases where a proposed action will have significant international environmental effects, the Department of State should be consulted, and should be sent a copy of any draft and final impact statement which covers such action.

National Oceanic and Atmospheric Administration
Department of Defense–
Department of the Air Force
Department of the Interior–
Bureau of Reclamation

WATER RESOURCES COUNCIL

WATER

Water Quality

Department of Agriculture–
Soil Conservation Service
Forest Service
Atomic Energy Commission (radioactive substances)
Department of the Interior–
Bureau of Reclamation
Bureau of Land Management (public lands)
Bureau of Indian Affairs (Indian lands)
Bureau of Sports Fisheries and Wildlife
Bureau of Outdoor Recreation
Geological Survey
Office of Saline Water
Environmental Protection Agency
Department of Health, Education, and Welfare
Department of Defense–
Army Corps of Engineers
Department of the Navy (ship pollution control)
National Aeronautics and Space Administration (remote sensing)
Department of Transportation–
Coast Guard (oil spills, ship sanitation)
Department of Commerce–
National Oceanic and Atmospheric Administration
Water Resources Council
River Basin Commissions (as geographically appropriate)

Marine Pollution, Commercial Fishery Conservation, and Shellfish Sanitation

Department of Commerce–
National Oceanic and Atmospheric Administration
Department of Defense–
Army Corps of Engineers
Office of the Oceanographer of the Navy
Department of Health, Education, and Welfare
Department of the Interior–
Bureau of Sport Fisheries and Wildlife
Bureau of Outdoor Recreation
Bureau of Land Management (outer continental shelf)
Geological Survey (outer continental shelf)

Department of Transportation–
  Coast Guard
Environmental Protection Agency
National Aeronautics and Space Administration (remote sensing)
Water Resources Council
River Basin Commissions (as geographically appropriate)

Waterway Regulation and Stream Modification

Department of Agriculture–
  Soil Conservation Service
Department of Defense–
  Army Corps of Engineers
Department of the Interior–
  Bureau of Reclamation
  Bureau of Sport Fisheries and Wildlife
  Bureau of Outdoor Recreation
  Geological Survey
Department of Transportation–
  Coast Guard
Environmental Protection Agency
National Aeronautics and Space Administration (remote sensing)
Water Resources Council
River Basin Commissions (as geographically appropriate)

FISH AND WILDLIFE

Department of Agriculture–
  Forest Service
  Soil Conservation Service
Department of Commerce–
  National Oceanic and Atmospheric Administration (marine species)
Department of the Interior–
  Bureau of Sport Fisheries and Wildlife
  Bureau of Land Management
  Bureau of Outdoor Recreation
Environmental Protection Agency

SOLID WASTE

Atomic Energy Commission (radioactive waste)
Department of Defense–
  Army Corps of Engineers
Department of Health, Education, and Welfare
Department of the Interior–
  Bureau of Mines (mineral waste, mine acid waste, municipal solid waste, recycling)
  Bureau of Land Management (public lands)
  Bureau of Indian Affairs (Indian lands)
  Geological Survey (geologic and hydrologic effects)
  Office of Saline Water (demineralization)

Department of Transportation–
    Coast Guard (ship sanitation)
Environmental Protection Agency
River Basin Commissions (as geographically appropriate)
Water Resources Council

NOISE

Department of Commerce–
    National Bureau of Standards
Department of Health, Education, and Welfare
Department of Housing and Urban Development (land use and building materials aspects)
Department of Labor–
    Occupational Safety and Health Administration
Department of Transportation–
    Assistant Secretary for Systems Development and Technology
    Federal Aviation Administration, Office of Noise Abatement
Environmental Protection Agency
National Aeronautics and Space Administration

RADIATION

Atomic Energy Commission
Department of Commerce–
    National Bureau of Standards
Department of Health, Education, and Welfare
Department of the Interior–
    Bureau of Mines (uranium mines)
    Mining Enforcement and Safety Administration (uranium mines)
Environmental Protection Agency

HAZARDOUS SUBSTANCES

Toxic Materials

Atomic Energy Commission (radioactive substances)
Department of Agriculture–
    Agricultural Research Service
    Consumer and Marketing Service
Department of Commerce–
    National Oceanic and Atmospheric Administration
Department of Defense
Department of Health, Education, and Welfare
Environmental Protection Agency

Food Additives and Contamination of Foodstuffs

Department of Agriculture–
    Consumer and Marketing Service (meat and poultry products)

Department of Health, Education, and Welfare
Environmental Protection Agency

Pesticides

Department of Agriculture—
- Agricultural Research Service (biological controls, food and fiber production)
- Consumer and Marketing Service
- Forest Service

Department of Commerce—
- National Oceanic and Atmospheric Administration

Department of Health, Education, and Welfare
Department of the Interior—
- Bureau of Sport Fisheries and Wildlife (fish and wildlife effects)
- Bureau of Land Management (public lands)
- Bureau of Indian Affairs (Indian lands)
- Bureau of Reclamation (irrigated lands)

Environmental Protection Agency

Transportation and Handling of Hazardous Materials

Atomic Energy Commission (radioactive substances)
Department of Commerce—
- Maritime Administration
- National Oceanic and Atmospheric Administration (effects on marine life and the coastal zone)

Department of Defense—
- Armed Services Explosive Safety Board
- Army Corps of Engineers (navigable waterways)

Department of Transportation—
- Federal Highway Administration, Bureau of Motor Carrier Safety
- Coast Guard
- Federal Railroad Administration
- Federal Aviation Administration
- Assistant Secretary for Systems Development and Technology
- Office of Hazardous Materials
- Office of Pipeline Safety

Environmental Protection Agency

ENERGY SUPPLY AND NATURAL RESOURCES DEVELOPMENT

Electric Energy Development, Generation, and Transmission, and Use

Atomic Energy Commission (nuclear)
Department of Agriculture—
- Rural Electrification Administration (rural areas)

Department of Defense—
- Army Corps of Engineers (hydro)

Department of Health, Education, and Welfare (radiation effects)
Department of Housing and Urban Development (urban areas)

Department of the Interior–
- Bureau of Indian Affairs (Indian lands)
- Bureau of Land Management (public lands)
- Bureau of Reclamation
- Power Marketing Administrations
- Geological Survey
- Bureau of Sport Fisheries and Wildlife
- Bureau of Outdoor Recreation
- National Park Service

Environmental Protection Agency
Federal Power Commission (hydro, transmission, and supply)
River Basin Commissions (as geographically appropriate)
Tennessee Valley Authority
Water Resources Council

Petroleum Development, Extraction, Refining, Transport, and Use

Department of the Interior–
- Office of Oil and Gas
- Bureau of Mines
- Geological Survey
- Bureau of Land Management (public lands and outer continental shelf)
- Bureau of Indian Affairs (Indian lands)
- Bureau of Sport Fisheries and Wildlife (effects on fish and wildlife)
- Bureau of Outdoor Recreation
- National Park Service
- Department of Transportation (Transport and Pipeline Safety)
- Environmental Protection Agency
- Interstate Commerce Commission

Natural Gas Development, Production, Transmission, and Use

Department of Housing and Urban Development (urban areas)
Department of the Interior–
- Office of Oil and Gas
- Geological Survey
- Bureau of Mines
- Bureau of Land Management (public lands)
- Bureau of Indian Affairs (Indian lands)
- Bureau of Sport Fisheries and Wildlife
- Bureau of Outdoor Recreation
- National Park Service

Department of Transportation (transport and safety)
Environmental Protection Agency
Federal Power Commission (production, transmission, and supply)
Interstate Commerce Commission

Coal and Minerals Development, Mining, Conversion, Processing, Transport, and Use

Appalachian Regional Commission
Department of Agriculture–
- Forest Service

Department of Commerce
Department of the Interior–
- Office of Coal Research
- Mining Enforcement and Safety Administration
- Bureau of Mines
- Geological Survey
- Bureau of Indian Affairs (Indian lands)
- Bureau of Land Management (public lands)
- Bureau of Sport Fisheries and Wildlife
- Bureau of Outdoor Recreation
- National Park Service

Department of Labor–
- Occupational Safety and Health Administration

Department of Transportation
Environmental Protection Agency
Interstate Commerce Commission
Tennessee Valley Authority

Renewable Resource Development, Production, Management, Harvest, Transport, and Use

Department of Agriculture–
- Forest Service
- Soil Conservation Service

Department of Commerce
Department of Housing and Urban Development (building materials)
Department of the Interior–
- Geological Survey
- Bureau of Land Management (public lands)
- Bureau of Indian Affairs (Indian lands)
- Bureau of Sport Fisheries and Wildlife
- Bureau of Outdoor Recreation
- National Park Service

Department of Transportation
Environmental Protection Agency
Interstate Commerce Commission (freight rates)

Energy and Natural Resources Conservation

Department of Agriculture–
- Forest Service
- Soil Conservation Service

Department of Commerce–
- National Bureau of Standards (energy efficiency)

Department of Housing and Urban Development–
- Federal Housing Administration (housing standards)

Department of the Interior–
- Office of Energy Conservation
- Bureau of Mines
- Bureau of Reclamation
- Geological Survey
- Power Marketing Administration

Department of Transportation
Environmental Protection Agency
Federal Power Commission
General Services Administration (design and operation of buildings)
Tennessee Valley Authority

LAND USE AND MANAGEMENT

Land Use Changes, Planning and Regulation of Land Development

Department of Agriculture–
 Forest Service (forest lands)
 Agricultural Research Service (agricultural lands)
Department of Housing and Urban Development
Department of the Interior–
 Office of Land Use and Water Planning
 Bureau of Land Management (public lands)
 Bureau of Indian Affairs (Indian lands
 Bureau of Sport Fisheries and Wildlife (wildlife refuges)
 Bureau of Outdoor Recreation (recreation lands)
 National Park Service (NPS units)
Department of Transportation
Environmental Protection Agency (pollution effects)
National Aeronautics and Space Administration (remote sensing)
River Basins Commissions (as geographically appropriate)

Public Land Management

Department of Agriculture–
 Forest Service (forests)
Department of Defense
Department of the Interior–
 Bureau of Land Management
 Bureau of Indian Affairs (Indian lands)
 Bureau of Sport Fisheries and Wildlife (wildlife refuges)
 Bureau of Outdoor Recreation (recreation lands)
 National Park Service (NPS units)
Federal Power Commission (project lands)
General Services Administration
National Aeronautics and Space Administration (remote sensing)
Tennessee Valley Authority (project lands)

PROTECTION OF ENVIRONMENTALLY CRITICAL AREAS– FLOODPLAINS, WETLANDS, BEACHES AND DUNES, UNSTABLE SOILS, STEEP SLOPES, AQUIFER RECHARGE AREAS, ETC.

Department of Agriculture–
 Agricultural Stabilization and Conservation Service

Soil Conservation Service
Forest Service
Department of Commerce–
National Oceanic and Atmospheric Administration (coastal areas)
Department of Defense–
Army Corps of Engineers
Department of Housing and Urban Development (urban and floodplain areas)
Department of the Interior–
Office of Land Use and Water Planning
Bureau of Outdoor Recreation
Bureau of Reclamation
Bureau of Sport Fisheries and Wildlife
Bureau of Land Management
Geological Survey
Environmental Protection Agency (pollution effects)
National Aeronautics and Space Administration (remote sensing)
River Basins Commissions (as geographically appropriate)
Water Resources Council

LAND USE IN COASTAL AREAS

Department of Agriculture–
Soil Conservation Service (soil stability, hydrology)
Department of Commerce–
National Oceanic and Atmospheric Administration (impact on marine life and coastal zone management)
Department of Defense–
Army Corps of Engineers (beaches, dredge and fill permits, Refuse Act permits)
Department of Housing and Urban Development (urban areas)
Department of the Interior–
Office of Land Use and Water Planning
Bureau of Sport Fisheries and Wildlife
National Park Service
Geological Survey
Bureau of Outdoor Recreation
Bureau of Land Management (public lands)
Department of Transportation–
Coast Guard (bridges, navigation)
Environmental Protection Agency (pollution effects)
National Aeronautics and Space Administration (remote sensing)

REDEVELOPMENT AND CONSTRUCTION IN BUILT-UP AREAS

Department of Commerce–
Economic Development Administration (designated areas)
Department of Housing and Urban Development
Department of the Interior–
Office of Land Use and Water Planning

Department of Transportation
Environmental Protection Agency
General Services Administration
Office of Economic Opportunity

DENSITY AND CONGESTION MITIGATION

Department of Health, Education, and Welfare
Department of Housing and Urban Development
Department of the Interior—
    Office of Land Use and Water Planning
    Bureau of Outdoor Recreation
Department of Transportation
Environmental Protection Agency

NEIGHBORHOOD CHARACTER AND CONTINUITY

Department of Health, Education, and Welfare
Department of Housing and Urban Development
National Endowment for the Arts
Office of Economic Opportunity

IMPACTS ON LOW-INCOME POPULATIONS

Department of Commerce—
    Economic Development Administration (designated areas)
Department of Health, Education, and Welfare
Department of Housing and Urban Development
Office of Economic Opportunity

HISTORIC, ARCHITECTURAL, AND ARCHEOLOGICAL PRESERVATION

Advisory Council on Historic Preservation
Department of Housing and Urban Development
Department of the Interior—
    National Park Service
    Bureau of Land Management (public lands)
    Bureau of Indian Affairs (Indian lands)
General Services Administration
National Endowment for the Arts

SOIL AND PLANT CONSERVATION AND HYDROLOGY

Department of Agriculture—
    Soil Conservation Service
    Agricultural Service
    Forest Service

Department of Commerce–
  National Oceanic and Atmospheric Administration
Department of Defense–
  Army Corps of Engineers (dredging, aquatic plants)
Department of Health, Education, and Welfare
Department of the Interior–
  Bureau of Land Management
  Bureau of Sport Fisheries and Wildlife
  Geological Survey
  Bureau of Reclamation
Environmental Protection Agency
National Aeronautics and Space Administration (remote sensing)
River Basin Commissions (as geographically appropriate)
Water Resources Council

OUTDOOR RECREATION

Department of Agriculture–
  Forest Service
  Soil Conservation Service
Department of Defense–
  Army Corps of Engineers
Department of Housing and Urban Development (urban areas)
Department of the Interior–
  Bureau of Land Management
  National Park Service
  Bureau of Outdoor Recreation
  Bureau of Sport Fisheries and Wildlife
  Bureau of Indian Affairs
Environmental Protection Agency
National Aeronautics and Space Administration (remote sensing)
River Basin Commissions (as geographically appropriate)
Water Resources Council

APPENDIX III—OFFICES WITHIN FEDERAL AGENCIES AND FEDERAL-STATE AGENCIES FOR INFORMATION REGARDING THE AGENCIES' NEPA ACTIVITIES AND FOR RECEIVING OTHER AGENCIES' IMPACT STATEMENTS FOR WHICH COMMENTS ARE REQUESTED

ADVISORY COUNCIL ON HISTORIC PRESERVATION

Office of Architectural and Environmental Preservation, Advisory Council on Historic Preservation, Suite 430, 1522 K Street, N.W., Washington, D.C. 20005 254-3974

DEPARTMENT OF AGRICULTURE[3]

Office of the Secretary, Attn: Coordinator Environmental Quality Activities, U.S. Department of Agriculture, Washington, D.C. 20250 447-3965

APPALACHIAN REGIONAL COMMISSION

Office of the Alternate Federal Co-Chairman, Appalachian Regional Commission, 1666 Connecticut Avenue, N.W., Washington, D.C. 20235 967-4103

DEPARTMENT OF THE ARMY (CORPS OF ENGINEERS)

Executive Director of Civil Works, Office of the Chief of Engineers, U.S. Army Corps of Engineers, Washington, D.C. 20314 693-7168

ATOMIC ENERGY COMMISSION

For nonregulatory matters: Office of Assistant General Manager for Biomedical and Environmental Research and Safety Programs, Atomic Energy Commission, Washington, D.C. 20545 973-3208

For regulatory matters: Office of the Assistant Director for Environmental Projects, Atomic Energy Commission, Washington, D.C. 20545 973-7531

DEPARTMENT OF COMMERCE

Office of the Deputy Assistant Secretary for Environmental Affairs, U.S. Department of Commerce, Washington, D.C. 20230 967-4335

DEPARTMENT OF DEFENSE

Office of the Assistant Secretary for Defense (Health and Environment), U.S. Department of Defense, Room 3E172, The Pentagon, Washington, D.C. 20301 697-2111

DELAWARE RIVER BASIN COMMISSION

Office of the Secretary, Delaware River Basin Commission, Post Office Box 360, Trenton, N.J. 08603 (609) 883-9500

ENVIRONMENTAL PROTECTION AGENCY[4]

Director, Office of Federal Activities, Environmental Protection Agency, 401 M Street, S.W., Washington, D.C. 20460 755-0777

[3] Requests for comments or information from individual units of the Department of Agriculture, e.g., Soil Conservation Service, Forest Service, etc. should be sent to the Office of the Secretary, Department of Agriculture, at the address given above.

[4] Contact the Office of Federal Activities for environmental statements concerning legislation, regulations, national program proposals or other major policy issues.

For all other EPA consultation, contact the Regional Administrator in whose area the

FEDERAL POWER COMMISSION

Commission's Advisor on Environmental Quality, Federal Power Commission, 825 N. Capitol Street, N.E., Washington, D.C. 20426 386-6084

GENERAL SERVICES ADMINISTRATION

Office of Environmental Affairs, Office of the Deputy Administrator for Special Projects, General Services Administration, Washington, D.C. 20405 343-4161

GREAT LAKES BASIN COMMISSION

Office of the Chairman, Great Lakes Basin Commission, 3475 Plymouth Road, P.O. Box 999, Ann Arbor, Michigan 48105 (313) 769-7431

DEPARTMENT OF HEALTH, EDUCATION AND WELFARE[5]

Office of Environmental Affairs, Office of the Assistant Secretary for Administration and Management, Department of Health, Education and Welfare, Washington, D.C. 20202 963-4456

---

proposed action (e.g., highway or water resource construction projects) will take place. The Regional Administrators will coordinate the EPA review. Addresses of the Regional Administrators, and the areas covered by their regions are as follows: Regional Administrator, I, U.S. Environmental Protection Agency, Room 2303, John F. Kennedy Federal Bldg., Boston, Mass. 02203, (617) 223-7210 (Connecticut, Maine, Massachusetts, New Hampshire, Rhode Island, Vermont); Regional Administrator, II, U.S. Environmental Protection Agency, Room 908, 26 Federal Plaza, New York, New York 10007, (212) 264-2525 (New Jersey, New York, Puerto Rico, Virgin Islands); Regional Administrator, III, U.S. Environmental Protection Agency, Curtis Bldg., 6th & Walnut Sts., Philadelphia, Pa. 19106, (215) 597-9801 (Delaware, Maryland, Pennsylvania, Virginia, West Virginia, District of Columbia); Regional Administrator, IV, U.S. Environmental Protection Agency, 1421 Peachtree Street, N.E., Atlanta, Ga. 30309, (404) 526-5727 (Alabama, Florida, Georgia, Kentucky, Mississippi, North Carolina, South Carolina, Tennessee); Regional Administrator, V, U.S. Environmental Protection Agency, 1 N. Wacker Drive, Chicago, Illinois 60606, (312) 353-5250 (Illinois, Indiana, Michigan, Minnesota, Ohio, Wisconsin); Regional Administrator, VI, U.S. Environmental Protection Agency, 1600 Patterson Street, Suite 1100, Dallas, Texas 75201, (214) 749-1962 (Arkansas, Louisiana, New Mexico, Texas, Oklahoma); Regional Administrator, VII, U.S. Environmental Protection Agency, 1735 Baltimore Avenue, Kansas City, Missouri 64108, (816) 374-5493 (Iowa, Kansas, Missouri, Nebraska); Regional Administrator, VIII, U.S. Environmental Protection Agency, Suite 900, Lincoln Tower, 1860 Lincoln Street, Denver, Colorado 80203, (303) 837-3895 (Colorado, Montana, North Dakota, South Dakota, Utah, Wyoming); Regional Administrator, IX, U.S. Environmental Protection Agency, 100 California Street, San Francisco, California 94111, (415) 556-2320 (Arizona, California, Hawaii, Nevada, American Samoa, Guam, Trust Territories of Pacific Islands, Wake Island); Regional Administrator, X, U.S. Environmental Protection Agency, 1200 Sixth Avenue, Seattle, Washington 98101, (206) 442-1220 (Alaska, Idaho, Oregon, Washington).

[5]Contact the Office of Environmental Affairs for information on HEW's environmental statements concerning legislation, regulations, national program proposals or other major policy issues, and for all requests for HEW comment on impact statements of other agencies.

For information with respect to HEW actions occurring within the jurisdiction of the Departments' Regional Directors, contact the appropriate Regional Environmental Officer: Region I: Regional Environmental Officer, U.S. Department of Health, Education and Welfare, Room 2007B, John F. Kennedy Center, Boston, Massachusetts 02203 (617) 223-6837; Region II: Regional Environmental Officer, U.S. Department of Health, Education and Welfare, Federal

DEPARTMENT OF HOUSING AND URBAN DEVELOPMENT[6]

Director, Office of Community and Environmental Standards, Department of Housing and Urban Development, Room 7206, Washington, D.C. 20410 755-5980

DEPARTMENT OF THE INTERIOR[7]

Director, Office of Environmental Project Review, Department of the Interior, Interior Building, Washington, D.C. 20240 343-3891

Building, 26 Federal Plaza, New York, New York 10007, (212) 264-1308; Region III: Regional Environmental Officer, U.S. Department of Health, Education and Welfare, P.O. Box 13716, Philadelphia, Pennsylvania 19101, (215) 597-6498; Region IV: Regional Environmental Officer, U.S. Department of Health, Education and Welfare, Room 404, 50 Seventh Street, N.E., Atlanta, Georgia 30323, (404) 526-5817; Region V: Regional Environmental Officer, U.S. Department of Health, Education and Welfare, Room 712, New Post Office Building, 433 West Van Buren Street, Chicago, Illinois 60607, (312) 353-1644; Region VI: Regional Environmental Officer, U.S. Department of Health, Education and Welfare, 1114 Commerce Street, Dallas, Texas 75202, (214) 749-2236; Region VII: Regional Environmental Officer, U.S. Department of Health, Education and Welfare, 601 East 12th Street, Kansas City, Missouri 64106, (816) 374-3584; Region VIII: Regional Environmental Officer U.S. Department of Health, Education and Welfare, 9017 Federal Building, 19th and Stout Streets, Denver, Colorado 80202, (303) 837-4178; Region IX: Regional Environmental Officer, U.S. Department of Health, Education and Welfare, 50 Fulton Street, San Francisco, California 94102, (415) 556-1970; Region X: Regional Environmental Officer, U.S. Department of Health, Education and Welfare, Arcade Plaza Building, 1321 Second Street, Seattle, Washington 98101, (206) 442-0490.

[6] Contact the Director with regard to environmental impacts of legislation, policy statements, program regulations and procedures, and precedent-making project decisions. For all other HUD consultation, contact the HUD Regional Administrator in whose jurisdiction the project lies, as follows: Regional Administrator I, Environmental Clearance Officer, U.S. Department of Housing and Urban Development, Room 405, John F. Kennedy Federal Building, Boston, Mass. 02203, (617) 223-4066; Regional Administrator II, Environmental Clearance Officer, U.S. Department of Housing and Urban Development, 26 Federal Plaza, New York, New York 10007, (212) 264-8068; Regional Administrator III, Environmental Clearance Officer, U.S. Department of Housing and Urban Development, Curtis Building, Sixth and Walnut Street, Philadelphia, Pennsylvania 19106, (215) 597-2560; Regional Administrator IV, Environmental Clearance Officer, U.S. Department of Housing and Urban Development, Peachtree-Seventh Building, Atlanta, Georgia 30323, (404) 526-5585; Regional Administrator V, Environmental Clearance Officer, U.S. Department of Housing and Urban Development, 360 North Michigan Avenue, Chicago, Illinois 60601, (312) 353-5680; Regional Administrator VI, Environmental Clearance Officer, U.S. Department of Housing and Urban Development, Federal Office Building, 819 Taylor Street, Fort Worth, Texas 76102, (817) 334-2867; Regional Administrator VII, Environmental Clearance Officer, U.S. Department of Housing and Urban Development, 911 Walnut Street, Kansas City, Missouri 64106, (816) 374-2661; Regional Administrator VIII, Environmental Clearance Officer, U.S. Department of Housing and Urban Development, Samsonite Building, 1051 South Broadway, Denver, Colorado 80209, (303) 837-4061; Regional Administrator IX, Environmental Clearance Officer, U.S. Department of Housing and Urban Development, 450 Golden Gate Avenue, Post Office Box 36003, San Francisco, California 94102, (415) 556-4752; Regional Administrator X, Environmental Clearance Officer, U.S. Department of Housing and Urban Development, Room 226, Arcade Plaza Building, Seattle, Washington 98101, (206) 583-5415.

[7] Requests for comments or information from individual units of the Department of the Interior should be sent to the Office of Environmental Project Review at the address given above.

INTERSTATE COMMERCE COMMISSION

Office of Proceedings, Interstate Commerce Commission, Washington, D.C. 20423 343-6167

DEPARTMENT OF LABOR

Assistant Secretary for Occupational Safety and Health, Department of Labor, Washington, D.C. 20210 961-3405

MISSOURI RIVER BASINS COMMISSION

Office of the Chairman, Missouri River Basins Commission, 10050 Regency Circle, Omaha, Nebraska 68114 (402) 897-5714

NATIONAL AERONAUTICS AND SPACE ADMINISTRATION

Office of the Comptroller, National Aeronautics and Space Administration, Washington, D.C. 20546 755-8440

NATIONAL CAPITAL PLANNING COMMISSION

Office of Environmental Affairs, Office of the Executive Director, National Capital Planning Commission, Washington, D.C. 20576 382-7200

NATIONAL ENDOWMENT FOR THE ARTS

Office of Architecture and Environmental Arts Program, National Endowment for the Arts, Washington, D.C. 20506 382-5763

NEW ENGLAND RIVER BASINS COMMISSION

Office of the Chairman, New England River Basins Commission, 55 Court Street, Boston, Mass. 02108 (617) 223-6244

OFFICE OF ECONOMIC OPPORTUNITY

Office of the Director, Office of Economic Opportunity, 1200 19th Street, N.W., Washington, D.C. 20506 254-6000

OHIO RIVER BASIN COMMISSION

Office of the Chairman, Ohio River Basin Commission, 36 East 4th Street, Suite 20820, Cincinnati, Ohio 45202 (513) 684-3831

PACIFIC NORTHWEST RIVER BASINS COMMISSION

Office of the Chairman, Pacific Northwest River Basins Commission, 1 Columbia River, Vancouver, Washington 98660 (206) 695-3606

SOURIS-RED-RAINY RIVER BASINS COMMISSION

Office of the Chairman, Souris-Red-Rainy River Basins Commission, Suite 6, Professional Building, Holiday Mall, Moorhead, Minnesota 56560 (701) 237-5227

DEPARTMENT OF STATE

Office of the Special Assistant to the Secretary for Environmental Affairs, Department of State, Washington, D.C. 20520 632-7964

SUSQUEHANNA RIVER BASIN COMMISSION

Office of the Executive Director, Susquehanna River Basin Commission, 5012 Lenker Street, Mechanicsburg, Pa. 17055 (717) 737-0501

TENNESSEE VALLEY AUTHORITY

Office of the Director of Environmental Research and Development, Tennessee Valley Authority, 720 Edney Building, Chattanooga, Tennessee 37401 (615) 755-2002

DEPARTMENT OF TRANSPORTATION[8]

Director, Office of Environmental Quality, Office of the Assistant Secretary for Environment, Safety, and Consumer Affairs, Department of Transportation, Washington, D.C. 20590 426-4357

For other administrations not listed above, contact the Office of Environmental Quality, Department of Transportation, at the address given above.

For comments on other agencies' environmental statements, contact the appropriate administration's regional office. If more than one administration within the Department of Transportation is to be requested to comment, contact the Secretarial Representative in the appropriate Regional Office for coordination of the Department's comments:

[8] Contact the Office of Environmental Quality, Department of Transportation, for information on DOT's environmental statements concerning legislation, regulations, national program proposals, or other major policy issues.

For information regarding the Department of Transportation's other environmental statements, contact the national office for the appropriate administration: *U.S. Coast Guard*, Office of Marine Environment and Systems, U.S. Coast Guard, 400 7th Street, S.W., Washington, D.C. 20590, 426-2007; *Federal Aviation Administration*, Office of Environmental Quality, Federal Aviation Administration, 800 Independence Avenue, S.W., Washington, D.C. 20591, 426-8406; *Federal Highway Administration*, Office of Environmental Policy, Federal Highway Administration, 400 7th Street, S.W., Washington, D.C. 20590, 426-0351; *Federal Railroad Administration*, Office of Policy and Plans, Federal Railroad Administration, 400 7th Street, S.W., Washington, D.C. 20590, 426-1567; *Urban Mass Transportation Administration*, Office of Program Operations, Urban Mass Transportation Administration, 400 7th Street, S.W., Washington, D.C. 20590, 426-4020.

SECRETARIAL REPRESENTATIVE

Region I Secretarial Representative, U.S. Department of Transportation, Transportation Systems Center, 55 Broadway, Cambridge, Massachusetts 02142 (617) 494-2709

Region II Secretarial Representative, U.S. Department of Transportation, 26 Federal Plaza, Room 1811, New York, New York 10007 (212) 264-2672

Region III Secretarial Representative, U.S. Department of Transportation, Mall Building, Suite 1214, 325 Chestnut Street, Philadelphia, Pennsylvania 19108 (215) 597-0407

Region IV Secretarial Representative, U.S. Department of Transportation, Suite 515, 1720 Peachtree Rd., N.W. Atlanta, Georgia, 30309 (404) 526-3738

Region V Secretarial Representative, U.S. Department of Transportation, 17th Floor, 300 S. Wacker Dirve, Chicago, Illinois 60606 (312) 353-4000

Region VI Secretarial Representative, U.S. Department of Transportation, 9-C-18 Federal Center, 1100 Commerce Street, Dallas, Texas 75202 (214) 749-1851

Region VII Secretarial Representative, U.S. Department of Transportation, 601 E. 12th Street, Room 634, Kansas City, Missouri 64106 (816) 374-2761

Region VIII Secretarial Representative, U.S. Department of Transportation, Prudential Plaza, Suite 1822, 1050 17th Street, Denver, Colorado 80225 (303) 837-3242

Region IX Secretarial Representative, U.S. Department of Transportation, 450 Golden Gate Avenue, Box 36133, San Francisco, California 94102 (415) 556-5961

Region X Secretarial Representative, U.S. Department of Transportation, 1321 Second Avenue, Room 507, Seattle, Washington 98101 (206) 442-0590

FEDERAL AVIATION ADMINISTRATION

New England Region, Office of the Regional Director, Federal Aviation Administration, 154 Middlesex Street, Burlington, Massachusetts 01803 (617) 272-2350

Eastern Region, Office of the Regional Director, Federal Aviation Administration, Federal Building, JFK International Airport, Jamaica, New York 11430 (212) 995-3333

Southern Region, Office of the Regional Director, Federal Aviation Administration, P.O. Box 20636, Atlanta, Georgia 30320 (404) 526-7222

Great Lakes Region, Office of the Regional Director, Federal Aviation Administration, 2300 East Devon, Des Plaines, Illinois 60018 (312) 694-4500

Southwest Region, Office of the Regional Director, Federal Aviation Administration, P.O. Box 1689, Fort Worth, Texas 76101 (817) 624-4911

Central Region, Office of the Regional Director, Federal Aviation Administration, 601 E. 12th Street, Kansas City, Missouri 64106 (816) 374-5626

Rocky Mountain Region, Office of the Regional Director, Federal Aviation Administration, Park Hill Station, P.O. Box 7213, Denver, Colorado 80207 (303) 837-3646

Western Region, Office of the Regional Director, Federal Aviation Administration, P.O. Box 92007, WorldWay Postal Center, Los Angeles, California 90009 (213) 536-6427

Northwest Region, Office of the Regional Director, Federal Aviation Administration, FAA Building, Boeing Field, Seattle, Washington 98108 (206) 767-2780

FEDERAL HIGHWAY ADMINISTRATION

Region 1, Regional Administrator, Federal Highway Administration, 4 Normanskill Boulevard, Delmar, New York 12054 (518) 472-6476

Region 3, Regional Administrator, Federal Highway Administration, Room 1621, George H. Fallon Federal Office Building, 31 Hopkins Plaza, Baltimore, Maryland 21201 (301) 962-2361

Region 4, Regional Administrator, Federal Highway Administration, Suite 200, 1720 Peachtree Road, N.W., Atlanta, Georgia 30309 (404) 526-5078

Region 5, Regional Administrator, Federal Highway Administration, Dixie Highway, Homewood, Illinois 60430 (312) 799-6300

Region 6, Regional Administrator, Federal Highway Administration, 819 Taylor Street, Fort Worth, Texas 76102 (817) 334-3232

Region 7, Regional Administrator, Federal Highway Administration, P.O. Box 7186, Country Club Station, Kansas City, Missouri 64113 (816) 361-7563

Region 8, Regional Administrator, Federal Highway Administration, Room 242, Building 40, Denver Federal Center, Denver, Colorado 80225

Region 9, Regional Administrator, Federal Highway Administration, 450 Golden Gate Avenue, Box 36096, San Francisco, California 94102 (415) 556-3895

Region 10, Regional Administrator, Federal Highway Administration, Room 412, Mohawk Building, 222 S.W. Morrison Street, Portland, Oregon 97204 (503) 221-2065

URBAN MASS TRANSPORTATION ADMINISTRATION

Region I, Office of the UMTA Representative, Urban Mass Transportation Administration, Transportation Systems Center, Technology Building, Room 277, 55 Broadway, Boston Massachusetts 02142 (617) 494-2055

Region II, Office of the UMTA Representative, Urban Mass Transportation Administration, 26 Federal Plaza, Suite 1809, New York, New York 10007 (212) 264-8162

Region III, Office of the UMTA Representative, Urban Mass Transportation Administration, Mall Building, Suite 1214, 325 Chestnut Street, Philadelphia, Pennsylvania 19106 (215) 597-0407

Region IV, Office of UMTA Representative, Urban Mass Transportation Administration, 1720 Peachtree Road, Northwest, Suite 501, Atlanta, Georgia 30309 (404) 526-3948

Region V, Office of the UMTA Representative, Urban Mass Transportation Administration, 300 South Wacker Drive, Suite 700, Chicago, Illinois 60606 (312) 353-6005

Region VI, Office of the UMTA Representative, Urban Mass Transportation Administration, Federal Center, Suite 9E24, 1100 Commerce Street, Dallas, Texas 75202 (214) 749-7322

Region VII, Office of the UMTA Representative, Urban Mass Transportation Administration, c/o FAA Management Systems Division, Room 1564D, 601 East 12th Street, Kansas City, Missouri 64106 (816) 374-5567

Region VIII, Office of the UMTA Representative, Urban Mass Transportation Administration, Prudential Plaza, Suite 1822, 1050 17th Street, Denver, Colorado 80202 (303) 837-3242

Region IX, Office of the UMTA Representative, Urban Mass Transportation Administration, 450 Golden Gate Avenue, Box 36125, San Francisco, California 94102 (415) 556-2884

Region X, Office of the UMTA Representative, Urban Mass Transportation Administration, 1321 Second Avenue, Suite 5079, Seattle, Washington (206) 442-0590

DEPARTMENT OF THE TREASURY

Office of Assistant Secretary for Administration, Department of the Treasury, Washington, D.C. 20220 964-5391

UPPER MISSISSIPPI RIVER BASIN COMMISSION

Office of the Chairman, Upper Mississippi River Basin Commission, Federal Office Building, Fort Snelling, Twin Cities, Minnesota 55111 (612) 725-4690

WATER RESOURCES COUNCIL

Office of the Associate Director, Water Resources Council, 2120 L Street, N.W., Suite 800, Washington, D.C. 20037 254-6442

APPENDIX IV—STATE AND LOCAL AGENCY REVIEW OF IMPACT STATEMENTS

1. OMB Circular No. A-95 through its system of clearinghouses provides a means for securing the views of State and local environmental agencies, which can assist in the preparation of impact statements. Under A-95, review of the proposed project in the case of federally assisted projects (Part I of A-95) generally takes place prior to the preparation of the impact statement. Therefore, comments on the environmental effects of the proposed project that are secured during this stage of the A-95 process represent inputs to the environmental impact statement.

2. In the case of direct Federal development (Part II of A-95), Federal agencies are required to consult with clearinghouses at the earliest practicable time in the planning of the project or activity. Where such consultation occurs prior to completion of the draft impact statement, comments relating to the environmental effects of the proposed action would also represent inputs to the environmental impact statement.

3. In either case, whatever comments are made on environmental effects of proposed Federal or federally assisted projects by clearinghouses, or by State and local environmental agencies through clearinghouses, in the course of the A-95 review should be attached to the draft impact statement when it is circulated for review. Copies of the statement should be sent to the agencies making such comments. Whether those agencies then elect to comment again on the basis of the draft impact statement is a matter to be left to the discretion of the commenting

agency depending on its resources, the significance of the project, and the extent to which its earlier comments were considered in preparing the draft statement.

4. The clearinghouses may also be used, by mutual agreement, for securing reviews of the draft environmental impact statement. However, the Federal agency may wish to deal directly with appropriate State or local agencies in the review of impact statements because the clearinghouses may be unwilling or unable to handle this phase of the process. In some cases, the Governor may have designated a specific agency, other than the clearinghouse, for securing reviews of impact statements. In any case, the clearinghouses should be sent copies of the impact statement.

5. To aid clearinghouses in coordinating State and local comments, draft statements should include copies of State and local agency comments made earlier under the A-95 process and should indicate on the summary sheet those other agencies from which comments have been requested, as specified in Appendix I of the CEQ Guidelines.

[FR Doc. 73-15783 Filed 7-31-73; 8:45 am]

Appendix C

Data Sources for Environmental Factors

The sources of information for environmental factors will vary somewhat from state to state, but the listing provided herein can be adapted by substituting the appropriate agency or association (1). For convenience the listing contains the factors in the sequence in which they are listed in *U.S. Geological Survey Circular* 645. The sources include the federal agencies with jurisdiction by law or with special expertise to comment on various types of environmental impacts, as listed in Appendix II to the Council on Environmental Quality Guidelines. Other sources are state and local agencies, educational institutions, and private associations.

The sources may have published and/or unpublished information for the subjects listed, and in some instances they may have only limited data without analysis.

Subject	Sources of information
A Physical and chemical characteristics.	
1 Earth.	
a Mineral resources.	a U.S. Bureau of Mines, U.S. Geological Survey, state bureaus of economic geology, state railroad commissions (oil and gas), geology departments at state and private universities and

	colleges in the region, Mid-continent Oil & Gas Association, Geological Society of America chapters in the region.
b Construction material.	b U.S. Bureau of Mines, U.S. Geological Survey, Corps of Engineers, U.S. Soil Conservation Service, Bureau of Reclamation, state bureaus of economic geology, state highway departments, civil engineering departments at universities.
c Soils.	c U.S. Soil Conservation Service, U.S. Agricultural Research Service, state universities, state agricultural experiment stations.
d Land form.	d U.S. Geological Survey, university geology and geography departments.
e Force fields and background radiation.	e Energy Research and Development Administration Technical Information Center, Environmental Protection Agency Office of Radiation, Bureau of Mines (uranium mines), state departments of health.
f Unique physical features.	f National Park Service, U.S. Geological Survey, state historical survey committees, universities and colleges.
2 Water.	
a Surface.	a U.S. Geological Survey, Corps of Engineers, International Boundary and Water Commission, state water development boards, state water rights commissions.
b Ocean.	b U.S. Geological Survey, National Oceanic and Atmospheric Administration, Corps of Engineers, Environmental Protection Agency, universities and colleges.
c Underground.	c U.S. Geological Survey, state water development boards, underground water districts, university and college geology departments.
d Quality.	d U.S. Geological Survey, Environmental Protection Agency, Corps of Engineers, Coast Guard, state water development boards, state water quality boards, state health departments, universities and colleges.
e Temperature.	e Same as d.
f Recharge.	f U.S. Geological Survey, state water

Factor	Source
	development boards, underground water districts.
g Snow, ice, and permafrost.	g Weather bureaus.
3 Atmosphere.	
a Quality (gases, particulates).	a Environmental Protection Agency Air Pollution Control Office, National Oceanic and Atmospheric Administration, state air control boards, local or regional air control boards.
b Climate (micro, macro).	b National Oceanic and Atmospheric Administration, meteorology departments at universities and colleges.
c Temperature.	c National Oceanic and Atmospheric Administration, Corps of Engineers, state agricultural experiment stations, state water development boards.
4 Processes.	
a Floods.	a U.S. Geological Survey, Corps of Engineers, U.S. Soil Conservation Service, International Boundary and Water Commission, state water development boards.
b Erosion.	b U.S. Soil Conservation Service, U.S. Geological Survey, Corps of Engineers, International Boundary and Water Commission, state water development boards.
c Deposition (sedimentation, precipitation).	c Same as b.
d Solution.	d U.S. Geological Survey.
e Sorption (ion exchange, complexing).	e U.S. Geological Survey.
f Compaction and settling.	f U.S. Geological Survey, state water development boards.
g Stability (slides, slumps).	g U.S. Geological Survey, U.S. Soil Conservation Service, U.S. Agricultural Research Service, universities and colleges.
h Stress–strain (earthquake).	h U.S. Geological Survey, universities and colleges.
i Air movements.	i Same as 3b.
B Biological conditions.	
1 Flora.	
a Trees.	a U.S. Forest Service, U.S. Soil Conservation Service, state forest services, universities and colleges.
b Shrubs.	b Same as a, plus U.S. Agricultural Research Service.
c Grass.	c U.S. Soil Conservation Service, U.S. Agricultural Research Service, Bureau of Land Management, state

	agricultural extension services, universities and colleges.
d Crops.	d Same as c.
e Microflora.	e U.S. Agricultural Research Service, universities and colleges.
f Aquatic plants.	f Corps of Engineers, U.S. Soil Conservation Service, U.S. Agricultural Research Service, National Oceanic and Atmospheric Administration, universities and colleges.
g Endangered species.	g U.S. Agricultural Research Service, state agricultural research services, biology departments of universities and colleges.
h Barriers.	h U.S. Agricultural Research Service universities and colleges.
i Corridors.	i Same as h.
2 Fauna.	
a Birds.	a Bureau of Sport Fisheries and Wildlife, state parks and wildlife departments, biology departments of universities and colleges.
b Land animals, including reptiles.	b Same as a.
c Fish and shellfish.	c Bureau of Sport Fisheries and Wildlife, National Marine Fisheries Service, National Oceanic and Atmospheric Administration, state parks and wildlife departments, universities and colleges.
d Benthic organisms.	d Same as c.
e Insects.	e Bureau of Sport Fisheries and Wildlife, U.S. Agricultural Research Service, state parks and wildlife departments, state agricultural research services, biology departments of universities and colleges.
f Microfauna.	f U.S. Agricultural Research Service, state agricultural research services, universities and colleges.
g Endangered species.	g Bureau of Sport Fisheries and Wildlife, National Marine Fisheries Service, National Oceanic and Atmospheric Administration, state parks and wildlife departments, universities and colleges.
h Barriers.	h Same as g.
i Corridors.	i Same as g.
C Cultural factors.	
1 Land use.	

a Wilderness and open space.	a U.S. Forest Service, National Park Service, Bureau of Outdoor Recreation, Bureau of Land Management, state parks and wildlife departments.
b Wetlands.	b U.S. Soil Conservation Service, U.S. Forest Service, National Oceanic and Atmospheric Administration, Corps of Engineers, Bureau of Sport Fisheries and Wildlife, Bureau of Outdoor Recreation, state parks and wildlife departments, universities and colleges.
c Forestry.	c U.S. Forest Service, National Park Service, Bureau of Outdoor Recreation, Bureau of Land Management, state forest services, state parks and wildlife departments, universities and colleges.
d Grazing.	d Bureau of Land Management, U.S. Soil Conservation Service, U.S. Agricultural Research Service, universities and colleges.
e Agriculture.	e U.S. Soil Conservation Service, U.S. Agricultural Research Service, state commissioners of agriculture, state agricultural extension services, universities and colleges.
f Residential.	f Department of Housing and Urban Development.
g Commercial.	g Department of Housing and Urban Development, Economic Development Administration (Department of Commerce).
h Industrial.	h Economic Development Administration (Department of Commerce), state industrial commissions.
i Mining and quarrying.	i U.S. Bureau of Mines, state bureaus of economic geology.
2 Recreation.	
a Hunting.	a Bureau of Outdoor Recreation, Bureau of Sport Fisheries and Wildlife, U.S. Forest Service, state parks and wildlife departments, universities and colleges.
b Fishing.	b Bureau of Outdoor Recreation, Bureau of Sport Fisheries and Wildlife, state parks and wildlife departments, universities and colleges.
c Boating.	c Bureau of Outdoor Recreation, state parks and wildlife departments, state

	highway departments, universities and colleges.
d Swimming.	d Bureau of Outdoor Recreation, National Park Service, state parks and wildlife departments.
e Camping and hiking.	e Bureau of Outdoor Recreation, U.S. Forest Service, National Park Service, state parks and wildlife departments, universities and colleges.
f Picnicking.	f Same as e.
g Resorts.	g Same as e, plus state tourist development agencies.
3 Aesthetics and human interest.	
a Scenic views and vistas.	a Bureau of Outdoor Recreation, National Park Service, U.S. Forest Service, state parks and wildlife departments.
b Wilderness qualities.	b Same as a.
c Open space qualities.	c Same as a.
d Landscape design.	d Same as a.
e Unique physical features.	e Same as a, plus U.S. Geological Survey.
f Parks and reserves.	f Same as a.
g Monuments.	g National Park Service, Advisory Council on Historic Preservation, state historical survey committees.
h Rare and unique species or ecosystems.	h Bureau of Sport Fisheries and Wildlife, U.S. Agricultural Research Service, state parks and wildlife departments, state agricultural research services.
i Historical or archeological sites or objects.	i National Park Service, Advisory Council on Historic Preservation, state historical survey committees (state archeologist is in this agency), state historical societies, universities and colleges.
4 Cultural status.	
a Cultural patterns (life-style).	a Bureau of the Census, Department of Housing and Urban Development, state fine arts commissions.
b Health and safety.	b Department of Health, Education and Welfare, Department of Justice, state departments of health, state departments of public safety.
c Employment.	c Department of Labor, state employment commissions, Bureau of Business Research, universities and colleges.
d Population density.	d Bureau of the Census.
5 Constructed facilities and activities.	
a Structures.	a Corps of Engineers, Bureau of Reclamation, U.S. Soil Conservation

	Service, Department of Housing and Urban Development, International Boundary and Water Commission, state water development boards, state highway departments, state water quality boards, local political subdivisions.
b Transportation network (movement, access).	b Corps of Engineers, state highway departments, local political subdivisions.
c Utility networks.	c Department of Housing and Urban Development, state departments of health, local political subdivisions.
d Waste disposal.	d Environmental Protection Agency, state water quality boards, local political subdivisions.
e Barriers and corridors.	e Bureau of Land Management, Department of Housing and Urban Development, local political subdivisions, local power companies.
D Ecological relationships.	
1 Salinization of water resources.	1 Environmental Protection Agency Office of Saline Water, U.S. Geological Survey, state water quality boards.
2 Eutrophication.	2 Environmental Protection Agency, state water quality boards.
3 Disease–insect vectors.	3 Public Health Service, state departments of health, local health organizations.
4 Food chains.	4 Bureau of Sport Fisheries and Wildlife, U.S. Agricultural Research Service, Department of Agriculture, state parks and wildlife departments, state agricultural research services.
5 Salinization of surficial materials.	5 U.S. Soil Conservation Service, U.S. Geological Survey, Environmental Protection Agency, U.S. Agricultural Research Service, state water quality boards, state agricultural research services.
6 Brush encroachment.	6 U.S. Soil Conservation Service, Bureau of Land Management, Bureau of Reclamation, Corps of Engineers, state water development boards, state agricultural research services.

SELECTED REFERENCE

1 Vandertulip, J. A.: Informational Sources for Environmental Factors, unpublished report, International Boundary and Water Commission, El Paso, Tex., 1972.

Appendix D

Impact Methodologies from Warner and Preston Study (1973)

1 Adkins, W. G. and D. Burke, Jr.: "Interim Report: Social, Economic, and Environmental Factors in Highway Decision Making," research conducted for the Texas Highway Department in cooperation with the U.S. Department of Transportation, Texas Transportation Institute, Texas A&M University, College Station, Oct. 1971.

The Adkins methodology is a checklist using a plus five to a minus five rating system for evaluating impacts but providing no guidelines for measuring impacts. The approach was developed to deal specifically with the evaluation of highway route alternatives. Because the bulk of parameters used relate directly to highway transportation, the approach is not readily adaptable to other types of projects. The parameters are broken down into the categories transportation, environmental, sociological, and economic impacts. Environmental parameters are generally deficient in ecological considerations. Social parameters emphasize community facilities and services.

Route alternatives are scored plus five to minus five in comparison to the present state of the project area, not the expected future state without the project. Since the approach uses only subjective relative estimations of impacts, the data, manpower, and cost requirements are very flexible. An interesting feature is the summarizing of the number as well as the magnitude of plus and minus ratings for each impact category. The number of pluses and minuses may be a more reliable indicator for alternative comparison since it is less affected by the arbitrariness of subjective weighting. These summarizations are additive and thus implicitly weigh all impacts equally.

2 Dee, Norbert, et al.: "Environmental Evaluation System for Water Resource Planning," report prepared by Battelle-Columbus for Bureau of Reclamation, Jan. 31, 1972.

This methodology is a checklist procedure emphasizing quantitative impact assessment. It was designed for major water resource projects, but many parameters used are also appropriate for other types of projects. Seventy-eight specific environmental parameters are defined within the four categories ecology, environmental pollution, aesthetics, and human interest. The approach does not deal with economic or secondary impacts, and social impacts are only partially covered within the human interest category.

Parameter measurements are converted to a common base of environmental quality units through specified graphs or value functions. Impacts can be aggregated using a set of preassigned weights of relative importance. The resource requirements are rather high, particularly data requirements, and they probably restrict the use of the approach to major project assessments.

The approach emphasizes explicit procedures for impact measurement and evaluation and should therefore produce highly replicable results. Both spatial and temporal aspects of impacts are noted and explicitly weighted in the assessment. Public participation, uncertainty, and risk concepts are not considered. An important idea of the approach is to highlight key impacts via a "red flag" system.

3 Dee, N., et al.: "Planning Methodology for Water Quality Management: Environmental Evaluation System," Battelle-Columbus, Columbus, Ohio, July 1973.

This methodology of impact assessment defines ready classification since it contains elements of checklist, matrix, and network approaches. Areas of possible impacts are defined by a hierarchical system of 4 categories (ecology, physical/chemical, aesthetic, social), 19 components, and 64 parameters. An interaction matrix is presented to indicate which activities associated with water quality treatment projects generally impact which parameters. The range of parameters used is comprehensive, excluding only economic variables.

Impact measurement incorporates two important elements. A set of ranges is specified for each parameter to express impact magnitude on a scale from zero to one. The ranges assigned to each parameter within a component are then combined by means of an environmental assessment tree into a summary environmental impact score for that component. The significance of impacts on each component is quantified by a set of assigned weights. A net impact can be obtained for any alternative by multiplying each component score by its weight factor and summing across components.

The data, time, and cost requirements of the methodology when used for impact assessment are moderate, although a small amount of training would be required to familiarize users with the techniques. Because of its explicitness, the methodology possesses only minor ambiguities and should be highly replicable. Because the environmental assessment trees are developed specifically for water treatment facilities, the methodology cannot be adapted to other types of projects without reconstructing the trees, even though the parameters could be useful as a simple checklist.

4 Institute of Ecology, University of Georgia: Optimum Pathway Matrix Analysis Approach to the Environmental Decision Making Process: Test Case: Relative Impact of Proposed Highway Alternatives, Athens, 1971.

The Georgia methodology incorporates a checklist of 56 environmental components. Measurable indicators are specified for each component. The actual values of alternative plan impacts on a component are normalized and expressed as a decimal of the largest impact (on that one component). These normalized values are multiplied by a subjectively determined weighting factor, which is the sum of a weight for initial effects plus 10 times a weight for long-term effects. The methodology is used to evaluate highway project alternatives, and the components listed are not suitable for other types of projects. A wide range of impact types is analyzed including land-use, social, aesthetic, and economic impacts.

The lower replicability of the analysis produced by using subjectively determined weighting factors is compensated for by conducting several passes at the analysis and incorporating randomly generated error variation in both actual measurements and weights. This procedure provides a basis for testing the significance of differences in total impact scores between alternatives.

The procedures for normalizing or scaling measured impacts to obtain commensurability and testing of significant differences between alternatives are notable features of potential value to other impact analyses and methodologies.

5 Krauskopf, T. M., and D. C. Bunde: Evaluation of Environmental Impact Through a Computer Modelling Process, in Robert Ditton and Thomas Goodale (eds.), "Environmental Impact Analysis: Philosophy and Methods," pp. 107–125, University of Wisconsin Sea Grant Program, Madison, 1972.

This methodology employs an overlay technique via computer mapping. Data on a large number of environmental characteristics are collected and stored in the computer on a grid system of 1 km^2 cells. Either highway route alternatives can be evaluated by the computer (by noting the impacts on intersected cells) or new alternatives may be generated via a program identifying the route of least impact.

The environmental characteristics used are rather comprehensive, particularly as regards land-use and physiographic characteristics. Although the methodology was developed and applied to a highway setting, it is adaptable, with relatively small changes in characteristics examined, to other project types with geographically well-defined and concentrated impacts. Because the approach requires considerable amounts of data on the project region, it is not practical for the analysis of programs of broad geographical scope.

The estimation of impact importance is done through specification of subjective weights. Because the approach is computerized, the effects of several alternative weighting schemes can be readily analyzed.

6 Leopold, L. B., et al.: A Procedure for Evaluating Environmental Impact, *Geolog. Surv. Circ.* 645, 1971.

This is an open-cell matrix approach identifying 100 project actions and 88 environmental characteristics or conditions. For each action involved in a project, the analyst evaluates the impact on every impacted environmental characteristic in terms of impact magnitude and significance. These evaluations are subjectively determined by the analyst. Ecological and physical-chemical impacts are treated comprehensively, social and indirect impacts are less well handled, and economic and secondary impacts are not addressed.

Because the assessments made are subjective, resource requirements of the approach are very flexible. The approach was not developed in reference to any specific type of project and may be broadly applied with some alterations. Guidelines for use of the approach are minimal, and several important ambiguities are likely in the definition and separation of impacts. The reliance on subjective judgment, again without guidelines, reduces the replicability of the approach, which is chiefly valuable as a means of identifying project impacts and as a display format for communicating results of an analysis.

7 A. D. Little, Inc.: "Transportation and Environment: Synthesis for Action: Impact of National Environmental Policy Act of 1969 on the Department of Transportation," vol. 3, prepared for Office of the Secretary, Department of Transportation, July 1971.

This is less a complete methodology than an overview discussion of the kinds of impacts that may be expected to occur from highway projects and the measurement techniques that may be available to handle some of them. A quite comprehensive list of impact types and the stages of project development at which each may occur are presented. As broad categories, the impact types identified are useful for other projects as well as for highways.

The approach suggests the separate consideration of an impact's amount, effect (public response), and value. Some suggestions are offered for measuring the amount of impact within each of seven categories: noise, air quality, water quality, soil erosion, ecological, economic, and sociopolitical impacts.

8 McHarg, I.: A Comprehensive Highway Route-Selection Method, *Highway Res. Rec.*, no. 246, pp. 1–15, 1968; or McHarg, I.: "Design With Nature," pp. 31–41, Natural History Press, Garden City, N.Y., 1969.

The McHarg approach is a system employing transparencies of environmental characteristics overlaid on a regional base map. Eleven to sixteen environmental and land-use characteristics are mapped. The maps represent three levels of the characteristics based upon compatibility with the highway.

This approach is basically an earlier, noncomputerized version of the ideas presented in the Krauskopf and Bunde methodology. Its basic value is as a method for screening alternative project sites or routes. Within this limited use, it is applicable to a variety of project types. Limitations of the approach include its inability to quantify as well as identify possible impacts and its implicit weighting of all characteristics mapped.

9 Moore, J. L., et al.: "A Methodology for Evaluating Manufacturing Environmental Impact Statements for Delaware's Coastal Zone," report prepared by Battelle-Columbus for the State of Delaware, June 15, 1973.

This approach was not designed as a method for impact analysis, but its principles could be adapted for such use. It employs a network approach, linking a list of manufacturing-related activities to potential environmental alterations to major environmental effects and finally to human uses affected. The primary strength of the set of linked matrices is its utility in displaying cause–condition–effect networks and tracing secondary impact chains.

Such networks are useful primarily for identifying impacts, and the issues of impact magnitude and significance are addressed only in terms of high, moderate, low, or negligible damage. As a result of these subjective evaluations, the approach would have low replicability as an assessment technique. For such a use, guidelines would likely need to be proposed to define the evaluation categories.

10 Central New York Regional Planning and Development Board: Environmental Resources Management, *Natl. Tech. Infor. Svc.* PB 217-517, Oct. 1972.

This methodology employs a matrix approach to assess in simple terms the major, minor, direct, and indirect impacts of certain water-related construction activities. It is designed primarily to measure only the physical impacts of water resource projects in a watershed and is based on an identification of the specific, small-scale component activities that are included in any project. Restricted to physical impacts on nine different types of watershed areas (e.g., wetlands) and fourteen types of activities (e.g., tree removal), the procedure indicates four possible levels of impact–receptor interactions (major direct through minor indirect). Low to moderate resources in terms of time, money, or personnel are required for the methodology, due principally to its simple way of quantification (major versus minor impact).

11 Smith, W. L.: Quantifying the Environmental Impact of Transportation Systems, Van Doren-Hazard-Stallings-Schnacke, Topeka, Kan., undated.

The Smith approach, as developed for highway route selection, is a checklist system based on the concepts of probability and supply–demand. The approach attempts to identify the alternative with least social cost to environmental resources and maximum social benefit to system resources. Environmental resources elements are agriculture, wildlife conservation, interference, noise, physical features, and replacement. System resources elements are aesthetics, cost, mode interface, and travel desires. For each element, categories are defined and used to classify zones of the project area. Numerical probabilities of supply and demand are then assigned to each zone for each element. These are multiplied to produce a probability of least social cost (or maximum social benefit). These least social cost probabilities are then multiplied across the elements to produce a total for the route alternative under examination.

The approach is tailored and perhaps limited to project situations requiring comparison of siting alternatives. The range of environmental factors examined is very limited, but presumably it could be expanded to cover more adequately ecological, pollution, and social considerations.

12 Sorensen, J.: "A Framework for Identification and Control of Resource Degradation and Conflict in the Multiple Use of the Coastal Zone," University of California, Berkeley, 1971; and Sorensen, J., and James E. Pepper: "Procedures for Regional Clearing House Review of Environmental Impact Statements—Phase Two," report to the Association of Bay Area Governments, Apr. 1973.

These two publications present a network approach usable for environmental impact analysis. The approach is not a full methodology but rather a guide to the identification of impacts. Several potential uses of the California coastal zone are examined through networks relating uses to causal factors (project activities) to first-order conditions changes to second- and third-order condition changes and finally to effects. The major strength of the approach is its ability to identify the pathways by which both primary and secondary environmental impacts are produced.

The second reference also indicates types of data relevant to each effect identified, although no specific measurable indicators are suggested. In this reference some general criteria for identifying projects of regional significance are suggested, based on project size and types of impacts generated, particularly land-use impacts.

Because the preparation of the required detailed networks is a major undertaking, the approach is presently limited to some commercial, residential, and transportation uses of the California coastal zone for which networks have been prepared. An agency wishing to use the approach in other circumstances might develop the appropriate networks for reference in subsequent environmental impact assessments.

13 Stover, L. V.: "Environmental Impact Assessment: A Procedure," Sanders and Thomas, Miami, 1972.

This methodology is a checklist procedure for a general quantitative evaluation of environmental impacts from development activities. Fifty different impact parameters are sufficient to include most possible effects, thereby allowing much flexibility. Subparameters indicate specific impacts, but there is no description of how the individual measures are aggregated into a single parameter value. While spatial differences in impacts are not indicated, both initial and future impacts are included and explicitly compared.

14 Multiagency Task Force: Guidelines for Implementing Principles and Standards for Multiobjective Planning of Water Resources, review draft, U.S. Bureau of Reclamation, 1972.

The task force approach is an attempt to coordinate features of the Water Resources Council's (WRC) Proposed Principles and Standards for Planning Water and Related Land Resources with requirements of NEPA. It develops a checklist of environmental components and categories organized in the same manner as the WRC guidelines. The categories of potential impacts examined deal comprehensively with biological, physical, cultural, and historical resources and with pollution factors but do not treat social or economic impacts. Impacts are measured in quantitative terms where possible and also are rated subjectively on quality and human influence. In addition, uniqueness and irreversibility considerations are included where appropriate. Several suggestions for summary tables and bar graphs are offered as communications aids.

The approach is general enough to have wide applicability to various types of projects, even though its impact categories are perhaps better tailored to rural than urban environments. No specific data or other resources are required to conduct an analysis, although an interdisciplinary project team is specified to assign the subjective weightings. Key ideas incorporated in the approach include explicit identification of the without-project environment as distinct from present conditions and use of a uniqueness rating system for evaluating quality and human influence (worst known, average, best known, etc.).

15 Tulsa District, U.S. Army Corps of Engineers: Matrix Analysis of Alternatives for Water Resource Development, draft technical paper July 31, 1972.

Despite the title, this methodology can be considered a checklist under the definitions used here since, although a display matrix is used to summarize and compare the impacts of project alternatives, impacts are not linked to specific project actions. The approach was developed to deal specifically with reservoir construction projects but could be readily adapted to other project types.

Potential impacts are identified within three broad objectives: environmental quality, human life quality, and economics. For each impact type identified, a series of factors is described, indicating possible measurable indicators. Impact magnitude is not measured in physical units but by a relative impact system. This system assigns the future state of an environmental characteristic without the project a score of zero; it then assigns the project alternative possessing the greatest positive impact on that characteristic a score of plus five or for negative impact, minus five. All other alternatives are assigned scores between zero and five by comparison. The raw scores thus obtained are multiplied by weights determined subjectively by the impact analysis team.

Like the Georgia approach, the Tulsa methodology tests for the significance of differences between alternatives by introducing error factors and conducting repeated runs. The major limitations of the approach, aside from the required computerization, are the lack of clear guidelines on exactly how to measure impacts and the lack of guidance on how the future no-project state is to be defined and described in the analysis.

The key ideas in the Tulsa approach include reliance on relative rather than absolute impact measurement, statistical tests of significance with error introduction, and specific use of the no-project conditions, as a base line for impact evaluation.

16 Walton, L. E., Jr., and J. E. Lewis: A Manual for Conducting Environmental Impact Studies, *Natl. Tech. Infor. Svc.* PB 210-222, Jan. 1971.

The Walton methodology is a checklist, unique in its almost total reliance on social impact categories and strong public participation. The approach was developed for the evaluation of highway alternatives and identifies different impact analysis procedures for the conceptual, corridor, and design states of highway planning. All impacts are measured by either their dollar value or a weighted function of the number of persons affected. (The weights used are to be determined subjectively by the study team.) The basis for most measurements is a personal interview with a representative of each facility or service impacted.

Analyses produced by the approach may have very poor replicability due to the lack of specific data used and the criticality of the decision regarding boundaries of the analysis since many impacts are measured in numbers of people affected. The key ideas in the approach are the use of only social impacts without direct consideration of other impacts (pollution, ecology, etc.), the heavy dependence on public involvement and specific suggestions on how the public may be involved, and the recognition of the need for different analyses of different stages of project development.

17 Western Systems Coordinating Council, Environmental Committee: "Environmental Guidelines," 1971.

These guidelines are intended primarily as a planning tool for siting power generation and transmission facilities. However, they address many of the concerns of environmental impact analysis and have been used in the preparation of impact statements. Viewed as an impact assessment methodology, the approach is an *ad hoc* procedure, suggesting general areas and types of impacts but not listing specific parameters to examine.

The approach considers a range of pollution, ecological, economic (business economics), and social impacts but does not address secondary impacts such as induced growth or energy-use patterns. The format of the approach is an outline of considerations important to the selection of sites for each of several types of facilities—for example, thermal generating plants, transmission lines, hydroelectric and pumped storage, and substations. An additional section offers suggestions for a public information program.

Appendix E

Impact Methodologies from Smith Study (1974) and Not Included in Appendix D

1 Eckenrode, R. T.: Weighting Multiple Criteria, *Management Sci.*, vol. 12, no. 3, 1965.

This methodology is not directed toward environmental impact per se but rather to the handling of multiple criteria in general. A team of six to twelve experts rates the importance of criteria using several different ranking and rating techniques on a subjective basis. The outcomes are compared for agreement or disagreement. This basic approach has been employed in developing weighting scores for various impact methodologies.

2 Lamanna, R. A.: Value Consensus Among Urban Residents, *J. Am. Inst. Planners*, vol. 30, no. 4, 1964.

Priority areas of concern are identified by means of responses to a sample survey. Persons interviewed are presented with a list of potential concerns and are asked to weight elements according to importance on a scale from one to three, three being most important. This method does not ensure a sound basis for indicated preferences and determines only probable areas of lesser or greater impacts, not the extent of impact itself.

3 McKenny, C. E. B., et al.: Interstate-75; Evaluation of Corridors Proposed for South Florida, report prepared by University of Miami Center for Urban Studies for Florida Department of Transportation, 1971.

An *ad hoc* interdisciplinary panel of experts is convened to consider the qualitative advantages and disadvantages of proposed routes, making a recommendation based on a consensus judgment. Subunits of the panel prepare written reports of probable impacts in their area of expertise. These reports are then discussed in a series of deliberative sessions, and agreement on a recommendation is reached.

This method has been widely employed. It is doubtful, however, whether extensive expertise could be brought to bear on all projects in a consistent manner and whether sufficient time would be available for the painstaking preparation of reports and deliberations on a case-by-case basis.

4 Lacate, D. S.: "The Role of the Resource Inventories and Landscape Ecology in the Highway Route Selection Process," Department of Conservation, Cornell University, Ithaca, N.Y., 1970.

Highway route alternatives are compared and the optimum route selected based on a subjective analysis of resource inventories. A resource inventory is simply a collection of data on a particular social, economic, or environmental factor of concern. The information contained in the resource inventories is aggregated in a series of overlay maps; thus the method is actually the same as the McHarg method. Unlike McHarg, however, specific features and details are not transformed into three-color shades for evaluation at a generalized, macroscale level. This technique, therefore, while avoiding the possible misinterpretation and unintentional neglect resulting from simplification, lacks ease and clarity in employment.

Only four overlay maps are prepared in a case study. The first displays the various land uses; the second, type and intensity of farming; the third, soils; and the last, localized historical, cultural, and environmental values. A large series of overlays would easily result in some features being buried in detail. The advantages, in fact, of this method over a general collection and analysis of available information are difficult to determine.

5 Baker, R. W., and J. D. Gruendler: A Case Study of the Milwaukee-Green Bay Interstate Corridor Location, paper presented at Highway Research Board Summer Meeting, 1972.

This method, developed at the Environmental Awareness Center of the University of Wisconsin's Department of Landscape Architecture, is a computer application of McHarg's overlay method. Considered is a broad range of factors including environmental, engineering, economic, and social aspects. Ten levels of shading are used to describe the total range of values varying from dark (more suitable) to light (least suitable) for each environmental parameter. The parameters are weighted according to relative importance (the weighting method is not described).

This technique, like the McHarg method, does not predict actual impact but rather only areas of greater or lesser impact. It does identify those locations for which more detailed studies might be conducted.

6 Turner, A. K., and I. Hausmanis: Computer-Aided Transportation Corridor Selection in the Guelph-Dundas Area, Ontario, Canada, paper presented at Highway Research Board Summer Meeting, 1972.

This is another application of the computer overlay method. The various parameters considered are weighted on a scale ranging from one to ten.

7 Manheim, M. L., et al.: Community Values in Highway Location and Design: A Procedural Guide, report prepared by Urban Systems Laboratory, MIT, for Highway Research Board, Sept. 1971.

Coordination between an interdisciplinary location study team and community groups is used in identifying impacts, the interests affected by the impacts, and appropriate representatives for those interests. An impact matrix is devised for each affected interest to describe each alternative and the corresponding impacts. The information contained in the matrix may be qualitative, pictorial, or numerical. It is the responsibility of the location team to use this information in assisting a politically responsible official to make a decision on the proper course of action.

Although more documentation and more detail might be required and a more comprehensive and expert input could be attained, this technique resembles the method developed by Leopold in both its application and its shortcomings.

8 Highway Research Section, Engineering Research Division, Washington State University: A Study of the Social, Economic and Environmental Impact of Highway Transportation Facilities on Urban Communities, report prepared for Washington State Department of Highways, 1968.

Three forms are to be completed for each alternative route for evaluation: one based on appearance considerations, one on sociological considerations, and one on economic considerations. Various parameters describing each of these three subject areas are listed on the respective form.

In evaluating a particular route on appearance, for example, a number between one and ten is assigned to each appearance parameter to describe the route's desirability for that parameter. The rating is to be done subjectively by the administrator and staff. The rating number is then multiplied by a weighting factor that has been established by administrators and interested citizens prior to the route rating process; the weights are to reflect the objectives the road is intended to serve. This process is repeated for each parameter three times, because the life of the project is divided into periods of 0–5, 6–25, and 26–50 yr—the weights and desirability rating may change with time. The combined 50-yr weighted ratings for each appearance factor are aggregated to indicate the overall appearance rating. If a route has similar conditions throughout its entire length, it can be rated as one section. If not, the overall rating for a portion is multiplied by the ratio of its length to that of the entire route. (This process, unfortunately, would make the importance of a particular portion dependent upon its length. Also, no criteria are presented to determine when similar conditions are or are not present.)

Finally, the total weighted-rating values on the appearance, sociological, and economic forms completed for each route are listed with construction cost and other monetary considerations on a route comparison form. This method is very subjective, and as is stated, its value depends on the skill and ability of the user. Quantification in this technique appears to be primarily for the purpose of making the decision process systematic.

9 Hill, M.: "A Method for Evaluating Alternative Plans: The Goals-Achievement Matrix Applied to Transportation Plans," doctoral dissertation, University of Pennsylvania, Philadelphia, Pennsylvania, 1966.

A matrix is prepared. Across one axis are specific environmental goals (decrease in air pollution, etc.); the other axis contains various land-use categories (residential districts, open space, etc.) subdivided into specific area, buildings, etc. that are affected. At each subdivision–goal intersection, a plus, minus, or equals sign is used to indicate for each alternative route whether there is an increase in goal attainment, a decrease in goal attainment, or no change. A comparison between the alternatives themselves is then made at each land category–goal intersection, using the results of the more specific intersections as a basis for judgment. In other words, each alternative is first compared against the attainment of a goal for a group of specific locations, then compared against another alternative for the broader land-use category, with the basis of the latter comparison being how each alternative fares in the first comparison (in terms of the number of plus, minus, and equal signs). By examining the matrix horizontally, a comparison can be made between alternative routes in terms of one specific location for all the goals. By examining the matrix vertically, a comparison can be made in terms of one goal and all the land-use categories.

Many subjective decisions are required with this method in determining both goal achievement for an alternative and the relative importance of each land-use category and each goal. Further, it is more suited to assessing the degree to which project objectives/goals are met than to assessing environmental consequences of meeting these goals/objectives.

10 Klein, G. E.: Evaluation of New Transportation Systems, in "Defining Transportation Requirements—Papers and Discussions," American Society of Mechanical Engineers, New York, New York, 1969.

This technique was designed to evaluate systems in terms of economic and social factors; the method could possibly be applied to environmental factors. A utility index similar to the value functions devised for the Battelle method is employed. Instead of one function defining the quality state, two functions representing the upper and lower limits are presented. The evaluator is allowed to make a decision somewhere between the two extremes. Specific criteria used in establishing the functions are placed on the horizontal axis, with the point chosen by the evaluator corresponding on the vertical axis to a zero to ten utility scale. Zero on the utility scale represents the best situation, and ten, the worst, and the utility index for each factor is translated directly into dollars. This dubious procedure is intended to put actual dollar values on intangibles, with the total evaluation based on relative costs and benefits among alternatives.

11 Oglesby, C. H., G. Bishop, and G. Willeke: Socio-Economic and Community Factors in Planning Urban Freeways, Stanford University research project for California Transportation Agency, Oct. 1969.

For those nonmonetary aspects of a highway project, a profile of alternatives is prepared and a list of environmental factors developed. The route with the most beneficial (or detrimental) effect for a particular factor is set at 100 percent (or −100 percent) for that particular factor. The effects of the alternatives routes are then expressed as a percentage of the effects of the best (or worst) for that factor. The results for all factors considered are shown diagrammatically on a scale from −100 (worst) to +100 (best). Any alternative routes that are clearly dominated are eliminated. Paired comparisons are made for the remaining alternatives. Subjective decisions are required, since each factor is considered separately (there is no categorization or grouping) and there is no ranking of factors. The conclusion is that an irreducible level exists at which subjective trade offs must be made and that a systematic, organized process of data presentation is most appropriate.

12 Southeastern Wisconsin Regional Planning Commission: Land Use Transportation Study–Forecast and Alternative Plans: 1990, vol. 2, *Plan Rept.* 7, June 1966.

Various environmental objectives are stated. These are ranked in order of importance for each situation, then weighted values are assigned on a direct reverse listing of the numerical importance rank. Each of the alternative projects or routes is then rated against the environmental objectives–no particular rating value system is specified. The two numbers for rating and ranking are multiplied together for each objective, and the resulting values for all the objectives are added for each alternative. These final values represent the evaluation of specific alternatives against their achievement of the objectives and can be used for comparison against each other.

In this method the assignment of weights is arbitrary rather than objective. Subjective judgments are required in ranking and in the rating of alternatives against the objectives. Some measure of the relative value of alternative plans is achieved, but impact itself is not determined.

13 Dearinger, J. A.: "Esthetic and Recreational Potential of Small Naturalistic Streams Near Urban Areas," Water Resources Institute, University of Kentucky, Lexington, Kentucky, Apr. 1968.

This technique was designed for evaluating aesthetic and recreational aspects of small streams. Environmental factors, such as water quality, wildlife, and scenic views, are weighted on a scale from one to five depending on their importance for types of potential recreational uses: hiking trails, canoeing areas, etc. In evaluating a location, a rating number is established for each particular environmental factor in each use category. This number is between one and ten and is determined through a set value function (e.g., BOD and turbidity for water quality). The weighting and rating numbers are multiplied for each factor and the results from all the factors added for each type of use. This final number is made a percentage of the number that would apply to the suitability of the location for a use if all conditions were optimum (all rating numbers equal to 10).

14 Orlob, G. T., et al.: Wild Rivers: Methods for Evaluation, report prepared by Water Resources Engineers for the U.S. Department of the Interior, Oct. 1970.

Nonmonetary and intangible values are expressed in dollars, on the premises that such values are at least equal to the economic development benefits that are foregone in favor of preservation and that nonmonetary benefits equal between 0.25 and 2.0 times monetary benefits. Environmental values can then be subjected to the benefit-cost analysis. The assumptions made in this method are arbitrary, and no provision is made for analyzing the nonmonetary environmental impacts of development. The evaluations required are very lengthy mathematical computations.

Appendix F

Impact Methodologies from Viohl and Mason Study (1974) and Not Included in Appendices D or E

1 Fischer, D. W., and G. S. Davies: An Approach to Assessing Environmental Impact, *J. Environ. Management*, vol. 1, no. 3, pp. 207–227, July 1973.

A method of environmental analysis is proposed that expands the concept of environmental matrices. Three steps are used to identify and evaluate environmental feasibility: (1) a matrix for environmental base-line evaluation in terms of importance, present condition, and management, (2) an environmental compatibility matrix in terms of introduced activities, and (3) a decision matrix for evaluating alternatives available (no project, structural and nonstructural alternatives, locational alternatives). This methodology relies heavily on the use of multidisciplinary teams to provide value scales and to perform the environmental analysis.

2 Pikul, R.: Development of Environmental Indices, in J. W. Pratt (ed.), "Statistical and Mathematical Aspects of Pollution Problems," pp. 103–121, Marcel Dekker, New York, 1974.

An index approach to environmental assessment is used. The author presents an initial formulation of environmental indices in 14 environmental classes. A method of ranking indices while taking cost and value considerations into account is discussed. The assessment of impact is facilitated by tracking and analyzing the behavior of these indices. Each mathematical index is analyzed according to factors of impact, utility, value, cost, and importance.

3 Bureau of Land Management: "Environmental Analysis," Washington, D.C., Aug. 1973.

This is an approach to environmental assessment that, in keeping with Section 102(2), A and B, of NEPA, uses an environmental analysis worksheet that is both flexible and open. The suggested environmental elements to be considered are listed in an environmental digest. Impacts are evaluated on a scale of low–medium–high, positive–negative, as they relate to previously predicted bureau standards. Accompanying remarks are an integral part of the ongoing worksheet process.

4 Commonwealth Associates, Inc.: "Environmental Analysis System," Rept. R-1447, prepared for the Northern States Power Company, Jackson, Mich., Sept. 1972.

The use of a matrix system in evaluating environmental impact is further developed by adding the adaptability and flexibility of a computer. The approach starts with the hierarchical cataloging of all factors, then quantifies impacts by assigning numerical values for magnitude and importance, and combines multiple considerations through the use of algebraic equations and weighting operations. The weighting of the individual elements, that is, air versus water, is handled as an external consideration. The computer can modify the weighting operations as often as necessary in response to new thinking. This system bases environmental factor comparisons on a relative rather than an absolute basis.

5 Alden, H. R.: Environmental Impact Assessment: A Procedure for Coordinating and Organizing Environmental Planning, Thorne Ecological Institute, Boulder, Col.

Many of the stumbling blocks one faces in trying to integrate diverse ecological elements into a single end product are removed by a recommended procedure of seven steps, from initial coordination and communication to final recommendations. This comprehensive procedure employs an environmental resource inventory, composite strength of relationship and impact relationship matrices, map graphics, nonmonetary cost-benefit studies, and ranking operations, in evaluating both linear and more complex environmental interrelationships.

6 Jain, R. K., L. V. Urban, and G. S. Stacey: "Handbook for Environmental Impact Analysis," Construction Engineering Research Laboratory, Department of the Army, Champaign, Ill., Apr. 1974.

Recommended procedures for use by army personnel in preparing and processing environmental impact assessments (EIAs) and EISs are presented. A method for evaluating environmental impacts is given that employs an attribute descriptor package—an environmental factor index and a matrix analysis with summary comparison of individual attributes (environmental characteristics) by bar charts. The problems of determining relative importance and severity of impacts and of providing a single number to indicate total environmental impact (thereby masking the distribution of the impact among its attributes) are avoided by assuming that each attribute is of equal importance.

7 University of Wisconsin: "Handbook: Applications of Remote Sensing and Computer Techniques for Recreation Planning," vols. 1–4, Bureau of Outdoor Recreation, Madison, Wisconsin, Mar. 1974.

This environmental assessment system is unique in its remote sensing data gathering and its computer-generated overlays. Statistical methods for sampling the project area (divided into cells for study by random sampling) are documented. A total impact index (II) is calculated for each of the cells and compared to the expected impact (EI) on each. An imitation data bank is proposed as an aid in the testing of other methodologies.

8 Environmental Protection Agency and R. Rea of Resource Planning Associates: "Handbook for Assessing the Social and Economic Impacts of Water Quality Management Plans," Washington, D.C., July 1973.

These studies propose a system for environmental assessment that is definable in two discrete components–impact identification and impact evaluation. The methodology relies heavily on input from regional planning agencies and community groups. In the impact identification process objectives, by-products, short-term requirements, and long-term requirements are considered. The authors state that due to the large number of impact cells to be scrutinized, priorities for in-depth analysis should be resolved by the contractor with major input from resource planning agencies. Tables and charts present a summary view of the findings in readily understandable form.

9 Environmental Impact Center, Inc.: A Methodology for Assessing Environmental Impact of Water Resources Development, final report for Department of the Interior, Nov. 1973.

A new method of analysis is recommended that integrates existing analytical techniques into a single unified methodology. The model structure is modular with each of the six sectors–water, biology, population, industry, land use, and recreation–analyzed separately using various cost-benefit, matrix, and computer modeling subsystems. Interrelationships between sectors are outlined.

10 Kane, J.: "A Methodology for Interactive Resource Policy Simulation," pp. 65–79, Water Resources Research, 1973.

A paramathematical technique of environmental assessment is presented. The authors feel that computerization is a needed facet of assessment, yet comprehensiveness should not encompass the whole decision process. The study develops a simulation procedure (KSIM) that describes complex interaction of both hard and soft variables while not requiring mathematical simulation in its use. Further, KSIM is touted as easily modified and totally flexible. The KSIM approach utilizes geometric concepts and an interaction matrix. The authors have two purposes for introduction of the system: (1) to iterate that high-order complex systems do not respond continuously to small changes but rather have a critical threshold, and (2) to impart a new pattern of holistic thinking.

11 Tabors, J. G.: Model for Landscape Resource Assessment, Water Resources Research Center, University of Massachusetts at Amherst, Feb. 1973.

The development and application of a landscape resource assessment model designed to estimate landscape change, both positive and negative, caused by urbanization in the Boston metropolitan region since 1945 are described. The model incorporates the quantitative assessment techniques for estimating water quality, the use of questionnaires and matrices, mapping of value change isopleths, and other submodels for parameter evaluations. This is only the first step; a second step is planned that will provide planners with a model for land-use allocation and land-use activity management.

12 Heuting, R.: A Statistical System for Estimating the Deterioration of the Human Environment, in J. W. Pratt (ed.), "Statistical and Mathematical Aspects of Pollution Problems," pp. 123–132, Marcel Dekker, New York, 1974.

An innovative approach to environmental impact is presented. The author measures deterioration in the environment as decreased availability of functions of an environmental component. Such a decrease would be the manifest result of competition of functions. Hence in assessment, one should analyze where competition of functions takes place, thus tagging a deleterious effect. The author sees this competition taking on quantitative, spatial, and qualitative aspects. In schematic representation indicators are used with statistical reflection in the compilation of cells, which serve as units in planning processes. This paper is concluded with a section on the shadow pricing of environmental functions.

13 Landscape Architecture Research Service: "Three Approaches to Environmental Resource Analysis," Harvard University, Cambridge, Massachusetts, Nov. 1967.

An overview is presented of three land-use planning analysis methods: those of G. Angus Hill, Phillip H. Lewis, Jr., and Ian L. McHarg. The three methodologies are compared on (1) their identification of spatial and environmental patterns beyond single-factor analysis and simple spatial location, (2) their identification and evaluation of key elements and forces that affect the quality of the environment, (3) their review of analysis material for incorporation in the planning process, and (4) their treatment of the environment as an entity.

14 Chen, C. W., and G. T. Orlob: Ecological Simulation for Aquatic Environments, final report prepared for Office of Water Resources Research, U.S. Department of the Interior, Dec. 1972.

A method of modeling ecosystems and predicting environmental impact is presented. This study assesses the impact of water resources development on aquatic ecosystems considering not only environmental but also attendant economic and social gains or losses. An ecological model is developed with a set of balanced equations to be solved over a time–space continuum appropriate to the project. As many as 22 parameters and their couplings are considered simultaneously by the model.

15 Schlesinger, B., and D. Daetz: A Conceptual Framework for Applying Environmental Assessment Matrix Techniques, *Environ. Sci.*, vol. XVI, no. 4, pp. 11–16, July–Aug. 1973.

The matrix approach is thoroughly reappraised and emerges as a newly finished and valuable assessment tool. The authors add to the matrix the concepts of interrelationship and probability of occurrence. A description of parameters and several necessary predications for matrix use are included. Concluding the paper is a discussion of the value of matrix analysis in all steps of the planning and decision-making process.

Appendix G

Environmental Factors to Be Used in Comparing Alternative Plant Systems

Primary impact	Population or resources affected	Description	Unit of measure[a]	Method of computation
1 Natural surface water body	(Specify natural water body affected)			
1.1 Impingement or entrapment by cooling water-intake structure	1.1.1 Fish[b]	Juveniles and adults subject to attrition	Percent of harvestable or adult population destroyed per year for each important species	Identify all important species as defined in Sec. 2.7. Estimate the annual weight and number of each species that will be destroyed. (For juveniles destroyed, only the expected population that would have survived naturally need be considered.) Compare with the estimated weight and number of the species population in the water body
1.2 Passage through or retention in cooling systems	1.2.1 Phytoplankton and zooplankton	Plankton population (excluding fish) changed due to mechanical, thermal, and chemical effects	Percent changes in production rates and species diversity	Field studies are required (1) to estimate the diversity and production rates of readily recognizable groups (e.g., diatoms, green algae, zooplankton) and (2) to estimate the mortality of organisms passing through the condenser and pumps. Include indirect[c] effects that affect mortality
	1.2.2 Fish	All life stages (eggs, larvae, etc.) that reach the condenser subject to attrition	Percent of harvestable or adult population destroyed per year for each important species	Identify all important species as defined in Sec. 2.7. Estimate the annual weight and number of each species that will be destroyed. (For larvae, eggs, and juveniles destroyed, only the expected population that would have survived naturally need be considered.) Compare with the estimated weight and number of the species population in the water body

1.3 Discharge area and thermal plume	1.3.1 Water quality, excess heat	Rate of dissipation of excess heat, primarily to the atmosphere, depends on both the method of discharge and the state of the receiving water in respect to ambient temperature and water currents	Acres and acre-feet	Estimate the average Btus per hour dissipated to the receiving water at full power. Estimate the water volume and surface areas within differential temperature isotherms of 2, 3, and 5°F under conditions that would tend, with respect to annual variations, to maximize the extent of the areas and volumes
	1.3.2 Water quality, oxygen availability	Dissolved oxygen concentration of receiving waters may be modified as a consequence of changes in the water temperature, the translocation of water of different quality, and aeration	Acre-feet	Estimate volumes of affected waters with concentrations below 5, 3, and 1 ppm under conditions that would tend to maximize the impact
	1.3.3 Fish (nonmigratory)	Fish[b] may be affected directly or indirectly[c] due to adverse conditions in the plume	Net effect in pounds per year (as harvestable or adult fish by species of interest)	Field measurements are required to establish the average number and weight (as harvestable or adults) of important species (as defined in Sec. 2.7). Estimate their mortality in the receiving water from direct and indirect[c] effects
	1.3.4 Fish (migratory)	Suitable habitats (wetland or water surface) may be affected	Acres of defined habitat or nesting area	Determine the areas impaired as habitats because of thermal discharges, including effects on food resources. Document estimates of affected population by species
	1.3.5 Wildlife (including birds, aquatic and amphibious mammals, and reptiles)	A thermal barrier may inhibit migration, both hampering spawning and diminishing the survival of returning fish	Pounds per year (as adult or harvestable fish by species of interest)	Estimate the fraction of the stock that is prevented from reaching spawning grounds because of plant operation. Prorate this directly to a reduction in current and long-term fishing effort supported by that stock. Justify esti-

Primary impact	Population or resources affected	Description	Unit of measure[a]	Method of computation
				mate on basis of local migration patterns, experience at other sites, and applicable state standards
1.4 Chemical effluents	1.4.1 Water quality, chemical	Water quality may be impaired	Acre-feet, percent	Volume of water required to dilute the average daily discharge of each chemical to meet applicable water quality standards should be calculated. Where suitable standards do not exist, use the volume required to dilute each chemical to a concentration equivalent to a selected lethal concentration for the most important species (as defined in Sec. 2.7) in the receiving waters. The ratio of this volume to the annual minimum value of the daily net flow, where applicable, of the receiving waters should be expressed as a percentage, and the largest such percentage reported. Include the total solids if this is a limiting factor. Include in this calculation the blowdown from cooling towers
	1.4.2 Fish	Aquatic populations may be affected by toxic levels of discharge chemicals or by reduced dissolved oxygen concentrations	Pounds per year (by species as fish)	Total chemical effect on important species of aquatic biota should be estimated. Biota exposed within the facility should be considered as well as biota in receiving waters. Supporting documentation should include reference to applicable standards, chemicals discharged, and their toxicity to the aquatic populations affected

	1.4.3 Wildlife (including birds, aquatic and amphibious mammals, and reptiles)	Suitable habitats for wildlife may be affected	Acres	Estimate the area of wetland or water surface impaired as a wildlife habitat because of chemical contamination including effects on food resources. Document estimates of affected population by species
	1.4.4 People	Recreational water uses (boating, fishing, swimming) may be inhibited	Lost annual user-days and area (acres) or shoreline miles for dilution	Volume of the net flow to the receiving waters required for dilution to reach accepted water quality standards must be determined on the basis of daily discharge and converted to either surface area or miles of shore. Cross-section and annual minimum flow characteristics should be incorporated where applicable. Annual number of visitors to the affected area or shoreline must be obtained. This permits estimation of lost user-days on an annual basis. Any possible eutrophication effects should be estimated and included as a degradation of quality
1.5 Radionuclides discharged to water body	1.5.1 Aquatic organisms	Radionuclide discharge may introduce a radiation level that adds to natural background radiation	Rad per year	Sum dose contributions from radionuclides expected to be released
	1.5.2 People, external	Radionuclide discharge may introduce a radiation level that adds to natural background radiation for water users	Rem per year for individual; man-rem per year for estimated population as of the first scheduled year of plant operation	Sum annual dose contributions from nuclides expected to be released. Calculate for above-water activities (skiing, fishing, boating), in-water activities (swimming), and shoreline activities

Primary impact	Population or resources affected	Description	Unit of measure[a]	Method of computation
	1.5.3 People, ingestion	Radionuclide discharge may introduce a radiation level that adds to natural background radiation for ingested food and water	Rem per year for individuals (whole body and organ); man-rem per year for population as of first scheduled year of plant opertion	Estimate biological accumulation in foods and intake by individuals and population. Calculate doses by summing results for expected radionuclides
1.6 Consumptive use	1.6.1 People	Drinking water supplies drawn from the water body may be diminished	Gallons per year	Where users withdraw drinking water supplies from the affected water body, lost water to users should be estimated. Relevant delivered costs of replacement drinking water should be included
	1.6.2 Agriculture	Water may be withdrawn from agricultural usage and use of remaining water may be degraded	Acre-feet per year	Where users withdrawing irrigation water are affected, the loss should be evaluated as the sum of two volumes: the volume of the water lost to agricultural users and the volume of dilution water required to reduce concentrations of dissolved solids in remaining water to an agriculturally acceptable level
	1.6.3 Industry	Water may be withdrawn for industrial use	Gallons per year	
1.7 Plant construction (including site preparation)	1.7.1 Water quality, physical	Turbidity, color, or temperature of natural water body may be altered	Acre-feet and acres	Volumes of dilution water required to meet applicable water quality standards should be calculated. The areal extent of the effect should be estimated
	1.7.2 Water quality, chemical	Water quality may be impaired	Acre-feet, percent	To the extent possible, the applicant should treat problems of spills and drainage during construction in the same manner as Sec. 1.4.1

1.8 Other impacts				Applicant should describe and quantify any other environmental effects of the proposed plant that are significant
1.9 Combined or interactive effects				Where evidence indicates that the combined effect of a number of impacts on a particular population or resource is not adequately indicated by measures of the separate impacts, the total combined effect should be described
1.10 Net effects				See discussion in Sec. 5.8
2 Groundwater				
2.1 Raising/lowering of groundwater levels	2.1.1 People	Availability or quality of drinking water may be decreased and the functioning of existing wells may be impaired	Gallons per year	Volume of replacement water for local wells actually affected must be estimated
	2.1.2 Plants	Trees and other deep-rooted vegetation may be affected	Acres	Estimate the area in which groundwater level change may have an adverse effect on local vegetation. Report this acreage on a separate schedule by land use. Specify such uses as recreational, agricultural, and residential
2.2 Chemical contamination of groundwater (excluding salt)	2.2.1 People	Drinking water of nearby communities may be affected	Gallons per year	Compute annual loss of potable water
	2.2.2 Plants	Trees and other deep-rooted vegetation may experience toxic effects	Acres	Estimate area affected and report separately by land use. Specify such uses as recreational, agricultural, and residential

Primary impact	Population or resources affected	Description	Unit of measure[a]	Method of computation
2.3 Radionuclide contamination of groundwater	2.3.1 People	Radionuclides that enter groundwater may add to natural background radiation level for water and food supplies	Rem per year for individuals (whole body and organ); man-rem per year for population as of year of first scheduled operation	Estimate intakes by individuals and populations. Sum dose contributions for nuclides expected to be released
	2.3.2 Plants and animals	Radionuclides that enter groundwater may add to natural background radiation level for local plant forms and animal population	Rad per year	Estimate uptake in plants and transfer to animals. Sum dose contributions for nuclides expected to be released
2.4 Other impacts on groundwater				Applicant should describe and quantify any other environmental effects of the proposed plant that are significant
3 Air				
3.1 Fogging and icing (caused by evaporation and drift)	3.1.1 Ground transportation	Safety hazards may be created in the nearby regions in all seasons	Vehicle-hours per year	Compute the number of hours per year that driving hazards will be increased on paved highways by fog and ice from cooling towers and ponds. Documentation should include the visibility criteria used for defining hazardous conditions on the highways actually affected
	3.1.2 Air transportation	Safety hazards may be created in the nearby regions in all seasons	Hours per year, flights delayed per year	Compute the number of hours per year that commercial airports will be closed to visual (VFR) and instrumental (IFR) air traffic because of fog and ice from cooling towers. Estimate number of flights delayed per year

	3.1.3 Water transportation	Safety hazards may be created in the nearby regions in all seasons	Hours per year, number of ships affected per year	Compute the number of hours per year ships will need to reduce speed because of fog from cooling towers or ponds or warm water added to the surface of the river, lake, or sea
	3.1.4 Plants	Damage to timber and crops may occur through introduction of adverse conditions	Acres by crop	Estimate the acreage of potential plant damage by crop
3.2 Chemical discharge to ambient air	3.2.1 Air quality, chemical	Pollutant emissions may diminish the quality of the local ambient air	Percent and pounds or tons	Actual concentration of each pollutant in ppm for maximum daily emission rate should be expressed as a percentage of the applicable emission standard. Report weight for expected annual emissions
	3.2.2 Air quality, odor	Odor in gaseous discharge or from effects on water body may be objectionable	Statement	Statement must be made as to whether odor originating in plant is perceptible at any point off site
3.3 Radionuclides discharged to ambient air and direct radiation from radioactive materials (in plant or being transported)	3.3.1 People, external	Radionuclide discharge or direct radiation may add to natural background radiation level	Rem per year for individuals (whole body and organ); man-rem per year for population as of year of first scheduled operation	Sum dose contributions from nuclides expected to be released
	3.3.2 People, ingestion	Radionuclide discharge may add to the natural radioactivity in vegetation and in soil	Rem per year for individuals (whole body and organ); man-rem per year for population as of year of first scheduled operation	For radionuclides expected to be released estimate deposit and accumulation in foods. Estimate intakes by individuals and populations and sum results for all expected radionuclides

Primary impact	Population or resources affected	Description	Unit of measure[a]	Method of computation
	3.3.3 Plants and animals	Radionuclide discharge may add to natural background radioactivity of local plant and animal life	Rad per year	Estimate deposit of radionuclides on, and uptake in, plants and animals. Sum dose contributions for radionuclides expected to be released
3.4 Other impacts on air				Applicant should describe and quantify any other environmental effects of the proposed plant that are significant
4 Land				
4.1 Site selection	4.1.1 Land, amount	Land will be preempted for construction of nuclear power plant, plant facilities, and exclusion zone	Acres	State number of acres preempted for plant, exclusion zone, and accessory facilities such as cooling towers and ponds. By separate schedule state the type and class of land preempted (e.g., scenic shoreline, wetland, forest land)
4.2 Construction activities (including site preparation)	4.2.1 People (amenities)	There will be a loss of desirable qualities in the environment due to the noise and movement of people, material, and machines	Total population affected, years	Disruption of community life (or alternatively the degree of community isolation from such irritations) should be estimated. Estimate the number of residences, schools, hospitals, etc., within area of visual and audio impacts. Estimate the duration of impacts and total population affected
	4.2.2 People (accessibility of historical sites)	Historical sites may be affected by construction	Visitors per year	Determine historical sites that might be displaced by generation facilities. Estimate effect on any other sites in plant environs. Express net impact in terms of annual number of visitors

	4.2.3 People (accessibility of archeological sites)	Construction activity may impinge upon sites of archeological value	Qualified opinion	Summarize evaluation of impact on archeological resources in terms of remaining potential value of the site. Referenced documentation should include statements from responsible county, state, or federal agencies, if available
	4.2.4 Wildlife	Wildlife may be affected	Qualified opinion	Summarize qualified opinion including views of cognizant local and state wildlife agencies when available, taking into account both beneficial and adverse effects
	4.2.5 Land (erosion)	Site preparation and plant construction will involve cut and fill operations with accompanying erosion potential	Cubic yards and acres	Estimate soil displaced by construction activity and erosion. Beneficial and detrimental effects should be reported separately
4.3 Plant operation	4.3.1 People (amenities)	Noise may induce stress	Number of residents, school populations, hospital beds	Use the Proposed HUD Criterion Guideline for Non-Aircraft Noise to establish areas receiving noise in the categories of Clearly Unacceptable, Normally Unacceptable, and Normally Acceptable. For each area report separately the number of residences, the total school population, and the total number of hospital beds
	4.3.2 People (aesthetics)	Local landscape as viewed from adjacent residential areas and neighboring historical, scenic, and recreational sites may be rendered aesthetically objectionable by the plant facility	Qualified opinion	Summarize qualified opinion including views of cognizant local and regional authorities when available

Primary impact	Population or resources affected	Description	Unit of measure[a]	Method of computation
	4.3.3 Wildlife	Wildlife may be affected	Qualified opinion	Summarize qualified opinion including views of cognizant local and state wildlife agencies when available, taking into account both beneficial and adverse effects
	4.3.4 Land,flood control	Health and safety near the water body may be affected by flood control	Reference to Flood Control District approval	Reference must be made to regulations of cognizant Flood Control Agency by use of one of the following terms: No Implications for flood control, Complies with flood control regulation
4.4 Salts discharged from cooling towers	4.4.1 People	Intrusion of salts into groundwater may affect water supply	Pounds per square foot per year	Estimate the amount of salts discharged as drift and particulates. Report maximum deposition. Supporting documentation should include patterns of deposition and projection of possible effect on water supplies
	4.4.2 Plants and animals	Deposition of entrained salts may be detrimental in some nearby regions	Acres	Salt tolerance of local affected area vegetation must be determined. That area, if any, receiving salt deposition in excess of tolerance (after allowance for dilution) must be estimated. Report separately an appropriate tabulation of acreage by land use. Specify such uses as recreational, agricultural, and residential. Where wildlife habitat is affected identify populations
	4.4.3 Property resources	Structures and movable property may suffer degradation from corrosive effects	Dollars per year	If salt spray impinges upon a local community, then property damage may be estimated by applying to the local value

				of buildings, machinery, and vehicles a differential in average depreciation rates between this and a comparable seacoast community
4.5 Transmission route selection	4.5.1 Land, amount	Land will be preempted for construction of transmission line systems	Miles, acres	State total length and area of new rights-of-way. Estimate current market value of land involved
	4.5.2 Land use and land value	Lines may pass through visually sensitive (that is sensitive to presence of transmission lines and towers) areas, thus impinging on their present and potential use and value of neighboring property	Miles, acres, dollars	Total length of new transmission lines and area of right-of-way through various categories of visually sensitive land. Estimate minimum loss in current property values of adjacent areas
	4.5.3 People (aesthetics)	Lines may present visually undesirable features	Number of such features	Estimate total number of visually undesirable features, such as number of major road crossings in vicinity of intersection or interchanges; number of major waterway crossings; number of crest, ridge, or other high point crossings; number of long views of transmission lines perpendicular to highways and waterways
4.6 Transmission facilities construction	4.6.1 Land adjacent to right-of-way	Constructing new roads for access to right-of-way may have environmental impact	Miles	Estimate length of new access and service roads required for alternative routes
	4.6.2 Land, erosion	Soil erosion may result from construction activities	Tons per year	Estimate area with increased erosion potential traceable to construction activities
	4.6.3 Wildlife	Wildlife habitat and access to habitat may be affected	Number of important species affected	Identify important (Sec. 2.7) species that may be disturbed

Primary impact	Population or resources affected	Description	Unit of measure[a]	Method of computation
	4.6.4 Flora	Flora may be affected		
4.7 Transmission line operation	4.7.1 Land use	Land preempted by right-of-way may be used for additional beneficial purposes such as orchards, picnic areas, nurseries, hiking, and riding trails	Percent, dollars	Estimate percent of right-of-way for which no multiple-use activities are planned. Annual value of multiple-use activities less cost of improvements
	4.7.2 Wildlife	Modified wildlife habitat may result in changes	Qualified opinion	Summarize qualified opinion including views of cognizant local and state wildlife agencies when available
4.8 Other land impacts				Applicant should describe and quantify any other environmental effects of the proposed plant that are significant
4.9 Combined or interactive effects				Where evidence indicates that the combined effects of a number of impacts on a particular population or resource are not adequately indicated by measures of the separate impacts, the total combined effect should be described. Both beneficial and adverse interactions should be indicated
4.10 Net effects				See discussion in Sec. 5.8

[a] Applicant may substitute an alternative unit of measure where convenient. Such a measure should be related quantitatively to the unit of measure shown in this table.

[b] "Fish" as used in this table includes shellfish and other aquatic invertebrates harvested by humans.

[c] Indirect effects could include increased disease incidence, increased predation, interference with spawning, changed metabolic rates, and hatching of fish out of phase with food organisms.

Source: U.S. Atomic Energy Commission: "Preparation of Environmental Reports for Nuclear Power Plants," Regulatory Guide 4.2, Washington, D.C., Mar. 2, 1973.

Index